Instabilities Modeling in Geomechanics

SCIENCES

Mechanics, Field Director – Gilles Pijaudier-Cabot

Geomechanics, Subject Head – Cino Viggiani

Instabilities Modeling in Geomechanics

Coordinated by

Ioannis Stefanou

Jean Sulem

iSTE

WILEY

First published 2020 in Great Britain and the United States by ISTE Ltd and John Wiley & Sons, Inc.

ISTE Ltd
27-37 St George's Road
London SW19 4EU
UK

www.iste.co.uk

John Wiley & Sons, Inc.
111 River Street
Hoboken, NJ 07030
USA

www.wiley.com

Library of Congress Control Number: 2019955403

British Library Cataloguing-in-Publication Data
A CIP record for this book is available from the British Library
ISBN 978-1-78945-000-2

ERC code:
PE1 Mathematics
PE1_7 Topology
PE10 Earth System Science
PE10_5 Geology, tectonics, volcanology

MIX
Paper from responsible sources
FSC® C013604

Contents

Introduction

Ioannis Stefanou[1] **and Jean Sulem**[2]
[1]*GeM, Ecole Centrale de Nantes, France*
[2]*Navier, Ecole des Ponts ParisTech, Université Paris-Est, Marne-la-Vallée, France*

Nuclear waste disposal, petroleum engineering, CO_2 sequestration, geothermal energy, tunneling, slope stability, geotechnics, borehole stability, drying cracking and earthquake nucleation are important applications of geomechanics with short- and long-term environmental and societal impacts. Geomechanical systems involve various multiphysical and non-linear processes at several length and time scales. These complex mechanisms are described by non-linear differential equations that express the evolution of the various state variables of a system (e.g. displacements, temperature, pore pressure etc.). In order to study the evolution of the system and the possible occurrence of instabilities, it is necessary to explore the mathematical properties of the governing equations. Therefore, questions about the existence, uniqueness and stability of solutions arise naturally. Bifurcation theory and stability analysis are robust and rigorous tools for qualitatively and quantitatively investigating various instabilities such as strain localization, thermal runaway and unstable pressure increase, without explicitly determining the solutions of the governing non-linear equations of a geomechanical system.

The purpose of this book is to present the basic ideas of bifurcation theory and its application to classical problems of geomechanics. The book is organized into nine chapters.

The first chapter, "Multiphysics Role in Instabilities in Geomaterials: A Review", provides *in situ* and laboratory evidence of instabilities in geomechanical applications, mostly induced by multiphysical phenomena at various scales, such as heat generated during pre-cursor creep in the development of landslides, geo-

Instabilities Modeling in Geomechanics,
coordinated by Ioannis Stefanou and Jean Sulem.

chemical reactions, the effect of heat on inducing possible failure through pressurization of pore water, the effect of evaporation-induced suction and air entry during drying and subsequent cracking of soils.

The second chapter, "Fundamentals of Bifurcation Theory and Stability Analysis", aims to provide the basic ideas of bifurcation theory and stability analysis. It focuses on giving the necessary vocabulary for the classification of common bifurcations that are often met in applications and, finally, it presents the application of the theory for studying strain localization in solids. Some aspects related to shear band thickness, mesh dependency and higher order continua are also briefly discussed.

The third chapter, "Material Instability and Strain Localization Analysis", covers the principles of strain localization analysis as applied to geomaterials. The conditions for the formation of different types of deformation bands are given and an extension of the analysis to fluid-saturated porous media is presented.

The fourth chapter, "Experimental Investigation of the Emergence of Strain Localization in Geomaterials", focuses on strain field measurements of strain localization. Full field methods, imaging tools and experimental loading apparatus have evolved considerably over the past 15 years. This chapter presents recent developments on the characterization of the strain localization process and introduces the methods that are most frequently used.

The fifth chapter, "Numerical Modeling of Strain Localization", gives the basic concepts of numerical modeling of the post-localization regime of strain softening geomaterials. For this purpose, two higher-order continua with microstructure are presented. This type of continuum is used to regularize the ill-posed mathematical problem of strain-softening materials and to enable the modeling of progressive localization of deformation in zones of intense shearing that eventually leads to failure.

The sixth chapter, "Numerical Modeling of Bifurcation: Applications to Borehole Stability, Multilayer Buckling and Rock Bursting", presents typical boundary value problems of bifurcation theory in applications related to the petroleum industry, mining and structural geology. The formulation of the bifurcation problem is described and the governing equations are integrated numerically using higher-order continua with microstructure, such as the Cosserat continuum.

The seventh chapter, "Numerical Modeling of Multiphysics Couplings and Strain Localization", focuses on the numerical modeling of localized phenomena induced by multiphysical couplings. To deal with interactions that occur between the different phases of porous media, a regularization technique based on the second gradient model is used.

The next two chapters, "Multiphysics Couplings and Strain Localization in Geomaterials" and "On the Thermo-poro-mechanics of Chemically Active Faults", provide a review of research regarding the effects of temperature, pore pressure, chemical reactions and microstructure on strain localization in geomaterials. Examples have been given in relation to seismic slip in outcrops, core drillings on active faults and compaction banding.

The last chapter, "Theoretical and Numerical Analysis of Softening and Instabilities in Faults", focuses on the Cosserat continuum, strain-localization and thermo-hydro-mechanical couplings. The interplay of different length and time scales is shown and the role of the size of the microstructure is emphasized.

We would like to thank all of the contributors of this volume and we hope that the chapters provide a valuable introduction to bifurcation theory and stability in geomechanics, covering the state of the art of theoretical, experimental and numerical developments in the field.

This volume is dedicated to the memory of late Professor Ioannis Vardoulakis (1949–2009) on the 10th anniversary of his death, in recognition of his contributions and achievements in mechanics, geomechanics and bifurcation theory.

1

Multiphysics Role in Instabilities in Geomaterials: a Review

Tomasz HUECKEL
Duke University, Durham, USA

Multiphysical nature is characterized by complex phenomena involved in instabilities, strain localization and bifurcation in landslides, borehole instability in nuclear waste disposal and drying cracking, as seen *in situ* and in the laboratory. The multiphysics includes the effect of heat generated during precursor creep in the development of landslides, as well as the effect of geochemical reactions, the effect of heat on inducing possible failure through pressurization of pore water, the effect of evaporation-induced suction and air entry during drying and subsequent cracking of soils. The phenomena illustrated with specific natural or engineered events are interpreted as scenarios of processes that are either simultaneous or sequential, and that are coupled or result from an accumulation of dissipative processes. As pointed out by Terzaghi (1950), the causes of the instabilities are often long-term phenomena, rather than single events, such as major rainfalls, which are contributing factors. The need for a proper description of these long-term phenomena and their coupling with variable mechanical properties of soil and rock is emphasized.

1.1. Introduction

The engineering practice in all branches of geomechanics is now at an interesting stage of development, when the customary tools of evaluation of the margin of safety, such as "admissible stress" and "factors of safety", are felt to lead to an oversimplification of what we are capable of saying about a sample, soil/rock mass or structure. This is mainly because of the developed computational capabilities of

Instabilities Modeling in Geomechanics,
coordinated by Ioannis STEFANOU and Jean SULEM. © ISTE Ltd 2020.

contemporary engineering, as well as experimentally supported modeling capabilities, including coupled fields, through which soil and rock behavior is mathematically described. In the statement above I have adapted the words of Giulio Maier, with which he opens the foreword to a fascinating book by Davide Bigoni (2012) on bifurcation and material instabilities. While the book refers to bifurcations in a larger class of materials than just geomaterials, the above pronouncement catches the situation exactly: we can predict much more in detail than we could a few years ago: the stress field evolution, together with strain and/or damage progress along a process of loading following multiple scenarios of coupling with temperature; concentration of ions; salts or reaction progress field. It potentially includes patterns related to failure/instability and their precursors. However, how this information could/should be utilized to quantify "the distance from failure" or "factor of safety" (FOS) often remains an open question.

The purpose of this chapter is to provide an overview of a series of phenomena in geomechanics which qualify as instabilities/failures of various kinds. The use of this less-than-strictly defined term is intentional, as we want to encompass the widest possible class of phenomena for which the criteria are not necessarily within a single type of definition, but which correspond in loose terms to Lyapunov's definition: an unlimited response to a limited solicitation. Solicitation refers to a trigger of any sort: mechanical, hydraulic, thermal or chemical. We shall start with classical phenomena associated with purely mechanical loading induced instabilities and their criteria and implications to expand into an array of non-classical multiphysics instability phenomena. Current observations and understanding of geomechanical processes indicate a critical role of non-mechanical variables, whereas the conceptual base is lagging behind. Material (local) and field (global) instabilities based on the actual instability events leading to failure are both discussed.

1.2. General remarks

As we started with a promise to be wide open and inclusive, we have to issue several warnings in order to try to wave off an inevitable confusion that the subject brings, despite the appearance of a strictly rigorous approach.

To start with, in a geoengineering/geophysics context, instability or, better, loss of stability may mean instability of a material per se (at a point), instability of a soil/rock mass, or speaking mathematically of a boundary value problem. In other terms, we speak of a local or global stability. A local loss of stability at at least a single point of the continuum is considered a necessary (but far from sufficient) condition for global instability. Similarly, a local instability in a volume around a tunnel opening may be critical for a highway authority, but of no relevance to the stability of the mountain in which it is built. On the other end of the spectrum, local fault instability *may* induce global slope instability, or trigger an earthquake. It depends on the geometrical constraints that the considered boundary value problem implies.

Local stability is usually tested in a materials laboratory on a uniform specimen, or in a mathematical model, for a single material point. Global stability can rarely be tested on a large scale, but attempts have been made to monitor known landslide sites or earthquake source sites, or to conduct large liquefaction experiments.

Physically, instability may mean many things depending on the type of material and on the geometrical scale of consideration. In the plainest case, a macroscopically homogeneous material element in a laboratory under sufficiently low stress deforms in a homogeneous manner when a uniform traction is applied at its boundary. However, for unspecified physical reasons, at a certain stress level it responds with an unconstrained strain to, for example, a small stress perturbation. Often, the homogeneous strain is associated with a diffuse dilatancy (increased volume). This is a classical representation of instability. The key point is the homogeneity of the response maintained during the unstable phase.

Alternatively, we perceive as critical a loss of uniqueness of response, which means that a repetition of what is theoretically the same experiment would yield a different, still homogeneous, response. An additional option is to treat as unstable a response in which the increase in internal energy over a virtual displacement is less than the work of the external forces. Each of the above critical conditions, in principle, leads to a different criterion, both locally and globally.

In addition, when in a homogeneous sample the straining becomes more advanced, this strain homogeneity may be spontaneously lost. It is inevitable that, due to the actual irregularities in the distribution of minerals with a different stiffness, interface properties, pre-existence of structural defects, etc., high concentrations of stress and/or strain occur. Such concentrations may result in various forms of local damage, like mineral decohesion, micro-fissures, nucleation of pores, and intermineral or intergrain slips, depending on the structure, mineralogy and history of the material. The interaction of such local singularities gives rise to a variety of failure mechanisms. Despite their differences, their common feature is the possibility of microdefects developing into large-scale mechanisms of failure, such as slip-surface, fault or compaction band, leading to landslides, sinkholes, faulting, subsidence, etc. Such mechanisms are often characterized as macroscopic strain localization, when certain features of continuity abruptly cease to persist, like intergranular or intermineral phase contact. Commonly, more than one mechanism occur simultaneously, as in Figure 1.1, which was obtained for a sample of sand, and in which two separate slip surfaces and a substantial volume of diffuse dilatancy are observed after a triaxial compression test.

A mathematical representation of the physics above is equally complex. Often, Lyapunov-type instability (defined as an unconstrained response to a limited perturbation) is implied to result from the solution of a system of differential equations describing the nonlinear material behavior. The instability consists of bifurcation of the solution of the system of equations, which clearly implies that the solution is not unique.

Figure 1.1. *Multiple shear bands and a volume increase due to diffuse dilatancy*

The criteria for exclusion of instability may be local (or for uniformly deformed systems) or global for a piece of continuum. There are several different criteria, expressing very similar, but not identical conditions.

The first type of criteria refers to certain characteristics of material energy or work required to deform in a unique manner (meaning one uniform solution) under the conditions of equilibrium, for a specific stress–strain relationship and geometrical strain–displacement relationship. The second type allows for strain localization, with a particular geometrical form of deformed configuration that satisfies the conditions of equilibrium that allow certain forms of discontinuity, that is, it admits a certain type of discontinuity (shear or compaction bands). Global criteria consist of the same requirements but integrated over the entire volume of the considered body. The very definition of "bifurcation point" varies from author to author, although in any specific situation there is usually no doubt about what should be called a bifurcation point. The above types of instabilities and their implications have been studied in the context of geomaterials for nearly 50 years, and an example of their implications and their relationships is reproduced in Figure 1.2 from an old paper (Bigoni and Hueckel 1991a). The most interesting (to this writer) recent developments concern geomaterials in their multiphysics context.

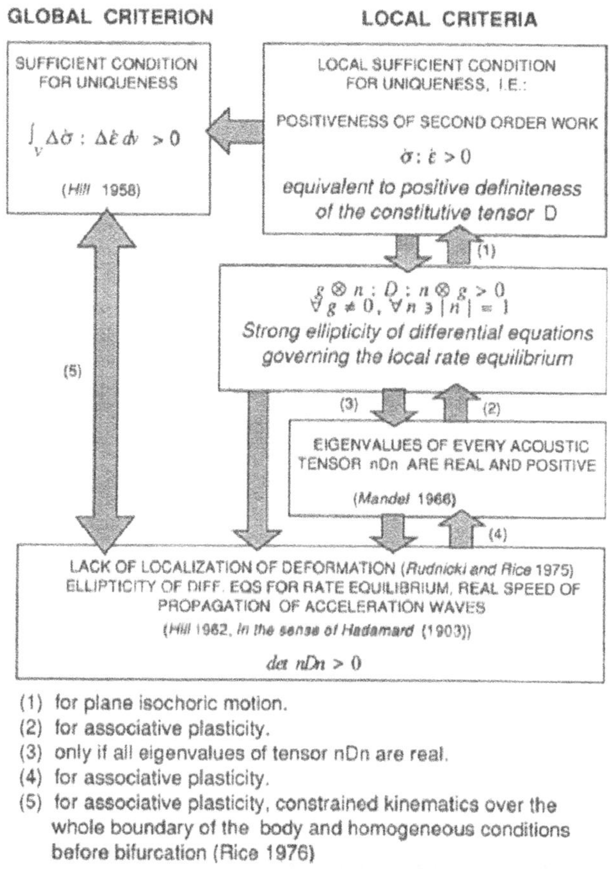

Figure 1.2. *Relationships between criteria for uniqueness, second-order work, strong ellipticity, Mandel's stability and localization*

1.3. Solid phase material criteria

To start with we will limit ourselves to considering exclusively incipient elasto-plastic straining at small strains. At the moment, we will not deal with materials that exhibit strong viscous effects.

In what follows, we will outline the basic formulae of elasto-plasticity (equation [1.1]) at small strains to introduce basic concepts that will be referred to (Roscoe and Burland 1968). The given set of equations implies the existence of a yield limit, $f = f(p',q,r,p_c')$ in the effective stress invariant space, which means yielding if $f = 0$, and means an elastic unloading state if $f < 0$. The case of $f > 0$ is deemed as statically inadmissible. The core of the multiphysical aspects of soil/sediment/rock behavior is embedded in the last of the equations, which describes the evolution of the principal mechanical property of the material, which is the size of the yield locus $p_c' = p_c'\left(\varepsilon^{pl}, \mu\left(\Delta T, \xi, c...\right)\right)$ with changes in a series of the environmental (non-mechanical) variables (see, for example Hueckel 1992, 2002).

$$
\begin{aligned}
& f = f(p',q,r,p_c') \leq 0 \\
& p' = \tfrac{1}{3}\sigma_{kk}'; \; q = \sqrt{3}\sqrt{J_2''}; \; J_2'' = \tfrac{1}{2} s_{ij} s_{ij}; \\
& f = \left(\frac{2p'}{p'_c} - 1\right)^2 + \left(\frac{2q}{gMp_c'}\right)^2 - 1 \\
& M = const. \\
& g = 2k \Big/ \left[1 + k + \tfrac{27}{2}(1-k)\left(\tfrac{r}{q}\right)^3\right]; \; k = \tfrac{3}{3+M} \\
& r = \sqrt[3]{J_3'}; \; J_3' = J_3'' = \tfrac{1}{3} s_{ik} s_{ij} s_{kj}; \\
& p_c' = p_c'\left(\varepsilon^{pl}, \mu\left(\Delta T, \xi, c...\right)\right)
\end{aligned}
\qquad [1.1]
$$

That implies that variables such as temperature, chemical mass removal (accretion) and ion concentration of pore fluid essentially affect the geomaterial strength, apparent preconsolidation pressure and elasto-plastic stiffness.

As is customary in inelasticity theories, the deformation can only be uniquely determined for an incremental stress, rather than total stress, as in elasticity. Therefore, the plastic strain rate is defined as

$$
d\varepsilon_{ij}^{pl} = d\lambda P_{ij}; \text{ under condition that } d\lambda df \leq 0, \text{ while } d\lambda \geq 0 \text{ and } df \leq 0 \qquad [1.2]
$$

where P_{ij} is the gradient of the plastic potential, or a resultant direction tensor of all inelastic strain components of various origins, including the irreversible change in elasticity due to plastic straining or damage (Hueckel 1975, 1976; Maier and Hueckel 1979; Bigoni and Hueckel 1991a, 1991b; Bigoni 2012). Most of the physical evidence suggests that $P_{ij} \neq Q_{ij}$, where $Q_{ij} = \partial f / \partial \sigma'_{ij}$ is the yield locus gradient, but for a number of reasons it is often seen as convenient to ignore the difference.

Non-associativity, i.e. when $P_{ij} \neq Q_{ij}$ (for instance via an elasto-plastic coupling), or any other forms of irreversible (not necessarily mechanical) straining leading to a non-symmetrical stress–strain incremental relationship, are notorious for inducing a premature loss of stability and/or strain localization.

Among several criteria for (local) material stability, the most commonly accepted is an energetic one: stating the positiveness of an increment of internal energy, or in other terms, of second-order work, in the sense of Hill (1958) or Drucker (1964). Following Drucker (1964), the continuum considered will be said to be stable in a given equilibrium configuration, if and only if an external agency imposing a kinematically admissible (compatible) infinitesimal geometric disturbance, by preserving equilibrium, performs non-negative (second-order) work regardless of what the disturbance is, i.e.

$$\delta\sigma'_{ij}\,\delta\varepsilon_{ij} \geq 0 \qquad [1.3]$$

Let us consider the set of constitutive equations [1.1] and express the incremental effective stress as a function of an incremental strain (Maier and Hueckel 1979) and properties of the material behavior: elasticity tensor E_{ijkl}, plastic hardening modulus H (positive during plastic strain hardening and negative during strain softening) and its critical value H_c, which may depend on stress and plastic strain $H = -\dfrac{\partial f}{\partial \varepsilon^p_{ij}} P_{ij}$; $H_c = -P_{ij} E_{ijkl} Q_{kl}$, where Q_{ij} and P_{ij} are, respectively, yield locus gradient and a second-order tensor defining the mode of incremental irreversible strain. We arrive at the following stress–strain incremental relationship characterized by elastoplastic stiffness tensor D_{ijkl}, as follows:

$$\begin{gathered}\delta\sigma'_{ij} = D_{ijkl}\,\delta\varepsilon_{kl};\ \text{with } D_{ijkl} = E_{ijkl} - \frac{1}{H - Hc} U_{ij} M_{kl} \\ and\ U_{ij} = E_{ijkl} Q_{kl} \text{ and } M_{ij} = E_{ijkl} P_{kl}\end{gathered} \qquad [1.4]$$

Employing the material properties introduced above in a local condition of stability [1.2], we conclude that the criterion is articulated through the hardening modulus H, which, if it is larger than the first value or lower than the second one (Maier and Hueckel 1979; Bigoni and Hueckel 1991a),

$$H > H_1 = \tfrac{1}{2}\left(H_c + \sqrt{P_{ij}E_{ijkl}P_{kl}Q_{mn}E_{mnrs}Q_{rs}}\right)$$
$$H < H_2 = \tfrac{1}{2}\left(H_c - \sqrt{P_{ij}E_{ijkl}P_{kl}Q_{mn}E_{mnrs}Q_{rs}}\right), \quad [1.5]$$

ensures stability.

For the associative flow rules ($P_{ij} = Q_{ij}$), the hardening modulus at loss of stability is $H = H_1 = 0$, while for materials with so-called subcritical softening, the hardening modulus at the point of re-gaining stability is $H = H_2 = H_c$. Hence, the stability range coincides with that of hardening and the postcritical range, while the softening range is all unstable (Figure 5.2). The most important departure from this rule is for all kinds of non-associative flow rules, for which it can be seen that $H_1 > 0$ (Bigoni and Hueckel 1991a, 1991b). Notably, an earlier, particular version of such conditions for a non-associated flow rule for a Mohr–Coulomb material was provided by Mróz (1963).

A separate issue is the uniqueness of material response. It appears that in certain situations, a certain type of incremental solicitation is not admissible, in the sense that the response is not unique, which means that two possible or infinite possible responses can be expected. The classical result is that in the softening range, i.e. $H \leq 0$, $d\sigma_{ij}$ such that $d\sigma_{ij}Q_{ij} \geq 0$ is statically not admissible, as it violates the flow rules of equation [1.3]. In addition, $d\sigma_{ij}$ such that $d\sigma_{ij}Q_{ij} < 0$ is admissible, but generates two possible types of response: an elastic unloading and an elastoplastic softening. In other terms, the range of $H \leq 0$ is not stress controllable. Analogous ranges of static admissibility may be established for $d\varepsilon_{ij}$, where a range of the hardening modulus where $H \leq H_c$ is not strain controllable. Consequently, the range $0 \leq H \leq H_c$ is strain controllable. The issues of controllability are relevant in experimental studies of material behavior.

Buscarnera and Nova (2011) have generalized the question of controllability to experiments with a mixed stress–strain control of loading programs. These are relevant, for instance, in undrained tests in which (under the assumption of incompressibility of solids and water) volumetric strain rate is imposed as zero.

An alternative form of instability is the one with strain localization. Originally identified in the seminal paper by Rudnicki and Rice (1975), the condition is less restrictive than [1.5]; nevertheless it does admit strain localization of a particular form of strain tensor in inequality [1.2], which is a product of a unit vector n_i normal to the planar band and a vector g_j defining the jump in the velocity derivative. This condition is equivalent to positive definiteness of all possible acoustic tensors $n_i D_{ijkl} n_l$. A special case of the strain localization into a planar band takes place when the differential equations describing the behavior of the material in equilibrium cease to be elliptic. Two modes of strain rate discontinuity are possible depending on directions of vectors n_i and g_i, which are normal modes of split and compaction (Castellanza *et al.* 2009): when the two vectors are coaxial, and a shear band mode, when they are not.

The criterium for localization in a particular direction specified by versor n_i orthogonal to the discontinuity shear band is given in the form of a hardening modulus

$$H^{sb}(n_j)=2G\left[2n_jP_{ij}Q_{jk}n_k-\left(n_jP_{ij}n_j\right)\left(n_mQ_{mn}n_n\right)-P_{ij}Q_{ij}-\tfrac{v}{1-v}\left(\left(n_jP_{ij}n_j\right)-P_{rr}\right)\left(\left(n_sQ_{st}n_t\right)-Q_{nn}\right)\right] \quad [1.6]$$

with G and v being the elastic shear modulus and Poisson coefficient. To determine a critical modulus at a given point, constrained maximization of equation [1.2] needs to be performed over all possible directions of n_i.

A global condition of stability is guaranteed in the medium for the entire boundary value problem if

$$2W \doteq \int_S \dot{T}_l \dot{u}_l dS+\int_V \dot{X}_i \dot{u}_i dV=\int_V \dot{\sigma}'_{ij}(\dot{\varepsilon}_{kl})\dot{\varepsilon}_{ij}(\dot{u}_r)dV \geq, \text{ for any } \dot{u}_r \quad [1.7]$$

which is again stated as non-negativeness of the second-order work of tractions integrated over the whole boundary (plus that of body forces, over volume). Through a standard consideration, the condition is brought to the requirement of positiveness of the second-order work integrated over the entire volume of the body, under the assumption that both the constitutive laws and the geometric conditions are fulfilled. That condition clearly is much less restrictive than the local condition [1.5], which requires the same, but at every point of the body. Thus, [1.7] admits large areas of a negative second-order work, as long that there are areas that can outweigh the negative areas.

One other way to guarantee the stability of the solution of the boundary value problem in geomechanics is to ensure that the solution is unique, or in other terms, that there are no alternative solutions for the one obtained to snap through to. The solution uniqueness is guaranteed only under sufficient conditions of local or overall nature, the overall condition being much less stringent than the above sufficient conditions for stability (Hueckel and Maier 1977).

In geoengineering practice analyses, the assessment of stability is often made on the basis of the finite element result through detection of failure as a loss of global equilibrium seen as a lack of convergence of the solution identified by the lack of convergence within a certain iteration number (Griffiths and Lane 1999; Zienkiewicz *et al.* 2005). Alternatively, loss of stability, for instance of a slope, is identified as an onset of a kinematically admissible "sliding" mechanism through monitoring of the selected nodes as the solution evolves to detect a sudden increase in displacements (Hicks and Spencer 2010).

1.4. Material sample stability: experimental

To provide an experimental illustration of a local instability, under the assumption of the absence of localization (diffuse plastic strain), is almost impossible. Figure 1.3(a) shows a series of triaxial test results with increasing confining stress values, from Paterson (1958) for Wombeyan marble. All but those with the highest confining stress eventually exhibit unstable behavior, but for pressure values lower than 35 MPa the material exhibits localized instability, either as a vertical spalling (1) or single (2) or conjugate (3) shear bands, whereas for larger confining stress the behavior is qualified as ductile (4,5) for 70 and 100 MPa (3,4) (Figure 1.3(b)).

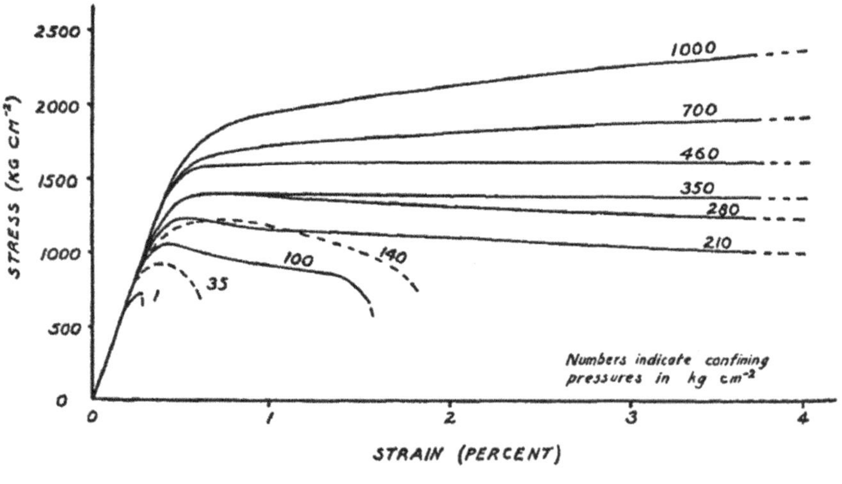

(a)

(1) (2) (3) (4) (5)

(b)

Figure 1.3. *Triaxial compression of Wombeyan marble. (a) Axial stress–strain curves; (b) localization and diffused damage modes, from Patterson (1958)*

Notably, for a higher level of confining stress in these tests (70 and 100 MPa), the deformation pattern, with a substantial inelastic component, can be viewed as uniform or non-localized. The stress–strain curves do not suggest an unstable or non-unique behavior. Nevertheless, the pattern formation of microcracks brings some concerns about the homogeneity of the strain and stress distribution across the sample, which in reality is not homogeneous. However, rigorous or at least somewhat codified measuring and understanding of what is "sufficiently homogeneous" are conspicuously missing.

In contrast, for all tests below 35 MPa of confining stress, one or more of stability criteria are failed, but the deformation is invariably localized at a certain point.

Acoustic emission recording techniques allow us to monitor sound-emitting micro-fracturing, which initiates long before the loss of stress–strain curve linearity, even before noticeable dilatancy onset, and long before approaching the peak stress in triaxial conditions for marble (Figure 1.4; Hallbauer *et al.* 1973). Indeed, what could qualify as an onset of non-uniqueness and instability coincides rather with the coalescence of microcracks into a shear band or macrocrack. Notably, the crack pattern evolution is a gradual process and does not suggest any threshold behavior or values. However, that is not always the case, as seen in comparisons of uniaxial failure in salt-rock, granite and marble (see Figure 1.7 from Zhang *et al.* (2015)).

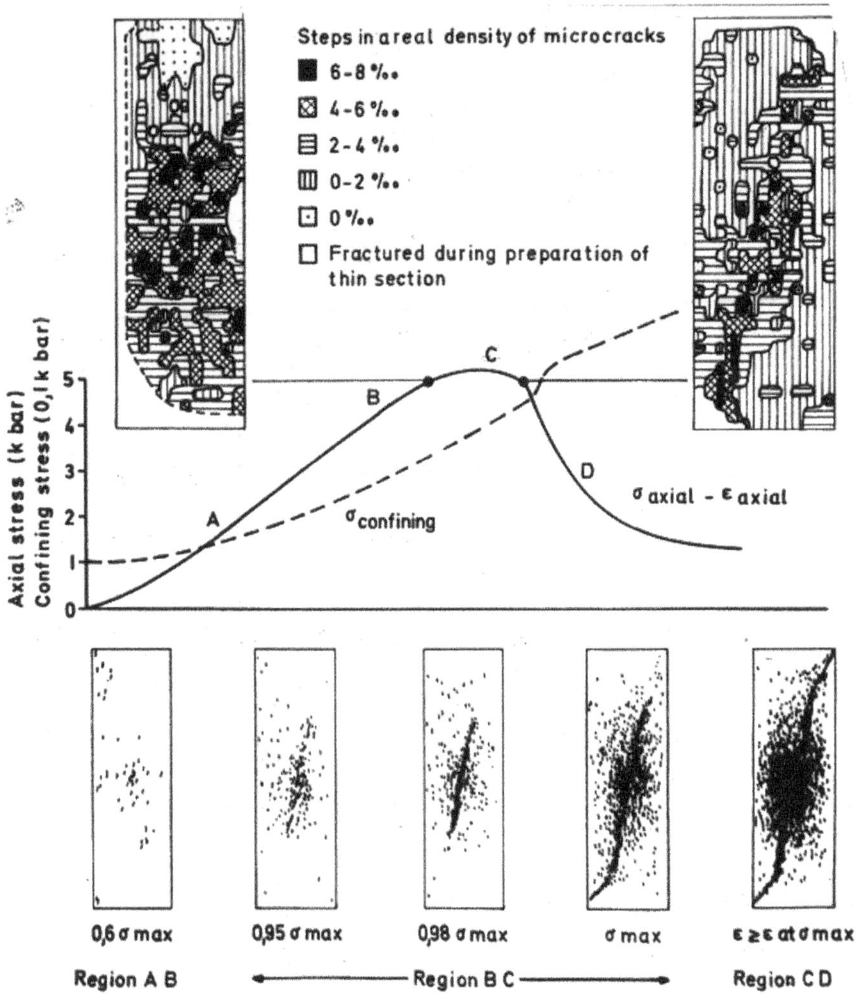

Figure 1.4. *Axial and lateral stress measured on a set of argillaceous quartzite with the corresponding evolution of the distribution of microcracking (Hallbauer* et al. *1973)*

For sand, the situation is quite similar. Unstable and non-unique behavior is seen in triaxial tests at low confining stress, and is invariably associated with a localized deformation into a shear band (Figure 1.5; Vardoulakis *et al.* 1978).

Figure 1.5. *Biaxial compression of sand with visible localized shear band (Vardoulakis* et al. *1978)*

Homogeneous behavior at higher confining stress is rarely seen in triaxial strain, and is usually associated with material stability and uniqueness of response to drained tests. There is no established database to support the claim that there may be unstable sand behavior with a uniform strain across the sample, at least in drained triaxial tests (Drescher 2016). Karner *et al.* (2008) report sound emission attributed to intense intergranular friction at low mean stress, but at higher stress (and sometimes elevated temperature), no strain localization is mentioned, while post-test observations indicate grain breaking and comminution. The associated stress–strain curve implies stable behavior at high confinement (Figure 1.6).

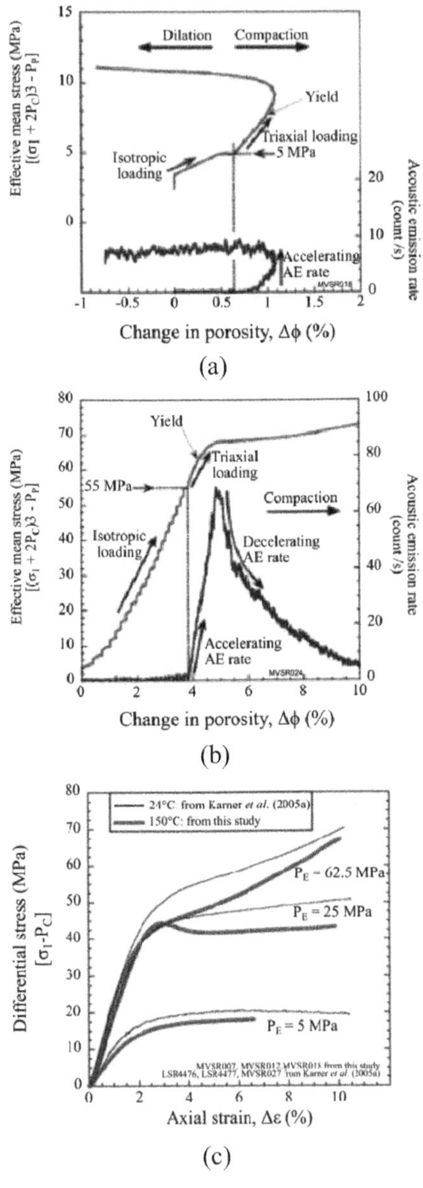

Figure 1.6. *(a) Low and (b) high confining stress compression of a quartz sand: isotropic effective stress versus porosity and acoustic emission decreasing after yielding at high confinement; (c) deviatoric stress–strain curves showing stable behavior at 24°C at high confining pressure (Karner* et al. *2008). For a color version of this figure, see www.iste.co.uk/stefanou/instabilities.zip*

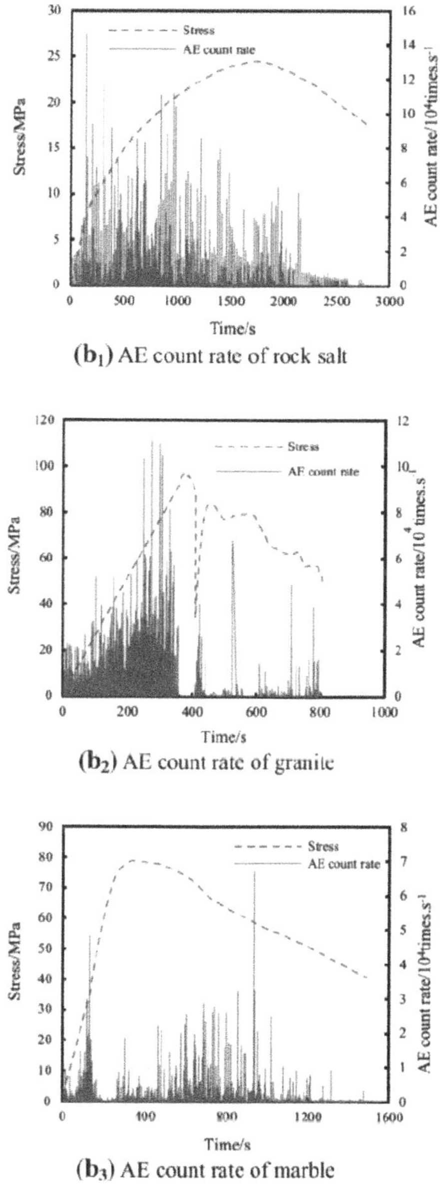

Figure 1.7. *Comparison of uniaxial compression of rock salt, granite and marble, with a different intensity of acoustic emission at different stages of loading (Zhang* et al. *2015). For a color version of this figure, see www.iste.co.uk/stefanou/instabilities.zip*

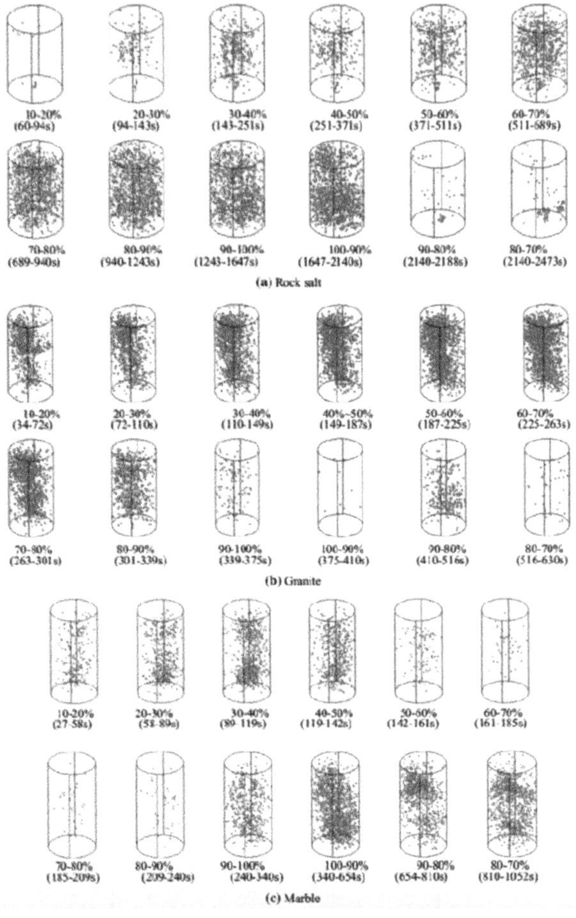

Figure 1.8. *Evolution of the distribution of acoustic emission during uniaxial compression of salt rock, granite and marble. For a color version of this figure, see www.iste.co.uk/stefanou/instabilities.zip*

A separate issue arises in undrained tests on sands, during which the volumetric strain is imposed to be constant. In such tests, the material exhibits unstable behavior; however, there is no indication of localization; in other words the deformation appears to be homogeneous or diffuse. The term "diffuse failure" has been adopted for this type of behavior (Daouadji *et al.* 2011). Figure 1.9 shows the corresponding stress–strain curve and the effective stress path for such a test on Hostun sand. Rightfully, Daouadji *et al.* indicate a restriction of the undrained or constant volume conditions, clearly imposing a peculiar deformation pattern.

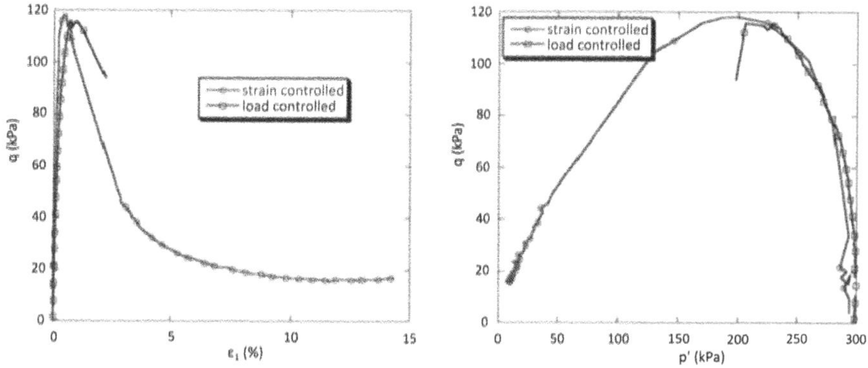

Figure 1.9. *Unstable behavior during an undrained test of Hostun sand (Daouadji* et al.*, 2011). For a color version of this figure, see www.iste.co.uk/stefanou/instabilities.zip*

Also interesting is the instability developing in sand during constant deviatoric stress drained tests. In these tests, failure coincides with the changing sign of the volumetric strain (or at a maximum of attained dilatancy) (see Figure 1.10). However, in such tests the condition of static admissibility of the local stress rate at the softening regime may be violated. As will be seen later in the paper, experimentation with thermal pressurization of clay at constant stress deviator undrained heating test leads to a similar response (Hueckel and Pellegrini 1991). However, in that test both the localized and the diffuse strains developed at failure.

There are several observations to be made concerning laboratory experiments on small triaxial samples. The underlying assumption for such experiments is that all the fields: stress, strain and plastic strain (and possibly microcracking) are uniform across the sample until a possible appearance of localization. This unfortunately is not necessarily true, as shown in Figure 1.4 (Hallbauer *et al.* 1973). Similarly, the evolution of local porosity monitored via CAT in sand suggests an early loss of uniformity prior to shear banding (Desrues *et al.* 1996). Stress is obviously not measurable directly in such experiments. It is usually considered as an average resulting from the measured force assumed as uniformly distributed across the area. A number of reasons for the non-uniformity are quoted, such as axial symmetry of specimens and a development of roughly planar shear bands, or stress concentration at piston boundaries, with a different intensity in sand, clay or rock. Remedies to the experimental techniques have been sought, by introducing truly triaxial testing (Muir Wood 1973), biaxial strain testing (e.g. Vardoulakis 1978), etc. Drescher and De Josselin de Jong (1972) conducted a series of tests in which they subjected a 2D photosensitive granular medium to shear under constant vertical load between two rigid smooth arms rotating around a pin with a controlled rate, causing a displacement of the medium between the arms with a globally unstable or stable force response, depending on the direction of the medium displacement, and dilative or contractile volumetric strain. These and many other subsequent similar experiments have shown that contact stresses lead to the formation of chains of compressed column forces within the medium, separated by lightly or completely unloaded grains. In addition, said compressed columns undergo periodic unstable buckling and rebuilding of such columns, so that the entire process – while apparently monotonic at the force–displacement level – is unstable, non-homogeneous and non-monotonic at the level of grain structures, which form vortices and other patterns (Kozicki and Tejchman 2016). Similar conclusions were derived from much later studies of discrete element methods (Iwashita and Oda 1998). This prompted the investigation of even small-scale experiments as boundary value problems and the treating of their stability and uniqueness through global criteria. Finally, there is a question of validity beyond the point of loss of stability or uniqueness of the stress–strain curves obtained in a single experiment. Indeed, each such curve in a non-uniqueness range is just one of an infinity of possible responses, as they go through a singularity point of $H = 0$.

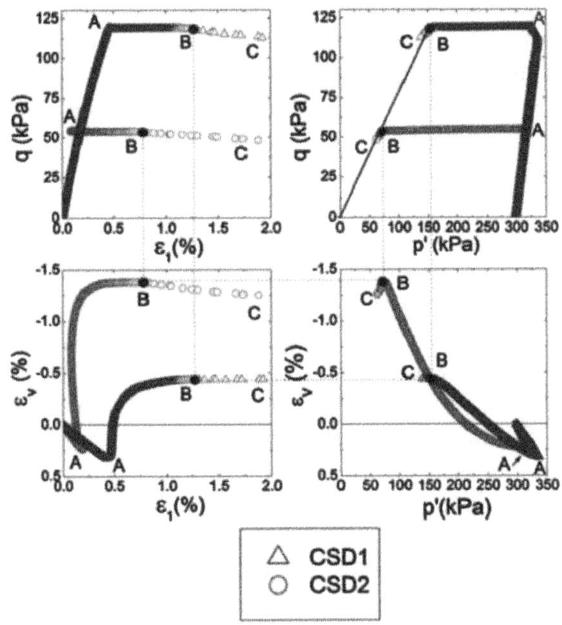

Figure 1.10. *Stress–strain curves in a q = cons. drained test with instability (Daouadji* et al. *2011). For a color version of this figure, see www.iste.co.uk/stefanou/instabilities.zip*

Some additional insight may be expected from multiscale analyses using discrete element method (DEM) computations, together with progress in a rigorous approach to interscale data interpretation.

1.5. Boundary value problems: uniqueness and stability at the field scale

1.5.1. *Landslides*

Landslides are the most common consequence of soil and rock instabilities at the field scale. In addition, they more common than earthquakes, and may claim many victims per occurrence in populated areas. Classically, shallow sliding is considered separately from deep rotational slope failures. Here, only the latter are considered. The traditional factor of safety (FOS) is understood as a ratio of resisting resultant moment to driving resultant moment the FOS approaching unity is considered as a condition of failure in classical geotechnics. However, as will be seen in what follows, in several historical landslides, failure occurred at a FOS substantially higher than 1, 2 or even 3. Incorrect evaluation of several variables, material properties or contributing processes led to misleading assessment of FOS.

A recent (February 2010) well-documented landslide is that of Maierato (Vibo Valentia), Calabria, which occurred at a site of paleo-landslides of 1783 and 1932. A rare, dramatic, but very instructive video by Patrizia Venturino (2010) has documented (https://www.youtube.com/watch?v=oWHjBsvmyLc) the event in detail. The most important thing to note in this video a succession of localized rotational collapses of individual scarps, intercalated with a mass flow. The general trigger of the landslide was a protracted period of rainfalls. As in so many similar cases, there were numerous precursors of the final event; in this case, a day to several hours prior to generalized failure consisting of minor local strikes, one of them toppling a high voltage powerline. Because of those warning signs, no fatalities occurred. The area affected was 0.3 km^2, with a front of about 800 m, depth of 60–70 m and involving about 10 Mm^3 (Figure 1.11). The geological structure, shown in a cross-section in Figure 1.12, points out to two particularities of relevance: the presence of two particularly weak permeable rocks: evaporitic limestone and Miopcene sandstone, intercalated by two layers of clay, roughly 10 m thick. Prior to the main collapse, a substantial swelling was observed, suggesting a rotational collapse of a part of mass. It is believed (Gattinoni *et al.* 2012) and confirmed by laboratory experiments that evaporitic limestone transformed from plastic to semifluid behavior. Such a transformation is suggested to explain the rapid change from a mode of sliding to the flowing debris mass observed in the videos.

Figure 1.11. *General view of the Maierato landslide site (Borrelli* et al. *2014). For a color version of this figure, see www.iste.co.uk/stefanou/instabilities.zip*

The most surprising finding from preliminary finite element (FE) calculations (Gattinoni *et al.* 2012) is that when dry the slope has FOS equal to 3.36, with slip surface within the evaporitic limestone, while with the water table up by a maximum of 10–15 m there is a reduction of FOS by 40% (about 2). Thus, the conventional understanding of slope stability highly overestimates FOS. Among the possible causes of failure not given proper consideration until now, the authors list penetration and pressurization of pore water along the contact between Miocene sandstone and Miocene clay, as well as the presence (and evolution upon inundation) of evaporitic limestone and Miocenic sandstone. They also suggest that autobrecciation induced by dissolution of halite and gypsum due to weathering and diagenesis might have played a role (Gattinoni *et al.* 2012; Borelli *et al.* 2014).

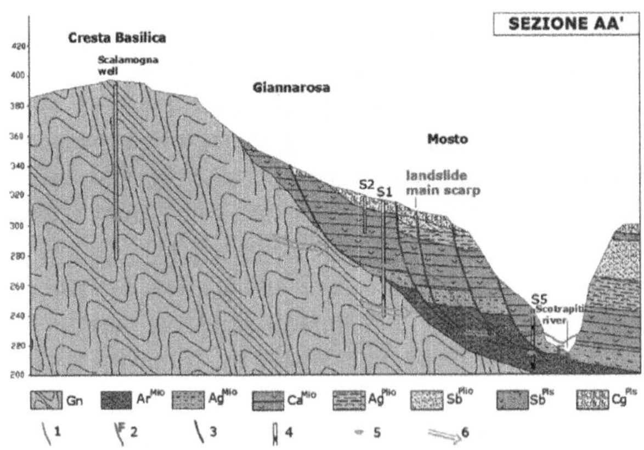

Figure 1.12. *A geological profile across the area prior to the landslide. The green line denotes post-event topography (from Gattinoni* et al. *2012). For a color version of this figure, see www.iste.co.uk/stefanou/instabilities.zip*

These findings from the preliminary analyses of the Maierato landslide (Gattinoni *et al.* 2012; Borrelli *et al.* 2014) interestingly connect to observations made about the instability of slopes made by Terzaghi (1950) in his paper on "Mechanism of landslides". Analyzing "landslide producing processes", he emphasizes the difference between the "causes" and the "contributing factors" of landslides. He defines the latter through an example as follows: "if a slope is old, heavy rainstorms (…) can hardly be the sole cause of a slope failure, because it is most unlikely that they are without any precedent in the history of the slope. They can only be considered contributing factors". In the same paper, in Table 1.1, Terzaghi (1950) introduces causes or "acting agents", among which there is water from heavy rainstorms or snow melting. Among several "modes of action of agent" linked to water, he lists in ninth place chemical weathering, which "weakens inter-granular bonds" and leads to a decrease of cohesion. For an "old slope" slide of Mount Turtle in Alberta (1903), he then shows after McConnell and Brock (2004) a progressive decrease in FOS over 15 years, with some variations corresponding to dry and wet spells, eventually reaching reaching 1, in conjunction with "a contributing factor" of high pore water pressure (Figure 1.13). He returns to this mechanism in the case of loess, where water from external reservoirs is listed as removing soluble binders, destroying intergranular bond(s), with the same macroscopic effect of a decreasing cohesion.

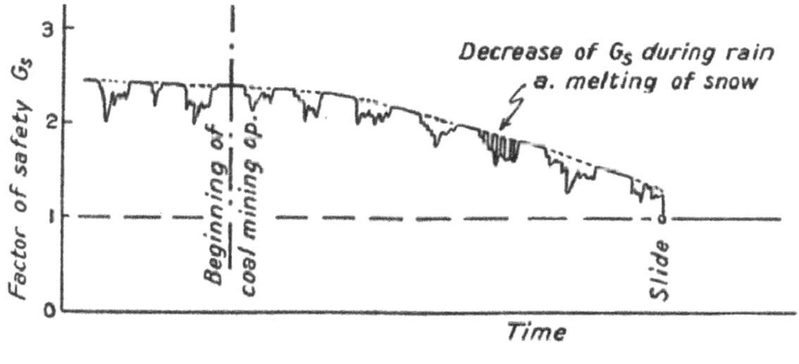

Figure 1.13. *Evolution of factor of safety of the slope at Mount Turtle along the process of soil weathering (from Terzaghi 1950)*

Sensitivity of basic mechanical properties to chemical processes in the environment has been seen as a critical factor of the stability of slopes and coastal structures due to several processes. These include: cyclic changes in the salinity and acidity of pore water, heat effect due to frictional slip during a creep phase of slope instability, oxidation and dissolution of sandstone, dissolution of calcite, dissolution of silica in aging sediments and dissolution of gypsum in abandoned mines, to mention just a few examples.

The conclusion from the above considerations, in slightly more modern terms, is that a proper approach to a realistic evaluation of FOS would be to view the stability as a scenario composed of several processes coupled between them. These processes would be solid and fluid mechanical as well as geochemical, coupled through both balance equations (including exchanges) and constitutive (phenomenological) coupling, i.e. reactions versus porosity (permeability) versus elasticity or softening. The processes may be either simultaneous or sequential. Hence, the evaluation of the instability of such processes or the differential equations describing them should be approached accordingly, as instability either of the whole process or of any element of the sequence.

An example of such an approach proposed by Veveakis *et al.* (2007) is an analysis of Vajont dam landslide of 1963 in Northern Italy (Figure 1.14). This was one of the largest disasters of this sort in the previous century, killing nearly 2,000 people in a town below the dam (which survived). In their analysis, Veveakis *et al.* postulated that 2–3 years prior to the collapse, creep of the rock mass localized in a clay-rich layer was self-alimented by shear-generated heat, pressurizing and possibly vaporizing water within the layer, and thus leading to thermal softening of clay and accelerating the creep, eventually leading to the massive landslide.

Figure 1.14. *Top left: view of the landslide scarp (2014). Top right: a crack (50–100 cm across, 2 km long) that developed parallel to the reservoir shore a few months before the collapse. Bottom: the town of Longarone 2 km below the dam, before (left) and after (right) the tragedy*

They calibrated their model against the data from the site at 35°C temperature and 2.5 MPa increase in pore pressure, which were sufficient to produce a creep of velocity 20 cm/day, prior to the failure (Figure 1.15).

The model has subsequently been generalized to a thermo-plasticity simulation by Cecinato *et al.* (2011).

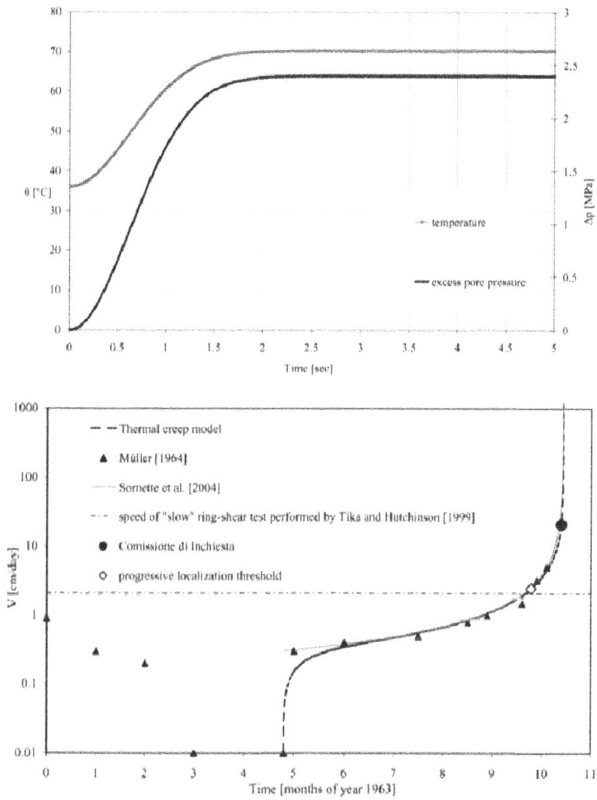

Figure 1.15. *Top: predicted tempearture and excess pore pressure arising in the shear band during sliding against time (seconds). Bottom: calculated velocity of creep against time in months, compared to other assessments (from Veveakis* et al. *2007)*

In a similar spirit, a case study of two landslide slip surfaces at Diao Jiao Zui and Qian Jiang Ping sites in the Three Gorges area, China, has been undertaken using a multiphysics approach. Geochemical testing revealed that due to acid rain (with pH between 5.4 and 3.45), potassium ions at the slide surface were released, the cementation was reduced and the ratio of interlayer clay minerals evolved.

Accordingly, illite or montmorillonite–illite mixtures were transformed first into kaolinite and subsequently into montmorillonite associated with a complex chemical softening evolution, including a 30% drop in shear strength at the slide surface, but not at a close vicinity (Zhao *et al.* 2011). Interestingly, both effective internal friction angle and cohesion appear for remolded smectite samples to be non-monotonic functions of time (Figure 1.16).

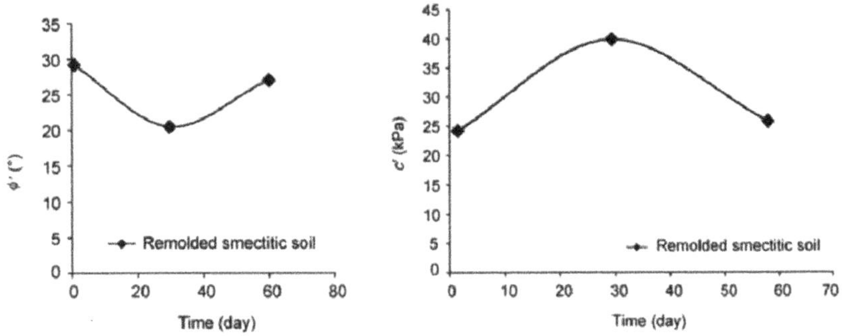

Figure 1.16. *Evolution with time of the effective internal friction angle and cohesion during laboratory-simulated chemical transformation due to increase in the pore water pH of a model smectite soil corresponding to the slip surface at Diao Jiao Zui and Qian Jiang Ping sites in the Three Gorges area, China (from Zhang* et al. *2015)*

In a similar direction of chemo-mechanical coupling at the slide surface, an interesting extension of their thermal runaway creep instability developed for Vajont simulation was proposed by Veveakis *et al.* (2010). They postulated that the aforementioned frictional motion is affected by not only the generated heat, but also the strain rate. It appears that steady-state heat diffusion across the shear band may be unstable and lead to localization of heat dissipation (and hence thermal softening) at the center of the band. Hence, the authors consider the effect of localized dissipation on endothermic chemical reactions, leading to a variety of possibilities depending on the nature of the specific chemical reaction.

In an analogous spirit, Stefanou and Sulem (2015) investigated the conditions of chemically induced compaction band instability via a chemo-plasticity model.

1.5.2. *Thermal pressurization problem*

One of concerns in the technology of nuclear waste disposal in clays is the effect of the heat of nuclear decay on the mechanical behavior of clay as a supporting medium surrounding the heat source. One of the multiphysics effects results from a huge difference between the thermal expansion of pore water and thermal expansion of clay (or thermal contraction, depending on whether it is in elastic or plastic range). Laboratory experiments have shown that undrained heating at constant total stress loading conditions relevant to the technology leads to instability at a temperature between 70 and 90°C, as can be seen from the effective stress path in Figure 1.17(a) (Hueckel and Pellegrini 1991).

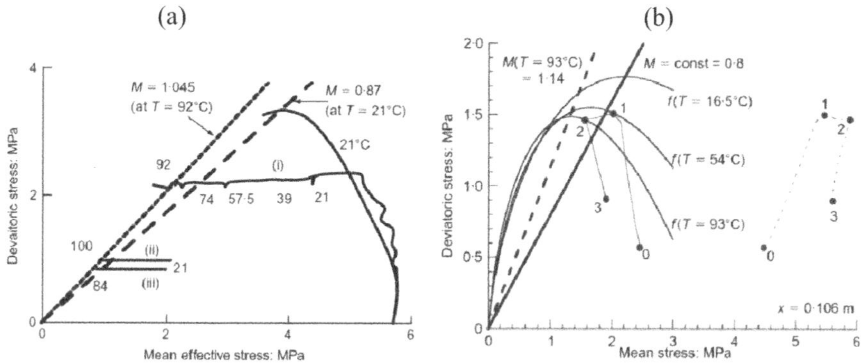

Figure 1.17. *(a) Effective stress paths during undrained heating of Boom clay (Hueckel and Pellegrini 1991). Notably the thermally generated pore pressure caused the effective stress to reach values where their further decrement is statically inadmissible (see section 1.3). This can be compared to Figure 1.10. (b) Simulation of the effective stress path and yield locus evolution for a centrally heated (from 16 to 93°C) Boom clay around an axisymmetric borehole with an increase in the critical state parameter of around 43% (from Hueckel* et al. *2011)*

In a borehole boundary value problem, the problem is exacerbated by the very low permeability of clay, compared to thermal conductivity. Indeed, the effective stress path up to the 0.5 m vicinity of the waste canister approaches the critical states in both considered cases of constant and variable friction angle (Figure 1.17(b); Hueckel *et al.* 2011). As the thermo-elasto-plastic deformation is coupled with heat flow and hydraulic flow, its stability and uniqueness should result from consideration of all three fields.

1.5.3. *Localization during drying of geomaterials*

Cracking of geomaterials during drying is a purely mechanical problem, but highly coupled with the pore fluid flow. As per the definition, drying is a multiphase phenomenon with quite a complex multiphysics, including phase change, capillarity, flow, deformability and (perhaps) water cavitation during the phenomenon of air entry, which per se is a fluid–gas interface instability. In addition, it requires considerations to be made at three scales: macroscale continuum, mesoscale of grain and pore clusters, and microscale of individual pore structure or grains with liquid bridges. Hueckel *et al.* (2014) postulate that drying cracking consists of a series of processes, starting with evaporation of water at the external surface, inducing a negative liquid pressure and flow out from the deformable soil undergoing shrinkage in response. According to Terzaghi (1950), the air invasion takes place when the menisci at the saturated external soil surface reach the size of the biggest pores, but when the soil pores are deformable, it affects air entry. As for the drying cracking, it is postulated that an air finger entering the soil as an instability of the water/air interface (Figure 1.18) constitutes a defect in the soil body, around which a stress concentration arises, when there are external constraints to shrinkage. An amplification of total tensile stress induces a localized high tensile effective stress, higher than the value of suction, resulting in a tensile failure, i.e. crack. Hence, a local value of air entry suction controls the local stress amplification at the defect. The mesoscale linear fracture mechanics analysis yields the stress values in the plausible range of tensile strength. In this scenario, the cracking needs to be considered at a continuum scale, while the air entry is a microscale phenomenon. In addition, the air entry requires a certain threshold suction to develop, which in turn changes the size of the pores. Hence, certain processes involved are sequential.

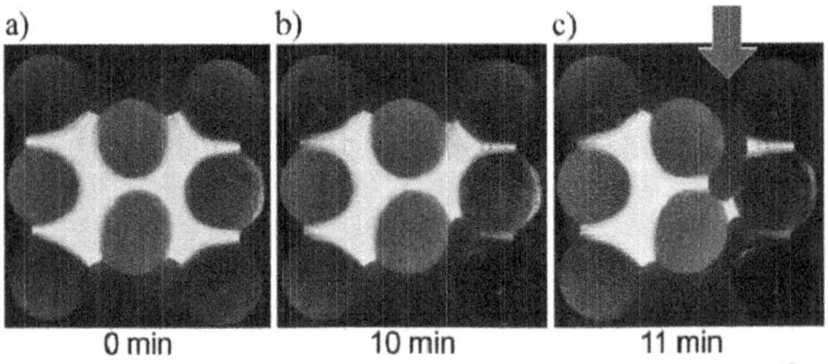

Figure 1.18. *Evolution of the water body between eight glass spheres subjected to evaporation at a constant temperature and constant ambient vapor pressure. The arrow indicates a localized non-symmetric unstable mode of the interface evolution (air entry finger) (from Hueckel* et al. *2014). For a color version of this figure, see www.iste.co.uk/stefanou/instabilities.zip*

There is a host of other similar problems in which multiphysical behavior leads to instabilities, which have not been addressed here due to limited space. They include liquefaction and instability of rock faults and mine pillars related to earthquakes, breakthrough flow due to dissolution of minerals affecting permeability, sinkholes and mine collapses.

1.6. Conclusion

A wide range of failure, instability, non-uniqueness and strain localization phenomena developing in geomaterials have been reviewed in this chapter. It was concluded that in many real life problems, such occurrences result from complex multiphysical fields, including flow of pore water, differential thermal expansion of soil constituents, generation of heat through friction, geochemical reactions, evaporation and air invasion. Mechanical instabilities of the solid skeleton, while no doubt an important part of the overall behavior of geomaterials, do not exhaust the complexity of the overall behavior of such materials. A call for a comprehensive approach to multiphysics instability is more than due. The phenomena involved can be modeled as scenarios of processes that occur either simultaneously or sequentially and that are either coupled or depend on the accumulation of dissipative variables. Hence, the stability of such processes should be investigated as those of coupled mechanical, hydraulic, thermal and chemical processes, or as single processes of a sequence. In the latter case, instability of a one-step process likely affects the formulation of the successor process model. As pointed out by Terzaghi (1950), the causes of the instabilities are often long-term phenomena rather than single events, such as major rainfalls, which are contributing factors. The need for a proper description of these long-term phenomena and their coupling to variable mechanical properties of soil and rock is emphasized.

1.7. References

Bigoni, D. (2012). *Nonlinear Solid Mechanics – Bifurcation Theory and Material Instability*. Cambridge University Press, Cambridge.

Bigoni, D. and Hueckel, T. (1991a). Uniqueness and localization I. Associative and non-associative elastoplasticity. *Int. J. Solids Struct.*, 28(2), 197–213.

Bigoni, D. and Hueckel, T. (1991b). Uniqueness and localization II. Coupled elastoplasticity. *Int. J. Solids Struct.*, 28(2), 215–224.

Borrelli L., Antronico, L., Gullà, G. and Sorriso-Valvo, G.M. (2014). Geology, geomorphology and dynamics of the 15 February 2010 Maierato landslide (Calabria, Italy). *Geomorphology*, 208, 50–73.

Buscarnera, G. and Nova, R. (2011). Loss of controllability in partially saturated soils. In *Bifurcations, Instabilities And Degradations in Geomaterials*, Wan, R., Alsaleh, M. and Labuz, J. (eds). Springer.

Castellanza, R., Gerolymatou, E. and Nova, R. (2009). Experimental observations and modelling of compaction bands in oedometric tests on high porosity rocks. *Strain*, 45(5), 410–423.

Cecinato, F., Zervos, A. and Veveakis, E. (2011). A thermo-mechanical model for the catastrophic collapse of large landslides. *International Journal for Numerical and Analytical Methods in Geomechanics*, 35(14), 1507–1535.

Daouadji, A., Darve, F., Al Gali, H., Hicher, P.Y., Laouafa, F., Lignon, S., Nicot, F., Nova, R., Pinheiro, M., Prunier, F., Sibille, L. and Wan, R. (2011). Diffuse failure in geomaterials: Experiments, theory and modelling. *Int. J. Numer. Anal. Meth. Geomech.*, 35, 1731–1773.

Desrues, J., Chambon, R., Mokni, M. and Mazerolle, F. (1996). Void ratio evolution inside shear bands in triaxial sand specimens studied by computed tomography. *Geotechnique*, 46(3), 529–546.

Drescher, A. (2016). Private communication.

Drescher, A. and De Josselin de Jong, D. (1972). Photoelastic verification of a mechanical model for the flow of a granular material. *Journal of the Mechanics and Physics of Solids*, 20(5), 337–340.

Drucker, D.C. (1964). On the postulate of stability of material in mechanics of continua. *J. Mécanique*, 3, 235–249.

Gattinoni, P., Scesi, L., Arieni, L. and Marco Canavesi, M. (2012). The February 2010 large landslide at Maierato, Vibo Valentia, Southern Italy. *Landslides*, 9, 255–261.

Griffiths, D.V. and Lane, P.A. (1999). Slope stability analysis by finite element. *Geotechnique*, 49(3), 387–403.

Hallbauer, D.K., Wagner, H. and Cook, N.G.W. (1973). Some observations concerning the microscopic and mechanical behaviour of quartzite specimens in stiff, triaxial compression tests. *Int. J. Rock Mech. Min. Sci. Geomech. Abs.*, 10(6), 713–726.

Hicks, M.A. and Spencer, W.A. (2010). Influence of heterogeneity on the reliability and failure of a long 3D slope. *Comput. Geotechnics*, 37(7–8), 948–955.

Hill, R. (1958). A general theory of uniqueness and stability in elastic-plastic solids. *J. Mech. Phys. Solids*, 6, 236.

Hill, R. (1962). Acceleration waves in solids. *J. Mech. Phys. Solids*, 10, 1–16.

Hueckel, T. (1975). On plastic flow of granular and rocklike materials with variable elasticity moduli. *Bulletin of the Polish Academy of Science, Engineering Sciences Series*, 23(8), 405–414.

Hueckel, T. (1976). Coupling of elastic and plastic deformations of bulk solids. *Meccanica*, 4(11), 227–235.

Hueckel, T. (1992). Water-mineral interaction in hygro-mechanics of clays exposed to environmental loads: A mixture theory approach. *Canadian Geotechnical Journal*, 29, 1071–1086.

Hueckel, T. (2002). Reactive plasticity for clays during dehydration and rehydration. Part I: Concepts and options. *International Journal of Plasticity*, 18(3), 281–312.

Hueckel, T. and Maier, G. (1977). Incremental boundary value problems in the presence of coupling of elastic and plastic deformations: A rock mechanics oriented theory. *International Journal of Solids and Structures*, 1, 1–15.

Hueckel, T. and Pellegrini, R. (1991). Thermoplastic modeling of undrained failure of saturated clay due to heating. *Soils & Foundations*, 31(3), 1–16.

Hueckel, T., Francois, B. and Laloui, L. (2009). Explaining thermal failure in saturated clays. *Geotechnique*, 59(3), 197–212.

Hueckel, T., Francois, B. and Laloui, L. (2011). Temperature dependent internal friction of clay in a heat source problem. *Geotechni*que, 61(14), 831–844.

Hueckel, T., Mielniczuk, B., El Youssoufi, M.S., Hu, L.B. and Laloui, L. (2014). A three-scale cracking criterion for drying soils. *Acta Geophysica*, 62(5), 1049–1059.

Iwashita, K. and Oda, M. (1998). Rolling resistance at contacts in simulation of shear band development by DEM. *Journal of Engineering Mechanics–ASCE*, 124(3), 285–292.

Jaeger, J.C. and Cook, N.G.W. (1971). *Fundamentals of Rock Mechanics*. Chapman and Hall, New York.

Karner, S.L., Kronenberg, A.K., Chester, F.M., Chester, J.S. and Hajash, A. Jr. (2008). Hydrothermal deformation of granular quartz sand. *J. Geophys. Res.*, 113, B05404.

Kozicki, J. and Tejchman, J. (2016). DEM investigations of two-dimensional granular vortex- and anti-vortex-structures during plane strain compression. *Granular Matter*, 18(2).

Le, T.M.H, Gallipoli, D., Sanchez, M. and Wheeler, S. (2015). Stability and failure mass of unsaturated heterogeneous slopes. *Canadian Geotechnical J.*, 52(11), 1747–1761.

Maier, G. and Hueckel, T. (1979). Non-associated and coupled flow rules of elastoplasticity for rock-like materials. *International Journal Rock Mechanics and Mining Science*, 16, 77–92.

Mandel, J. (1966). Conditions de stabilité et postulat de Drucker. In *Rheology and Soil Mechanics*, Kravtchenko, J. and Sirieys, P.M. (eds). Springer, Berlin.

McConnell, R.G. and Brock, R.W. (2004). Report on the great landslide at Frank, Alberta, Canada. Department Of Interior Annual Reports, 1902–1908.

Mroz, Z. (1963). Non-associated flow laws in plasticity. *J. de Mechanique*, 2, 21–42.

Muir Wood, D. (1973). Truly triaxial stress-strain behaviour of kaolin. In *Proceedings of the Symposium on the Role of Plasticity in Soil Mechanics*, Cambridge, September 13–15, Palmer, A.C. (ed.) Cambridge University Press, Cambridge.

Paterson, M.S. (1958). Experimental deformation and faulting in Wombeyan Marble. *Bull. Geol. Soc. Amer*, 60, 465–476.

Rice, J.R. (1976). The localization of plastic deformation. In *Theoretical and Applied Mechanics*, Koiter, W.T. (ed.). North-Holland, Amsterdam.

Roscoe, K.H. and Burland, J.B. (1968). The generalized stress-strain behaviour of soft clays. In *Engineering Plasticity*, Heyman, J. and Leckie, F. (eds). Cambridge University Press, Cambridge.

Rudnicki, J.W. and Rice, J.R. (1975). Conditions for localization of deformation in pressure-sensitive dilatant materials. *J. Mech. Ph. Solids*, 25, 371–394.

Stefanou, I. and Sulem, J. (2015). Chemically induced compaction bands: Triggering conditions and band thickness. *Journal of Geophysical Research–Solid Earth*, 119(2), 880–899.

Terzaghi, K. (1950). Mechanism of landslides. Harvard University, Department of Engineering, Cambridge, MA.

Vardoulakis, I., Goldscheider, M. and Gudehus, G. (1978). Formation of shear bands in sand bodies as a bifurcation problem. *International Journal for Numerical and Analytical Methods in Geomechanics*, 2(2), 99–128.

Venturino, P. (2010). Italy 16.02.2010: Impressive landslide caught on video–Italien Erdrutsch auf video. Available: https://www.youtube.com/watch?v=R__3DYQ CVnA.

Veveakis, E., Vardoulakis, I. and Di Toro, G. (2007). Thermoporomechanics of creeping landslides: The 1963 Vaiont slide, northern Italy. *Journal of Geophysical Research–Earth Surface*, 112(F3), F03026.

Veveakis, E., Alevizos, S. and Vardoulakis, I. (2010). Chemical reaction capping of thermal instabilities during shear of frictional faults. *Journal of the Mechanics and Physics of Solids*, 58(9), 1175–1194.

Zhang, Z., Zhang, R., Xie, H., Liu, J. and Were, P. (2015). Differences in the acoustic emission characteristics of rock salt compared with granite and marble during the damage evolution process. *Environmental Earth Sciences*, 73(11), 6987–6999.

Zhao, Y., Cui, P., Hu, L.B. and Hueckel, T. (2011). Multi-scale chemo-mechanical analysis of the slip surface of landslides in the Three Gorges, China. *Science China–Technological Sciences*, 54(7), 1757–1765.

Zienkiewicz, O.C., Taylor, R.L. and Zhu, J.L. (2005). *The Finite Element Method: Its Basis and Fundamentals*, 5th ed. McGraw-Hill, Oxford.

2

Fundamentals of Bifurcation Theory and Stability Analysis

Ioannis STEFANOU[1] and Sotiris ALEVIZOS[2]

[1]*GeM, Ecole Centrale de Nantes, France*

[2]*University of New South Wales, Sydney, Australia*

Geomechanical systems are of particular interest as they involve various multiphysical, nonlinear processes at several length and time scales. These complex mechanisms are described by nonlinear differential equations that express the evolution of the various state variables of a system (e.g. displacements, temperature, pore pressure, etc.). The solution of the governing equations, if it is possible to find, can provide complete information of the system and its behavior in time, but for specific initial and boundary conditions. Bifurcation theory and stability analysis are very useful tools for investigating qualitatively and quantitatively the behavior of complex systems without determining explicitly the solutions of its governing equations for various initial and boundary conditions. This chapter is an introduction to the corresponding mathematical theories. It aims to provide the basic ideas of bifurcation theory and stability analysis, focuses on giving the necessary vocabulary for the classification of equilibria and of common bifurcations that are often met in applications and, finally, presents the application of the theory for studying strain localization in solids. Some aspects related to shear band thickness, mesh dependency and generalized continua are also briefly discussed.

2.1. Introduction

Geomechanical systems are of particular interest as they involve various multiphysical, nonlinear processes that are characterized by several length and time

Instabilities Modeling in Geomechanics,
coordinated by Ioannis STEFANOU and Jean SULEM. © ISTE Ltd 2020.

scales. The inherent complex geomechanical procedures span from the terrestrial kilometric scale to the nanoscale of rock porosity, grain comminution and physicochemical activity, as well as from the geological time scale to the sudden formation of shear bands related to earthquake nucleation, landslides or failure of geotechnical sites.

These complex mechanisms are described by nonlinear differential equations that express the evolution of the state variables of a system in time (e.g. the evolution of displacements, temperature, pore fluid pressure and internal energy). The solution of the differential equations can provide complete information of the system and its behavior in time.

Ideally, we would like to compute directly and in analytical form all the solutions of a differential equation. Unfortunately, this is not possible except in the case of linear equations with constant coefficients or in the case of some special types of nonlinear differential equations. Numerical methods can help us go further and, by using fast computers, it is possible to approximate the solutions of specific initial and boundary value problems. Nevertheless, these are unique solutions, bound to the specific choice of numerical values for the initial and boundary conditions and no further information can be deduced for the spatiotemporal evolution of the system, even for small perturbations of these conditions. Moreover, a universal numerical method that can solve any problem (any system of nonlinear differential equations) does not exist yet. Numerical problems such as non-convergence of the numerical algorithm and inaccurate numerical results are common in practice. Finally, in most of the cases we are not interested in the exact evolution of the complete system, but just in the evolution of some critical state variables or its equilibrium.

It is, therefore, natural to ask if we can investigate the qualitative and quantitative properties of the solutions of a complex system without solving its governing equations analytically or numerically. Stability and bifurcation analyses are the main tools for that.

A complete list of references on bifurcation and stability analysis exceeds the scope of the present chapter. Here, we refer only to some that we find fundamental from a pedagogical point of view. For an introduction to bifurcation theory and stability analysis of general dynamical systems, we refer to Brauer and Nohel (1969), Crowford (1991) and Strogatz (1994). Of course, the pioneering work of Lyapunov (1966, 1892, 1992) is very interesting for further investigating dynamical systems and their stability. Concerning the application of bifurcation theory in solid mechanics and plasticity, we refer, among others, to Bigoni and Hueckel (1991a, 1991b), Lemaitre *et al.* (2009) and Rice (1976). Focusing on geomechanics and multiphysical couplings for classical and generalized continua such as Cosserat, we suggest the following references: Perrin and Leblond 1993; Vardoulakis and Sulem 1995; Benallal and Comi 2000; Bésuelle and Rudnicki 2000; Issen and Rudnicki 2000; Bésuelle 2001; Benallal and Comi 2003; Sulem *et al.* 2011; Veveakis *et al.* 2012, 2013; Stefanou and Sulem 2014.

Despite the various theoretical and mathematical complications related to constitutive modeling, one has to bear in mind that once the equations for the (dynamical) system are established, bifurcation (and stability) analysis is a standard methodology. One needs to identify and solve for certain types of solutions that act as attractors (or repellers) of the system of equations, i.e. "special" types of solutions that, irrespective of the initial data, all other solutions will tend toward (or move away from). These might be time-independent *equilibria* or periodic motions, for instance. In Figure 2.1, we attempt to illustrate this concept. The solid lines (constant solutions) depict the time-independent solutions and the dashed lines are the time-dependent ones. We observe that the latter can hover around (top), deviate (middle) or approach (bottom) the equilibria as time elapses. In this way, one might know where all the solutions will tend to in time without the need to compute them. Furthermore, the bifurcation theory can help us determine whether a particular attractor of a system, which we might not even know explicitly, is the only one, under which conditions (due to parameter variations) it might lose stability and become a repeller or whether other possible solutions can exist.

For instance, consider a homogeneously deformed solid under loading. Bifurcation analysis can indicate the existence of other solutions under a given load and their stability (for example, a localized zone inside the solid, such as a shear band, can develop). In other words, bifurcation theory can help us determine under which conditions a small perturbation of the reference solution (in this example, the homogeneous deformation of the solid) will grow in time (unstable solution leading to strain localization) or not (see Figure 2.1).

It is worth emphasizing that the notion of stability, well established and defined by the original work of Lyapunov (1892) at the end of 19th Century, is related with the time evolution of a system. Even if in common practice time is neglected (quasi-static conditions), the transition from a state (e.g. the homogeneous deformation state) to another one (e.g. the formation of shear bands) happens in a certain time scale, which might be very short (sudden failure of brittle materials) or very slow (geological phenomena). This is why time is central in stability theory, as will be seen in the following sections.

This chapter is organized as follows. In the beginning of section 2.2, we present a simple example of a dynamic mechanical system in order to introduce some basic notions of stability and bifurcation theory. Then, the necessary definitions of stable and unstable equilibria are given. The stability of general linear and nonlinear systems is investigated next. In section 2.3, the dynamics and stability of two-dimensional systems is described. A classification of the various equilibrium points is made. The dynamics can be surprisingly rich, allowing us even to represent Romeo and Juliet's affair (see Love mechanics, section 2.3.2). In section 2.4, we present the most common bifurcations and their classification. The notion of limit cycles is also introduced. All these sections focus on ordinary differential equations (ODEs), which is the key block for studying bifurcation and stability. The study of partial differential equations (PDEs), which is of main importance in geomechanics, is based on the same principles and techniques with ODEs. In section 2.5, we discuss how the study of ODEs is generalized in the case of PDEs. The condition for deformation band formation (such as compaction, shear, dilation bands and their combinations) is retrieved (acoustic tensor) with two different approaches and their stability is discussed. Finally, some aspects related to shear band thickness, mesh dependency and generalized continua are also briefly discussed.

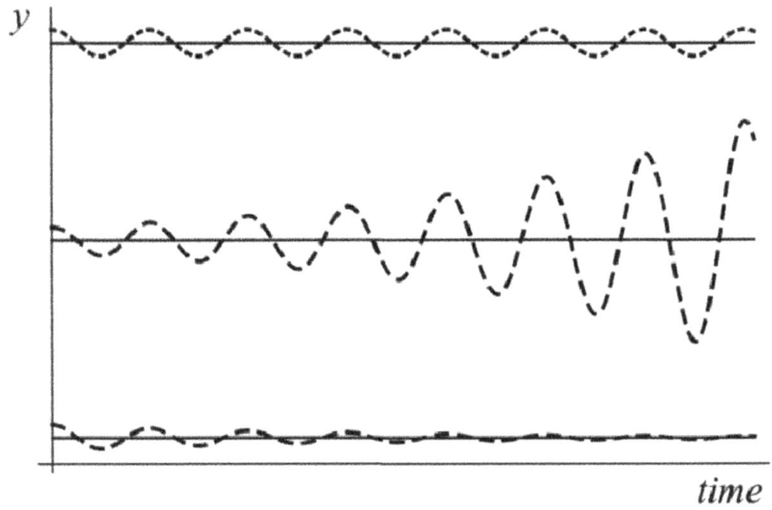

Figure 2.1. *Different types of stability. Solid lines depict the fixed points and dashed lines depict the time evolution of solutions starting from initial conditions near them. From bottom to top, the fixed points are asymptotically stable, unstable and (neutrally) stable*

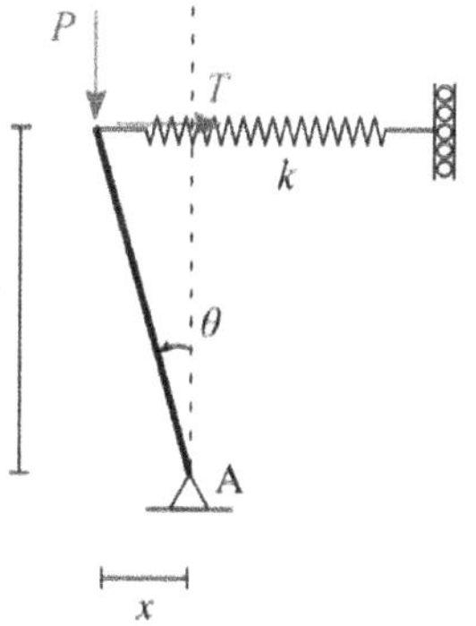

$$I_A \ddot{\theta} = \Sigma M_A = P x - T y$$
$$T = k x$$
$$x = \ell \sin \theta$$
$$y = \ell \cos \theta$$

Figure 2.2. *Spring – rigid beam system. For a color version of this figure, see www.iste.co.uk/stefanou/instabilities.zip*

2.2. Bifurcation and stability of dynamical systems

Let us start with the simple example of the mechanical system shown in Figure 2.2, which is subjected to a vertical force P. The spring coefficient is k, the length of the rigid beam is ℓ and its moment of inertia with respect to the out-of-plane axis passing through point A is I_A. The dynamic behavior of the system is described by the following nonlinear differential equation:

$$I_A \ddot{\theta} = k\ell^2 \sin\theta \left(\frac{P}{k\ell} - \cos\theta \right), \tag{2.1}$$

where θ is the rotation angle around A, as shown in Figure 2.2. The double dot represents the second derivative in time. By setting $\dot{\theta} = \omega$, the above equation can be written in the following equivalent form:

$$\begin{cases} \dot{\theta} = \omega \\ I_A \dot{\omega} = k\ell^2 \sin\theta \left(\dfrac{P}{k\ell} - \cos\theta \right) \end{cases} \tag{2.2}$$

The system is in equilibrium when $\ddot{\theta} = 0$ or equivalently when $\dot{\theta} = 0$ and $\dot{\omega} = 0$. Therefore, in order to be in equilibrium, either $\cos\theta_0 = P/k\ell$ or $\theta_0 = n\pi$, where $n \in \mathbb{Z}$. Figure 2.3 shows all the possible values of the angle θ for which equilibrium is possible for given $P^* = P/k\ell$. Points B_i are called bifurcation points of the (equilibrium) solutions. The diagram of Figure 2.3 is called a *bifurcation* *diagram* and P^* is the bifurcation parameter. Depending on the problem at hand, various bifurcation parameters can be selected. The bifurcation diagram is a very useful tool for presenting the possible equilibria or steady states of a system (mechanical, chemical, geomechanical, etc.). More details about bifurcation types and bifurcation diagrams are given in section 2.4. It is worth mentioning that for a given value of P^*, we may have several equilibrium solutions. However, some equilibria might be stable and some others might be unstable, in the sense that, if we are in a certain equilibrium and a tiny perturbation takes place (a fly that sits on the beam!), the system will stay close to its initial equilibrium (stable equilibrium) or it will diverge away from it (unstable equilibrium, see Figure 2.1). But how is stability rigorously defined and how can we assess it for a given system?

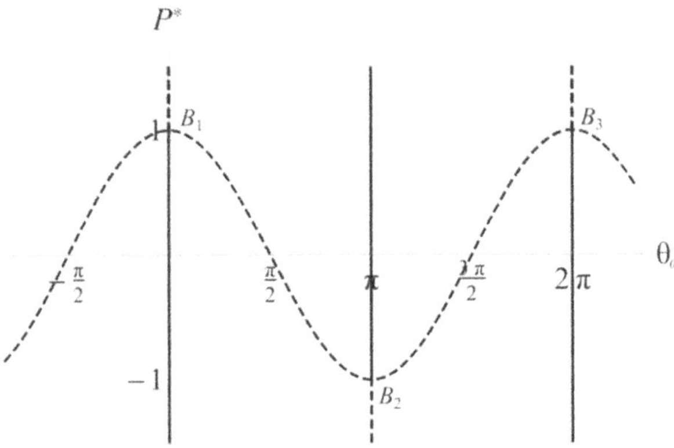

Figure 2.3. *Bifurcation diagram. Dashed lines represent unstable braches and solid stable*

2.2.1. *Definition of stability*

Let us consider a physical system, which is described by the following ODEs (set of first-order ordinary differential equations):

$$\dot{\underline{y}} = \underline{f}(\underline{y}) \qquad [2.3]$$

$\underline{y}$ is a vector of n components that contains the various quantities that determine the evolution of the physical system. The dot represents again the time derivative and $\underline{f}$ is a vector function that does not depend explicitly on the independent variable, which is the time t (autonomous system). $\underline{f}$ belongs to $C^1(D)$ ($\underline{f} \in C^1(D)$), which assures the existence and uniqueness of the initial value problem defined by equation [2.3]. D is the n-dimensional real Euclidean space over which $\underline{f}$ is defined and C^1 denotes that $\underline{f}$ and its derivatives, in terms of the components of $\underline{y}$, are continuous. The existence and uniqueness of solutions of the initial value problem does not mean that the system only has one equilibrium point. It means that for given initial conditions, the system follows a unique trajectory. In other words, it can be proven that the response of the initial value problem, even if it is very sensitive to initial conditions (chaotic behavior), has a unique evolution in time as long as $\underline{f} \in C^1(D)$, although various equilibria points (or steady states) might exist, as shown in the previous example (Figure 2.2 and Figure 2.3). In practice, when we use the term loss of uniqueness (see Chambon *et al.* (2004) for a discussion) of solutions, we refer to the existence of several different equilibrium solutions that satisfy $\underline{f}(\underline{y}_0) = 0$. $\underline{y}_0$ are called *fixed points*.

The important question, as far as applications are concerned, is if a certain equilibrium is stable or not. In other words, if at time t_0 we are in equilibrium ($\dot{\underline{y}}_0 = \underline{f}(\underline{y}_0) = 0$) and a tiny perturbation $\tilde{\underline{\psi}}$ takes place such as $\underline{\psi} = \underline{y}_0 + \tilde{\underline{\psi}}$, do we return to the initial equilibrium, $\underline{y}_0$, or does the system diverge to another state? Lyapunov (1966, 1892, 1992) introduced the following definitions of stability.

DEFINITION 2.1.– *The equilibrium solution $\underline{y}_0$ is said to be stable if for each number $\varepsilon > 0$ we can find a number $\delta > 0$ (depending on ε) such that if $\underline{\psi}(t)$ is any solution of equation [2.3] having $\left\| \underline{\psi}(t_0) - \underline{y}_0 \right\| < \delta$, then the solution $\underline{\psi}(t)$ exists for all $t \geq t_0$ and $\left\| \underline{\psi}(t) - \underline{y}_0 \right\| < \varepsilon$ for $t \geq t_0$ (see Figure 2.1, top).*

$\|\cdot\|$ denotes here the Euclidian norm ($\left\| \underline{a} \right\| = \sqrt{a_1^2 + a_2^2 + \ldots + a_n^2}$).

DEFINITION 2.2.– *The equilibrium solution $\underline{y}_0$ is said to be asymptotically stable if it is stable and if there exists a number $\delta_0 > 0$ such that if $\underline{\psi}(t)$ is any solution of equation [2.3] having $\left\| \underline{\psi}(t_0) - \underline{y}_0 \right\| < \delta_0$, then $\lim_{t \to +\infty} \underline{\psi}(t) = \underline{y}_0$ (see Figure 2.1, bottom).*

DEFINITION 2.3.– *The equilibrium solution $\underline{y}_0$ is said to be unstable if it is not stable (see Figure 2.1, middle).*

2.2.2. *Linear systems of ODEs*

The simplest dynamical system for which stability questions can be easily addressed is the following first-order linear system of ODEs:

$$\underline{\dot{y}} = \underline{\underline{A}}\ \underline{y}, \tag{2.4}$$

where $\underline{\underline{A}}$ is a real constant $n \times n$ matrix. $\underline{y}_0 = 0$ is the equilibrium solution.

We would like to determine its evolution in time. In other words, if initially the system was in equilibrium, how would a small, tiny perturbation evolve over time? To answer this question, we can determine the general solution of the system, a task that is possible because of its linearity.

A particular solution of the above system is $\underline{\psi}(t) = \underline{\eta} e^{st}$, where $\underline{\eta}$ is a vector of constants with $\underline{\eta} \neq 0$. Injecting this form into equation [2.4], we obtain:

$$\left(\underline{\underline{A}} - s\underline{\underline{I}}\right)\underline{\eta} = 0 \tag{2.5}$$

As $\underline{\eta}$ is not the zero vector, the above equation is satisfied for s such that the determinant $\det\left(\underline{\underline{A}} - s\underline{\underline{I}}\right) = 0$. Equation [2.5] defines an eigenvalue problem, which has n eigenvalues, $s^{(i)}$, and n associated eigenvectors $\underline{\eta}^{(i)}$. The calculation of the determinant results in a polynomial of n degree in terms of s, which is called a *characteristic polynomial* and whose roots are called *eigenvalues* $s^{(i)}$. If the eigenvalues of the system are distinct (no repeated eigenvalues, called *simple eigenvalues*), the general solution of this ODE system is:

$$\underline{y}(t) = \sum_{i=1}^{n} c_i \underline{\eta}^{(i)} e^{s^{(i)} t}, \tag{2.6}$$

where c_i are constants that are determined by the initial conditions of the problem. The eigenvalues of the system can be real or imaginary. The imaginary part is responsible for an oscillatory behavior of the system, while the real part is related to stability. If one of the eigenvalues is positive, then equation [2.6] indicates that the solution of the system will increase exponentially in time (monotonously increasing term).

If the characteristic polynomial has p distinct eigenvalues (roots) $s^{(i)}$ $(1 \leq i \leq p)$ with multiplicity $m^{(i)}$ each one (if the eigenvalue k is simple, then $m^{(k)} = 1$) and associated eigenvectors $\underline{\eta}^{(i)}$, then it can be shown that the general solution of the ODE system is:

$$\underline{y}(t) = \sum_{i=1}^{p} \sum_{j=1}^{m^{(i)}} c_{i,j} \underline{\eta}^{(i)} t^{j-1} e^{s^{(i)} t}, \tag{2.7}$$

where again $c_{i,j}$ are n, in total, constants that are determined by the initial conditions of the problem. For example, if the system consists of $n=3$ ODEs and it has only two distinct eigenvalues (one of the eigenvalues has multiplicity 2), then its general solution is: $\underline{y} = c_{1,1}\underline{\eta}^{(1)} e^{s^{(1)}t} + c_{2,1}\underline{\eta}^{(i)} e^{s^{(2)}t} + c_{2,2}\underline{\eta}^{(i)} t e^{s^{(2)}t}$. Note the term $te^{s^{(2)}t}$, which is strictly increasing in a region close to $t=0$, $0 \le t \le \varepsilon$, even if $s^{(2)} \le 0$.

By combining the aforementioned definitions of stability and the behavior of the solutions of equation [2.4], the following theorem can be proven (Brauer and Nohel 1969):

THEOREM 2.1.–

– If all eigenvalues of $\underline{\underline{A}}$ have non-positive real parts and all those eigenvalues with zero real parts are simple, then the zero solution $\underline{y}_0 = 0$ of equation [2.4] is stable.

– If (and only if) all eigenvalues of $\underline{\underline{A}}$ have negative real parts, the zero solution of equation [2.4] is asymptotically stable.

– If one or more eigenvalues of $\underline{\underline{A}}$ have a positive real part, the zero solution of equation [2.4] is unstable.

In other words, the stability of the equilibrium state of a linear system is investigated by simply studying the eigenvalues of the matrix $\underline{\underline{A}}$.

Can this theorem be extended for nonlinear systems like as the one presented at the beginning of this section?

2.2.3. *Nonlinear systems of ODEs*

The system described by equation [2.3], $\dot{\underline{y}} = \underline{f}(\underline{y})$, is nonlinear in the sense that $\underline{f}(\underline{y})$ is a nonlinear function of $\underline{y}$. Expressing its solution $\underline{\psi}(t)$ in the form:

$$\underline{\psi}(t) = \underline{y}_0 + \underline{\tilde{\psi}}(t), \tag{2.8}$$

where $\underline{y}_0$ is one of the equilibrium solutions (fixed point), we obtain:

$$\dot{\underline{\tilde{\psi}}}(t) = \underline{f}(\underline{y}_0 + \underline{\tilde{\psi}}(t)) = \underline{f}(\underline{y}_0 + \underline{\tilde{\psi}}(t)) - \underline{f}(\underline{y}_0) \quad [2.9]$$

If the difference at the right-hand side can be written in the following almost-linear form:

$$\dot{\underline{\tilde{\psi}}}(t) = \underline{\underline{A}}\,\underline{\tilde{\psi}} + \underline{p}(\underline{\tilde{\psi}}) \ , \quad [2.10]$$

where $\underline{\underline{A}} = \underline{\underline{J}}(\underline{y}_0) = \left\{ \left. \frac{\partial f_i}{\partial y_j} \right|_{\underline{y}=\underline{y}_0} \right\}$ the Jacobian of $\underline{f}(\underline{y})$ at point $\underline{y}_0$ ($\underline{\underline{A}}$ is a real constant $n \times n$ matrix), $\underline{p}$ a continuous function with $\underline{p}(\underline{0}) = \underline{0}$ and $\lim_{\underline{\psi} \to 0} \frac{\left\| \underline{p}(\underline{\tilde{\psi}}) \right\|}{\left\| \underline{\tilde{\psi}} \right\|} = 0$, then the following theorem can be proven (Brauer and Nohel 1969).

THEOREM 2.2.– Suppose that $\underline{p}$ is continuous, $\left\| \underline{\tilde{\psi}} \right\| < k$, where $k > 0$ is a constant, and $\underline{p}$ is small in the sense that $\lim_{\underline{\psi} \to 0} \frac{\left\| \underline{p}(\underline{\tilde{\psi}}) \right\|}{\left\| \underline{\tilde{\psi}} \right\|} = 0$, then:

– if all eigenvalues of $\underline{\underline{A}}$ have negative real parts, the solution $\underline{\tilde{\psi}} = 0$ of equation [2.10] is asymptotically stable;

– if one or more eigenvalues of $\underline{\underline{A}}$ have a positive real part, the solution $\underline{\tilde{\psi}} = 0$ of equation [2.10] is unstable.

Note that if the second derivative of $\underline{f}$ with respect to $\underline{y}$ exists, then the term $\underline{p}(\underline{\tilde{\psi}})$ is the remainder of a Taylor expansion of $\underline{f}$, which satisfies $\underline{p}(\underline{0}) = \underline{0}$ and $\lim_{\underline{\psi} \to 0} \frac{\left\| \underline{p}(\underline{\tilde{\psi}}) \right\|}{\left\| \underline{\tilde{\psi}} \right\|} = 0$. If all eigenvalues of $\underline{\underline{A}}$ have non-positive real parts and there exists at least one eigenvalue with zero real part, then the dynamics of the linearized system do not represent the dynamics of the nonlinear system and no conclusion can be safely derived for the stability of the nonlinear system. However, in the special case of conservative (systems where a conserved quantity exists, e.g. the total energy) or reversible systems (systems with time reversal symmetry), it can be proven that when all the eigenvalues of $\underline{\underline{A}}$ have non-positive real parts and there exists at least one eigenvalue with zero real part, then all orbits close to a fixed point are closed (see Strogatz 1994). In this case, the (isolated) fixed point is called the nonlinear center and is stable in the Lyapunov sense (but not asymptotically stable).

The above theorem gives the conditions for which any perturbation $\underline{\tilde{\psi}}$ is bounded, decays or grows exponentially with time. According to the definitions of stability, the system will be (*asymptotically*) *stable* or *unstable*. Therefore, the eigenvalues of the matrix $\underline{\underline{A}}$ can provide useful information about the stability of an equilibrium solution, even in the case of nonlinear ODEs. The investigation of stability by using the above theorem is called linear stability analysis (LSA), as it is based on the linearization of $\underline{f}(\underline{y})$.

2.2.4. *An example of LSA*

The system presented in the beginning of this section (equation [2.2]) is expressed in the form of equation [2.3] as follows:

$$\underline{y} = \begin{bmatrix} \theta \\ \omega \end{bmatrix} \text{ and } \underline{f} = \begin{bmatrix} \omega \\ \dfrac{k\ell^2}{I_A} \sin\theta \left(\dfrac{P}{k\ell} - \cos\theta \right) \end{bmatrix} \qquad [2.11]$$

At equilibrium, $\underline{\dot{y}} = 0$ and $\underline{y} = \underline{y}_0$. Perturbing the equilibrium solution, we replace $\underline{y}(t)$ by $\underline{\psi}(t) = \underline{y}_0 + \underline{\tilde{\psi}}(t)$ (equation [2.8]). Performing a Taylor expansion of $\underline{f}$ up to the first order around the point $\underline{y} = \underline{y}_0$, we retrieve a linear equation of the form of [2.10] where: $\underline{\underline{A}} = \underline{\underline{J}}(\underline{y}_0) = \left\{ \left. \dfrac{\partial f_i}{\partial y_j} \right|_{\underline{y}=\underline{y}_0} \right\} =$

$$\begin{bmatrix} 0 & 1 \\ \dfrac{k\ell^2}{I_A}\left(\dfrac{P}{k\ell} - \cos\theta_0 \right)\cos\theta_0 + \dfrac{k\ell^2}{I_A}\sin^2\theta_0 & 0 \end{bmatrix}$$

The characteristic polynomial of the eigenvalue problem is:

$$s^2 - \frac{k\ell^2}{I_A}\left(\frac{P}{k\ell} - \cos\theta_0 \right)\cos\theta_0 - \frac{k\ell^2}{I_A}\sin^2\theta_0 = 0 \qquad [2.12]$$

which leads to two eigenvalues:

$$s_{1,2} = \pm\sqrt{\frac{k\ell^2}{I_A}\left(\frac{P}{k\ell} - \cos\theta_0 \right)\cos\theta_0 + \frac{k\ell^2}{I_A}\sin^2\theta_0} \qquad [2.13]$$

Now we can investigate the stability of the various branches of the bifurcation diagram (Figure 2.3). When we are on the sinusoidal branch $\frac{P}{k\ell}-\cos\theta_0=0,$, and therefore $s_{1,2}=\pm|\sin\theta_0|\sqrt{\frac{k\ell^2}{I_A}}$, which means that there is always a positive eigenvalue (imaginary part is zero). When we are on the vertical branches $\sin\theta_0=0$ and $s_{1,2}=\pm\sqrt{\frac{k\ell^2}{I_A}\left(\frac{P}{k\ell}-1\right)}$ for $\theta_0=2n\pi$ or $s_{1,2}=\pm\sqrt{-\frac{k\ell^2}{I_A}\left(\frac{P}{k\ell}+1\right)}$ for $\theta_0=(2n+1)\pi$. In the first case, i.e. for $\theta_0=2n\pi$, if $P>k\ell$, then one of the eigenvalues is positive (the imaginary part is zero), which means that the system is unstable. If $P<k\ell$, then $s_{1,2}=\pm\mathbf{i}\sqrt{\frac{k\ell^2}{I_A}\left|\frac{P}{k\ell}-1\right|}$, which are two distinct imaginary eigenvalues ($\mathbf{i}=\sqrt{-1}$) and consequently, according to section 2.2.3, the equilibrium is (neutrally) stable (conservative and reversible system). In the second case, i.e. for $\theta_0=(2n+1)\pi$, if $P<-k\ell$, then one of the eigenvalues is positive (the imaginary part is zero), which means that the system is unstable. Figure 2.3 summarizes these results in the bifurcation diagram. If $P=\pm k\ell$ and $\theta_0=n\pi$, then $s_{1,2}=0$ and no conclusion can be drawn about the stability of these points.

2.3. Stability of two-dimensional linear dynamical systems

The general form of a two-dimensional linear system is:

$$\begin{aligned}\dot{y}_1&=a\,y_1+b\,y_2\\ \dot{y}_2&=c\,y_1+d\,y_2\end{aligned} \qquad [2.14]$$

The equilibrium solution of this system (fixed point) is obviously $\underline{y}_0=0$. The constants matrix is $\underline{\underline{A}}=\begin{bmatrix}a & b\\ c & d\end{bmatrix}$ and the characteristic polynomial:

$$s^2-\tau s+\Delta=0, \qquad [2.15]$$

where $\tau = a + d$ and $\Delta = ad - bc$. Let s_1 and s_2 be the roots of the characteristic polynomial (eigenvalues of $\underline{\underline{A}}$) :

$$s_1 = \frac{\tau + \sqrt{\tau^2 - 4\Delta}}{2}, \qquad s_2 = \frac{\tau - \sqrt{\tau^2 - 4\Delta}}{2} \qquad [2.16]$$

2.3.1. *Classification of fixed points*

Take, for instance, the following case for $\underline{\underline{A}} = \begin{bmatrix} a & 0 \\ 0 & -1 \end{bmatrix}$. The eigenvalues are $s_1 = a$ and $s_2 = -1$. Let a also be real. The solution of this linear system is $y_1 = c_1 e^{at}$, $y_2 = c_2 e^{-t}$ (see section 2.2.2). The initial conditions determine the constants c_1 and c_2. Plotting this solution in the (phase) space (y_1, y_2), we obtain the trajectories presented in Figure 2.4 for various initial conditions. Such a diagram is called a phase diagram and it depicts the dynamics or the so-called mathematical flow of the system. From the solution of this system, we get $\frac{dy_1}{dy_2} = -a\frac{c_1}{c_2}e^{(a-1)t}$, which illustrates that the dynamic behavior of the system evolves (is concentrated) in the direction with the slowest in absolute value eigenvalue (slow eigen-direction of the linearized system). In other words, for $a < -1$, y_2 reduces faster than y_1 and the solution approaches the equilibrium point having as an asymptote the axis y_2 (see Figure 2.4(a)). The direction of the slowest evolution of the system is called the *slow manifold* (slow manifold of an equilibrium point of a dynamical system). The contrary holds for $-1 < a < 0$ (see Figure 2.4(b)). In both cases, the equilibrium point is an *attractor* (y_1 and y_2 are stable manifolds) and the fixed point is called the *stable node*. If both eigenvalues are real and positive, then the fixed point is called the *unstable node*. If $a = 0$, then y_2 is constant ($y_1 = c_1$) and the system evolves as shown in Figure 2.4(c). In the case that $a > 0$, the system is unstable and the equilibrium point is a *saddle node* (Figure 2.4(d)). For initial conditions such that $c_1 = 0$, the system will evolve toward the equilibrium point. However, for tiny values of c_1, the system will diverge from the equilibrium point (y_1 is a stable manifold and y_2 unstable – saddle).

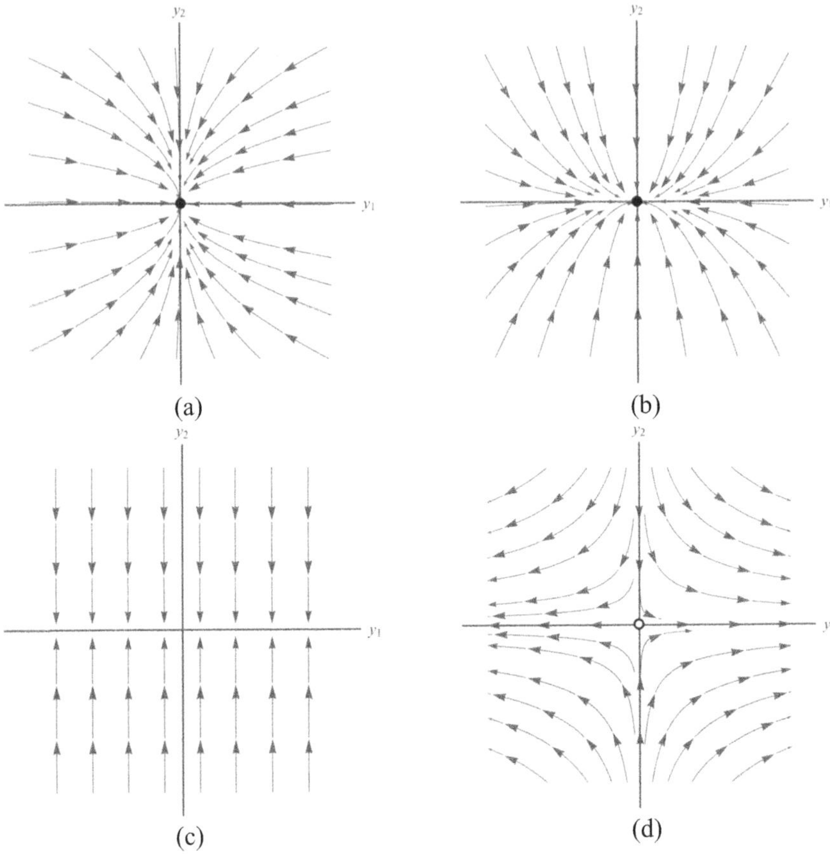

Figure 2.4. *Phase diagram for (a) $a<-1$, (b) $-1<a<0$, (c) $a=0$ and (d) $a>0$. A black dot indicates a stable fixed point (stable node, attractor), while an open circle indicates an unstable one (unstable node-repeller or saddle)*

In the general case (see Figure 2.5) of real eigenvalues, the manifolds (eigenvectors) are not perpendicular (except if $\underline{\underline{A}}$ is symmetric). If the real part of both eigenvalues of $\underline{\underline{A}}$ is zero and they have different non-zero imaginary parts, the equilibrium is neutrally stable (stable in the Lyapunov sense for a linear system; see definitions in sections 2.2.2 and 2.2.3). If the eigenvalues are complex with negative real part, then we have oscillations of reducing amplitude until equilibrium (Figure 2.6). If the real part is positive and the imaginary part is non-zero, then we diverge from equilibrium (oscillations with increasing amplitude, see also Figure 2.1). In the case of repeated but non-null eigenvalues, if the eigenvectors are distinct, then we have a star node, and when there is only one eigenvector the fixed point is called degenerated node (Figure 2.7).

Figure 2.8 summarizes the various types of fixed points in function of τ and Δ. In the bifurcation example of the previous section, $\tau = 0$. Therefore, we had either saddle points (unstable equilibria) or centers (neutrally stable equilibria).

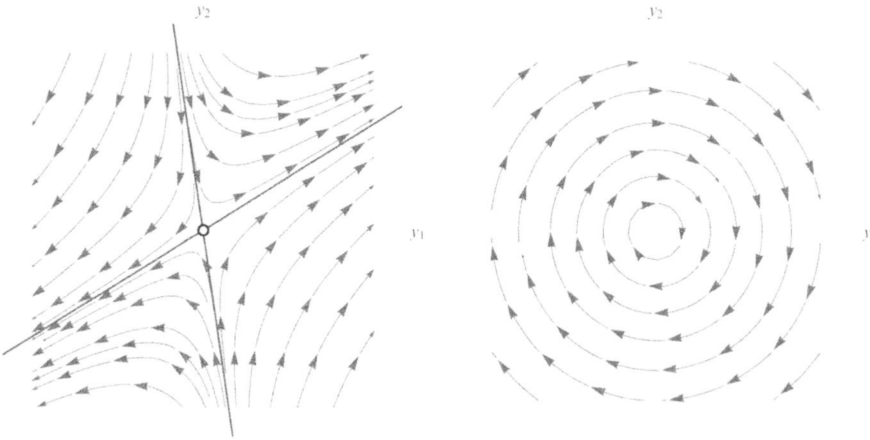

(a) (b)

Figure 2.5. *(a) Unstable fixed point with non-orthogonal eigenvectors (saddle), (b) neutrally stable fixed point (center)*

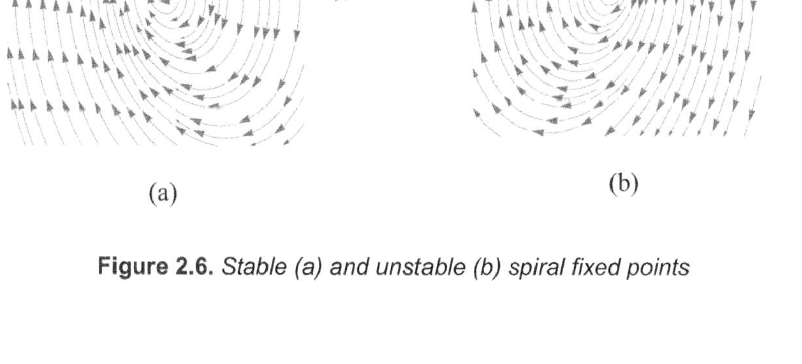

(a) (b)

Figure 2.6. *Stable (a) and unstable (b) spiral fixed points*

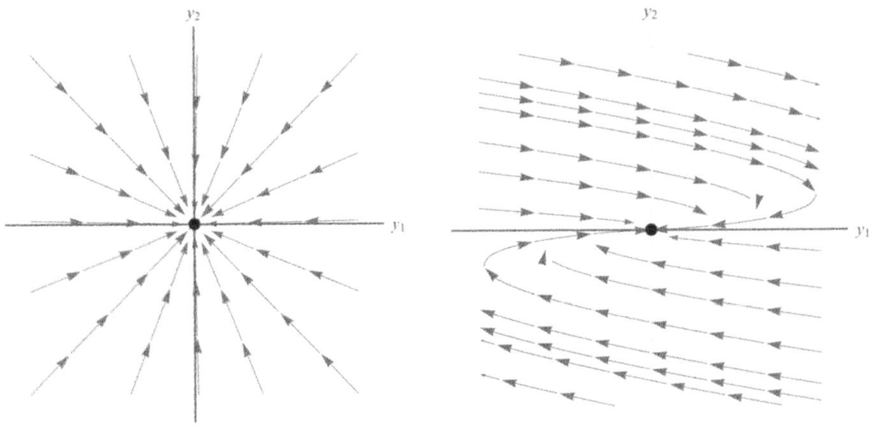

(a) (b)

Figure 2.7. *Degenerate cases: (a) star node and (b) degenerate node*

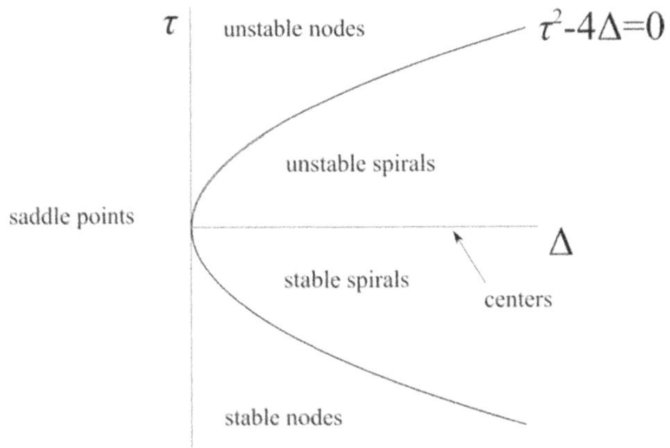

Figure 2.8. *Classification of fixed points of a two-dimensional dynamical systems*

2.3.2. *Love mechanics: Romeo and Juliet*

Two-dimensional systems are certainly more interesting than one dimensional systems. While one-dimensional systems can have nodes that are either stable or unstable and the solution might simply diverge or converge toward the equilibrium points, two-dimensional systems involve richer dynamics, such as oscillations.

Strogatz (see Strogatz 1988, 1994; Sprott 2004) used a simple linear two-dimensional system to describe the romantic affair between Romeo and Juliet. In his example, y_1 describes the love of Romeo for Juliet (R) and y_2 describes the love of Juliet for Romeo (J):

$$\begin{aligned} \dot{R} &= a\,R + b\,J \\ \dot{J} &= c\,R + d\,J \end{aligned} \qquad [2.17]$$

Positive values for J or R signify love and negative hate. $J=R=0$ signifies mutual indifference. The parameters a,b,c,d have a simple meaning as well. $a<0$ means that Romeo is a cautious lover. The more he realizes that he loves Juliet the more he is afraid, which reduces his affection. But if $b>0$ is great enough, the affection of Juliet makes his feelings stronger ($\dot{R}>0$). Similarly, Juliet's love is characterized by the parameters c,d.

Let us investigate the scenario with $a=d=0$, $b>0$ and $c<0$. Let us say that at the beginning Romeo is in love with Juliet and also Juliet with him. The more Romeo loves her ($R\uparrow$), the more Juliet is afraid and wants to run away ($\dot{J}=c\,R<0 \Rightarrow J<0$). Romeo gets disappointed and backs off ($\dot{R}=b\,J<0 \Rightarrow R<0$). But then Juliet begins to find him again attractive ($\dot{J}=c\,R>0 \Rightarrow J>0$) and after a while Romeo echoes her ($\dot{R}=b\,J>0 \Rightarrow R>0$). But then again Juliet loses her interest and so on. The sad outcome of this relation is a never ending cycle of love and hate (see Figure 2.5(b)). If Romeo was also a cautious lover ($a<0$), then their love would end soon following a stable spiral toward the fixed point of mutual indifference $J=R=0$. If Romeo was an enthusiastic lover ($a>0$), their relation would be an unstable spiral of increasing hate and love, but again if $a>0$ is very high, the more Romeo loves Juliet the more she is afraid and draws away. However, if both Romeo and Juliet were attracted in the beginning, their love would become a love fest provided that their love characteristics a,b,c,d were such that there is an unstable node or saddle in the phase space (see Figure 2.5(b)). In other words, love is instability.

Of course, in reality the dynamics of love are much more complicated than this simple model and certainly nonlinear. Nonlinearities and more degrees of freedom were taken into account in Sprott (2004) and interesting dynamics were observed.

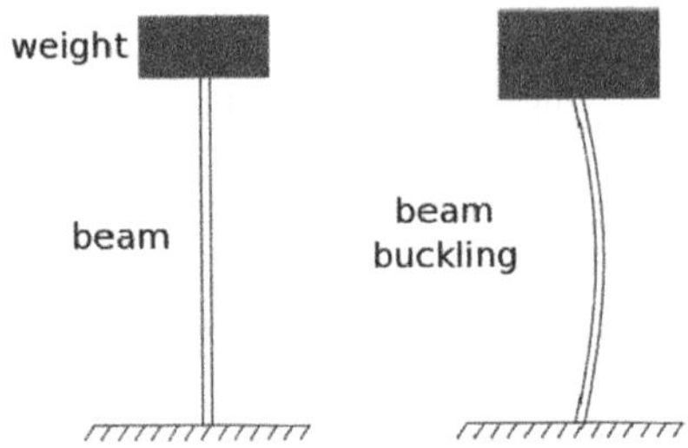

Figure 2.9. *Sketch of beam buckling due to high load. Here, the load acts as a control (bifurcation) parameter, which determines the transition from axial deformation to a buckled, flexural state*

2.4. Common types of bifurcations

The evolution laws that we commonly use in mechanics (and other scientific disciplines) can involve parameters that are unknown or non-constant (e.g. the loading of a beam or the elastic parameter of a spring). The dynamics of a system can significantly change with the variations of these parameters. For instance, in the example of section 2.2, the fixed point $\theta_0 = 0$ becomes unstable for values of the normalized vertical load $P^* > 1$ (see Figure 2.3). This is typical in many mechanical systems and central for the design of structures (see buckling of a beam due to high load, Figure 2.9).

In general, depending on the values of the parameters, fixed points can be created (appear) or destroyed (disappear) and/or their stability can change. These qualitative changes in the static and dynamic response of a system of equations are called *bifurcations* and the parameter values at which they first occur are called *bifurcation points*. The study of bifurcations is important since it provides the onset of instabilities and the transition across different states depending on the variation of the governing (bifurcation) parameters. In the following, we will go through some of the most well-known (mathematical) bifurcation types for ODEs.

2.4.1. *Saddle-node bifurcation*

The most fundamental bifurcation is the appearance and disappearance of equilibrium points for different values of a bifurcation parameter. This is the so-called *saddle-node* (or *fold*) *bifurcation* of equilibria. In this case, as the parameter varies, two fixed points of the underlying system of equations move toward each other, collide and mutually annihilate. The following differential equation is a classic example of this kind of bifurcation:

$$\dot{y} = \mu + y^2, \qquad [2.18]$$

where μ is a real number that can admit any real value. When μ is negative, the right-hand side of equation [2.18] (equilibrium solution) has two fixed points, one stable and one unstable (see Figure 2.10).

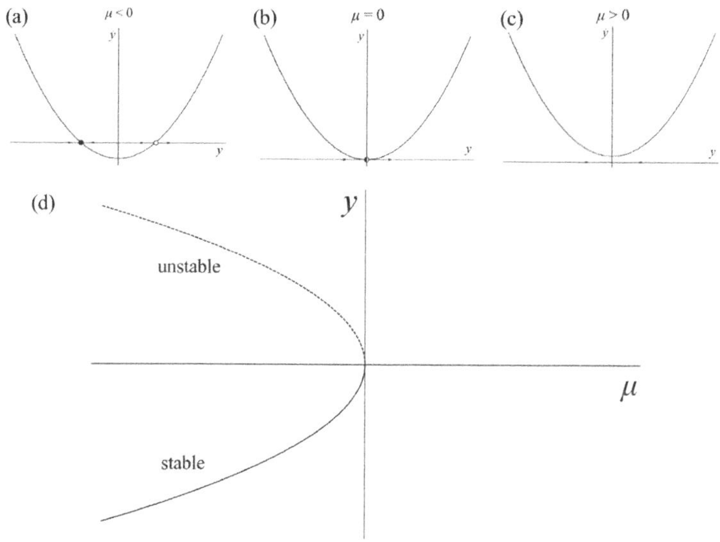

Figure 2.10. *Saddle-node bifurcation. A half full circle denotes half stable fixed points*

As $\mu \to 0$ from negative values, the two fixed points move toward each other and they collide to a half-stable point at $y_0 = 0$ for $\mu = 0$. This is the bifurcation point, since for $\mu > 0$ the dynamics change completely, equation [2.18] presents no fixed points and predicts infinite growth of y in time. This behavior is depicted in Figure 2.10(a–c). As μ plays the role of the independent variable, we can plot the steady-state solutions of equation [2.18] in a $\mu - y_0$ diagram (Figure 2.10(d)), where we can observe the number of steady-state solutions in function of the parameter μ, as well as their stability (calculated as described in section 2.2). As mentioned in section 2.2, such a diagram, where the fixed points of the equations (for a norm of the solution) are plotted against the bifurcation parameter is called a bifurcation diagram.

Note that equation [2.18] along with its symmetric $\dot{y} = \mu - y^2$ are representative of all saddle-node bifurcations. This means that close to a saddle bifurcation, the dynamics of a given system are qualitatively the same with $\dot{y}=\mu-y^2$ or $\dot{y}=\mu+y^2$ (see section 2.7). Equations like equation [2.18], which can characterize the dynamics of any system near a bifurcation point, are called *normal forms* of that bifurcation.

2.4.2. *Transcritical bifurcation*

The transcritical bifurcation happens when a pair of fixed points exchange stability as the bifurcation parameter varies. Its normal form is:

$$\dot{y}=\mu y-y^2 \quad [2.19]$$

As shown in Figure 2.11(a–c), the point $y_0=0$ is always a fixed point. Starting from negative values of μ, $y_0=0$ is stable and there exists a second fixed point, $y_0=\mu$, which is unstable. As the value of the parameter μ increases, the second fixed point moves toward $y_0=0$ and for $\mu=0$ (which is also the bifurcation point in this case) they collapse on a half-stable point. Upon further increasing of the parameter value, the two fixed points reappear but have opposite stability, $y_0=0$ is unstable and $y_0=\mu$ is stable. Thus, we can say that the transcritical bifurcation is a mechanism for exchanging stability between two fixed points. This is more apparent if one observes the corresponding bifurcation diagram (Figure 2.11(d)).

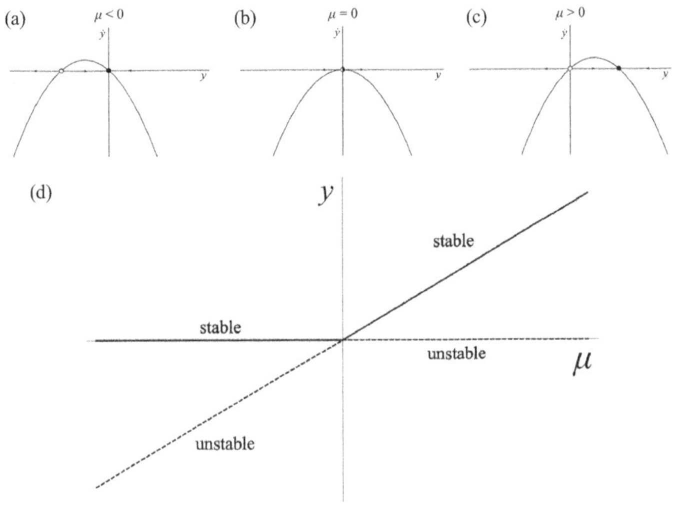

Figure 2.11. *Transcritical bifurcation*

2.4.3. *Supercritical and subcritical pitchfork bifurcation*

Pitchfork bifurcation (both the super- and the subcritical one) is common in problems that have symmetry and describes the appearance (or disappearance) of a symmetrical pair of fixed points after some critical value of the bifurcation parameter.

Revisiting the example of the buckling of a beam (see Figure 2.2 or Figure 2.9), after the load exceeds the critical threshold there exists no preferred direction for the deformation and only a defect in the mechanical problem or the perturbation itself can lead the beam to "choose" one direction or another, thus breaking its symmetry.

The normal form of the supercritical pitchfork bifurcation is:

$$\dot{y} = \mu y - y^3 \qquad [2.20]$$

Note that changing the variable $y \to -y$ does not change the equation of the system. This symmetry justifies mathematically the aforementioned existence of a symmetrical pair of fixed points.

The fixed point $y_0 = 0$ exists for all $\mu \in \mathbb{R}$. For $\mu \leq 0$, it is stable. At $\mu = 0$, the pitchfork bifurcation occurs and for $\mu > 0$ a symmetric pair of stable fixed points appears ($y_0 = \pm\sqrt{\mu}$). Therefore, three fixed points exist ($y_0 = \pm\sqrt{\mu}$ and $y_0 = 0$) for $\mu > 0$.

The reason for which this type of bifurcation is called "pitchfork" becomes apparent upon observing the bifurcation diagram in Figure 2.12(d).

The subcritical pitchfork bifurcation has the following normal form:

$$\dot{y} = \mu y + y^3 \qquad [2.21]$$

As shown in Figure 2.13, the corresponding bifurcation diagram is similar to the one of Figure 2.12(d), but the pitchfork is inverted. The pair $y_0 = \pm\sqrt{-\mu}$ is unstable and it exists only for $\mu < 0$. In addition, even though the origin is a fixed point for all $\mu \in \mathbb{R}$, it is stable only for $\mu < 0$. For all $\mu > 0$, there is no stable equilibrium solution and the system *blows up*, i.e. $y \to \pm\infty$ as time elapses. Furthermore, one can show that the blow-up happens in finite time for all initial conditions. It is worth mentioning that this type of bifurcation is the normal form of the bifurcation of the example studied in sections 2.2 and 2.3 (see Figure 2.2 and Figure 2.3).

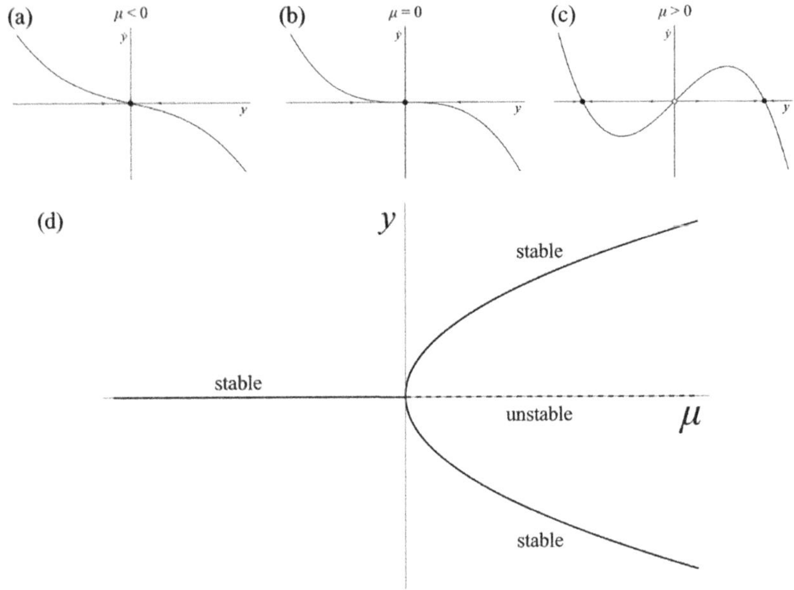

Figure 2.12. *Supercritical pitchfork bifurcation*

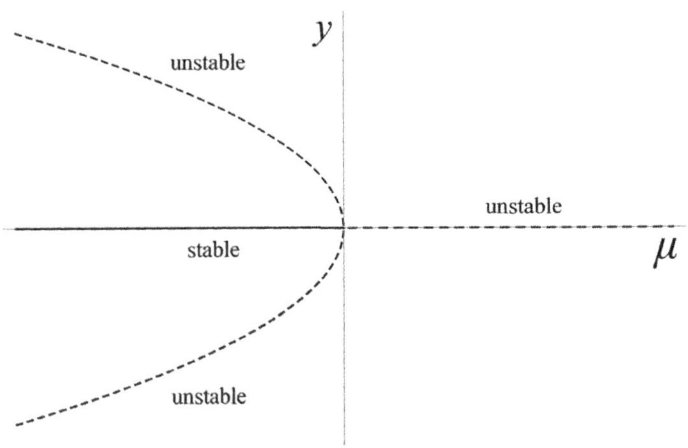

Figure 2.13. *Bifurcation diagram of the subcritical pitchfork bifurcation*

2.4.4. *From one to two dimensions – limit cycles*

All the cases considered in the previous sections concerned bifurcations of equilibrium solutions in one-dimensional problems. What about bifurcations in problems of high order?

Nonlinear dynamical systems of order higher than one can present perfectly periodic solutions. Such solutions appear on the phase space as isolated closed orbits, which can attract or repel all neighboring trajectories, much like the fixed points. These orbits are called *limit cycles*. Limit cycles are an inherent phenomenon of two or higher dimensional systems that are nonlinear. Although, linear systems can present closed orbits, when the fixed point is a stable center (neutral stability, see Figure 2.5(b)), such solutions are non-isolated, i.e. if $x(t)$ is a periodic solution, then $c\,x(t)$ is also a periodic solution for all $c \in \mathbb{R}^*$.

An illustrative example of a system with a stable limit cycle in polar coordinates is:

$$\begin{aligned} \dot{r} &= r\left(1-r^2\right), \\ \dot{\theta} &= 1 \end{aligned} \qquad [2.22]$$

where $r \geq 0$. It is easy to identify that the two equations are uncoupled and that the first one if treated alone. It has two fixed points, namely $r=0$ (unstable) and $r=1$ (stable). This means that all trajectories approach $r=1$. However, the system of two equations has no fixed points at all because $\dot{\theta} = 1 \neq 0$. $\dot{\theta} = 1$ describes the angular velocity, which is constant. Therefore, all trajectories on the phase plane are approaching the unit circle ($r=1$) monotonically. This can be visualized if we revert again to Cartesian coordinates (Figure 2.1(a)), i.e. $x(t) = r(t)\cos\left(\theta(t)\right)$ and $y(t) = r(t)\sin\left(\theta(t)\right)$. The evolution in time of the x-coordinate, for $r(0) = 0.01$ and $\theta(0) = 0$, is presented in Figure 2.1(b). As we can see, the amplitude of the oscillations is $r=1$ and the period is $T=2\pi$.

One of the most famous examples of equations that present limit cycles is the *van der Pol equation:*

$$\ddot{y} + \mu\left(y^2 - 1\right)\dot{y} + y = 0, \qquad [2.23]$$

where $\mu \geq 0$ is a parameter. In this equation, the nonlinear term $\mu\left(y^2 - 1\right)\dot{y}$ forces the oscillation. The limit cycle is no longer a circle (Figure 2.2(a)) and the waveform is not sinusoidal (Figure 2.2(b)).

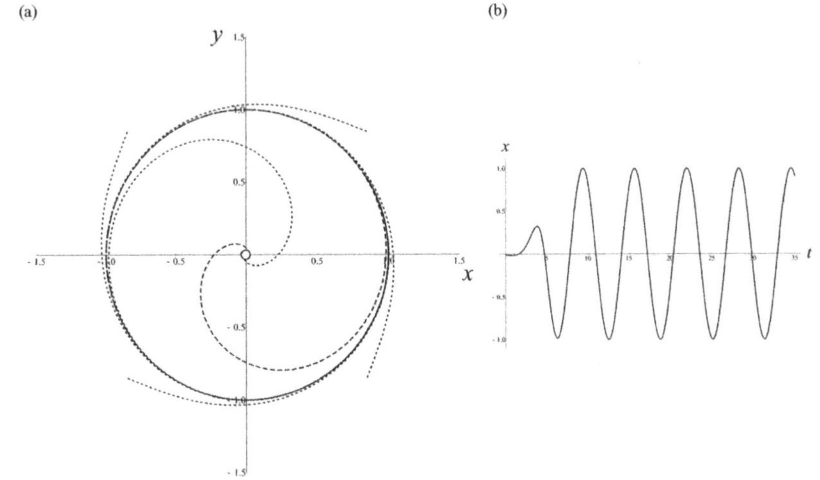

Figure 2.14. *(a) Phase diagram of the system of equation [2.22]. We observe the trajectories moving toward the limit cycle of the system. (b) Evolution in time of the system of equation [2.22] for* $r(0)=0.01$ *and* $\theta(0)=0$

A question that follows naturally is if fixed points and limit cycles are the only possible attractors (or repellers) of the trajectories of a system of ODEs. The answer to that is negative for higher dimensions. In two dimensions, the dimensionality of the system (and thus the corresponding trajectories on the phase space) is equal to the dimensionality of the limit cycles (both equal to two) and, hence, all trajectories on the phase space can be attracted to either points or closed orbits. On the contrary, dynamical systems of order $n\geq 3$ can have trajectories that might be in an open, bounded domain, yet, they can move freely inside it without settling into a fixed point or a closed orbit. They can be attracted to topological manifolds (called *stable manifolds*) or even to complex geometric objects that are called *strange attractors* or *fractals*. The study of such complex (or chaotic) dynamics is out of the scope of the present chapter and the reader should refer to Hale and Kocak (1991) and Strogatz (1994) for a first introduction to these phenomena.

2.4.5. *Bifurcations in two dimensions – supercritical and subcritical Hopf bifurcation*

We are now ready to answer the question about bifurcations in two-dimensional systems. In terms of bifurcations of fixed points, all the basic examples discussed in sections 2.4.1–2.4.3 have their analogs in two (and in higher) dimensions. The corresponding normal forms in two dimensions are:

$$\begin{aligned}\dot{y}_1 &= \mu + y_1^2 \\ \dot{y}_2 &= -y_2\end{aligned} \qquad [2.24]$$

$$\begin{aligned}\dot{y}_1 &= \mu y_1 - y_1^2 \\ \dot{y}_2 &= -y_2\end{aligned} \qquad [2.25]$$

$$\begin{aligned}\dot{y}_1 &= \mu y_1 - y_1^3 \\ \dot{y}_2 &= -y_2\end{aligned} \qquad [2.26]$$

$$\begin{aligned}\dot{y}_1 &= \mu y_1 + y_1^3 \\ \dot{y}_2 &= -y_2\end{aligned} \qquad [2.27]$$

for the saddle-node, transcritical, supercritical pitchfork and subcritical pitchfork, respectively. It is easy to prove that at the bifurcation point $(\mu = 0)$, the corresponding linearized problem has a zero eigenvalue. This means that they always involve the collision of fixed points. Furthermore, irrespectively of the dimensionality of the problem, these types of bifurcations are inherently one-dimensional phenomena in the sense that they occur on the one-dimensional unstable manifold of the unstable fixed point. There exists, however, another way for a fixed point to lose stability and it involves the creation or destruction of a limit cycle around it.

This case is the so-called *Hopf* (or *Andropov-Hopf*) bifurcation. Let us assume that the dynamical system at hand, $\dot{y} = f(y, \mu)$, has a stable fixed point. This means that the eigenvalues, λ_1, λ_2, of the Jacobian matrix of the system have negative real parts. The imaginary part is not necessarily zero. In other words, the eigenvalues lie on the left half-plane of the complex plane $\mathbb{C}$ (see Figure 2.16). For a two-dimensional system, there are only two possible cases for its eigenvalues, either $\lambda_1, \lambda_2 \in \mathbb{R}^-$ or they are complex conjugates. Let us then assume that there exists a value of the parameter $\mu = \mu_H$ for which the fixed point loses stability. In the first case, as we approach this bifurcation point by varying the parameter μ, one of the eigenvalues becomes zero. This corresponds to the cases of saddle-node, transcritical and pitchfork bifurcations. In the second case, the pair of complex conjugate eigenvalues crosses simultaneously the imaginary axis into the right half-plane (Figure 2.16). The latter is the fundamental mechanism described by the Hopf bifurcation.

As mentioned before, the Hopf bifurcation describes the creation or destruction of a limit cycle around a fixed point when the latter loses stability. The first potential scenario is the creation of a limit cycle from a fixed point. In this case, for all values

$\mu < \mu_H$, the system is stable and the fixed point is a stable spiral (Figure 2.6). As μ increases, it approaches and then surpasses the critical value μ_H for which the spiral becomes unstable. This is the case of the supercritical Hopf bifurcation and its normal form is given by:

$$\begin{aligned} \dot{x} &= \mu x - y - x\left(x^2 + y^2\right) \\ \dot{y} &= x + \mu y - y\left(x^2 + y^2\right) \end{aligned} \qquad [2.28]$$

or equivalently in polar coordinates:

$$\begin{aligned} \dot{r} &= \mu r - r^3 \\ \dot{\theta} &= 1 \end{aligned} \qquad [2.29]$$

It is worth mentioning that the unstable spiral is surrounded by a stable limit cycle. Note that the system [2.29] is just a generalization of the system [2.22]. For $\mu < 0$, it yields that $\dot{r}<0$, and thus all oscillations have decreasing amplitude. This means that the only attractor is the origin and it is a stable spiral (Figure 2.17(a)). For $\mu = 0$, , the origin becomes a center. For $\mu > 0$, as also shown in section 2.4.4, the origin becomes an unstable spiral and is surrounded by a stable limit cycle (Figure 2.17(b)). If we consider $x(t) = r(t)\cos(\theta(t))$, it is easy to show from the roots of $\mu r - r^3$ that the amplitude of the oscillations is $r = \sqrt{\mu}$ and the period $T = 2\pi$.

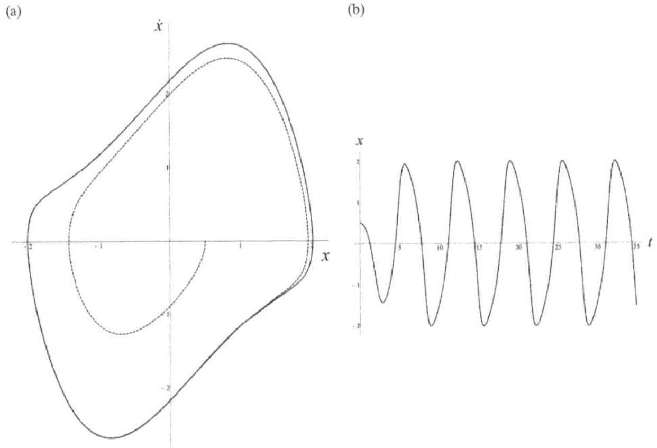

Figure 2.15. *(a) Phase diagram of the van der Pol equation for $\mu=1$. (b) Evolution of the solution of the van der Pol equation for $y(0)=0.5$ and $\mu=1$*

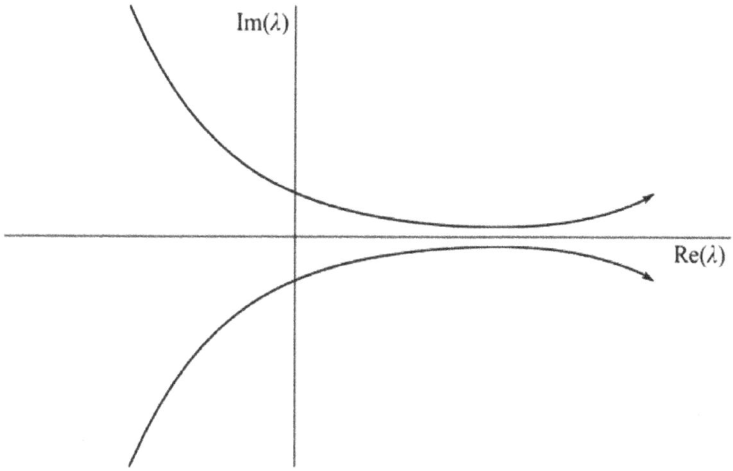

Figure 2.16. *Sketch of a pair of complex eigenvalues crossing the imaginary axis*

We should note here that the normal form represents the so-called *topological equivalent* of the Hopf bifurcation. This means that all limit cycles that are created by a Hopf bifurcation are equivalent (in a mathematical sense) to an oscillation of amplitude $\sqrt{\mu}$ and period 2π (in other words, of angular velocity $\omega=1$). Based on the system [2.29] one can construct more general systems of equations that can admit different modes of sinusoidal wave forms as solutions. Such a system is:

$$\begin{aligned}\dot{r} &= \mu r - ar^3 \\ \dot{\theta} &= \omega + br^2\end{aligned}, \qquad [2.30]$$

where ω is the frequency of the infinitesimal oscillations (near $\mu=0$) and b describes the dependency of the frequency (and of the angular velocity) on the amplitude. For $a>0$, it can be shown that the amplitude is $r=\sqrt{\mu/a}$ and the period is $T=2\pi/\left(\omega+br^2\right)$.

The second potential scenario is the destruction of a limit cycle and it is called *subcritical Hopf bifurcation*. Its normal form is as follows:

$$\begin{aligned}\dot{x} &= \mu x - y + x\left(x^2+y^2\right) \\ \dot{y} &= x + \mu y + y\left(x^2+y^2\right)\end{aligned} \qquad [2.31]$$

or equivalently in polar coordinates:

$$\dot{r} = \mu r + r^3$$
$$\dot{\theta} = 1 \qquad [2.32]$$

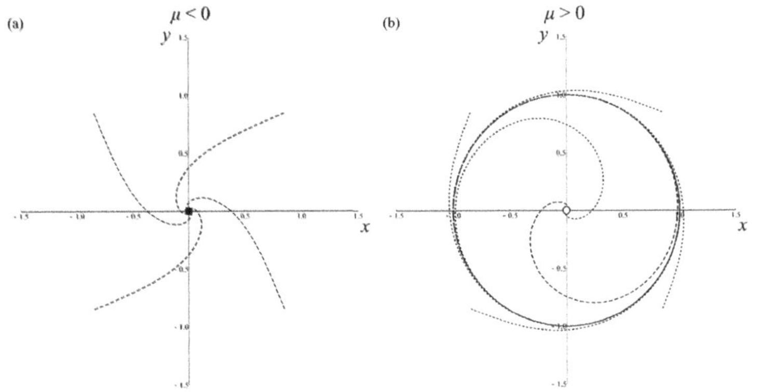

Figure 2.17. *Phase diagram of the supercritical Hopf bifurcation. We observe the transition from a stable spiral in (a) to an unstable one in (b) which is surrounded by a stable limit cycle*

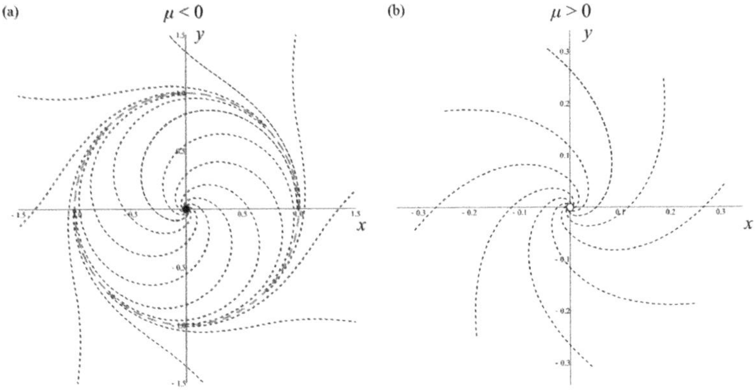

Figure 2.18. *Phase diagram of the subcritical Hopf bifurcation. In (a), $\mu < 0$, we observe that origin is a stable spiral that is surrounded by an unstable limit cycle (marked with a red dashed line). All trajectories starting from inside the cycle tend to the origin, while those starting out of it, diverge. The cycle radius decreases with increasing μ until it collapses to the fixed point for $\mu = 0$. In (b), $\mu > 0$, we observe that there exists only the origin as a fixed point and it is an unstable spiral. For a color version of this figure, see www.iste.co.uk/stefanou/instabilities.zip*

In this case, for $\mu < 0$ the right-hand side of the radial equation has two roots, $r_0 = 0$ and $r_0 = \mu$. It is easy to show that the origin is a stable fixed point (stable spiral). However, the second root represents an unstable limit cycle for the system. This means that all trajectories that start inside the cycle move toward the origin, whereas all the trajectories that start outside the cycle diverge since the cycle repels them (Figure 2.4(a)).

As the value of the parameter increases, the radius of the cycle decreases and collapses to the origin for $\mu = 0$. For $\mu > 0$, the limit cycle is destroyed and the origin exchanges stability with it becoming unstable (Figure 2.4(b)). This means that for $\mu > 0$, there is no stable solution (fixed or periodic) for the system [2.32].

2.4.6. *Mathematical bifurcations in PDEs*

The bifurcations presented in the previous sections are just indicative cases. As stated in section 2.4.4, the higher the order of the system, the more complex the behavior can be. However, change in the number of the equilibrium solutions as well as changes in the stability of equilibrium or periodic solutions are the most common and important bifurcations that can occur in systems of equations modeling problems in mechanics. Even though the analysis was so far restricted in ODEs, these mathematical bifurcations are observed in PDEs as well.

One famous example is the Bratu equation (Bratu 1914; Gelfand 1963) here written as a boundary value problem with symmetric boundary conditions and in one dimension:

$$\begin{aligned} &0 = \frac{d^2T(x)}{dx^2} + \lambda e^{T(x)}, \\ &T(-1) = T(1) = 1 \end{aligned} \qquad [2.33]$$

where x is the spatial coordinate and λ a bifurcation parameter. This equation can describe the time-independent behavior of an infinite layer under simple shear (Chen *et al.* 1989). This means that it provides the steady-state solutions ($\frac{\partial T(x,t)}{\partial t} = 0$), which is the equivalent of the fixed points of ODEs. The steady-state problem of equations [2.33] has two solutions for $\lambda < \lambda_c$, one for $\lambda = \lambda_c$ and none for all $\lambda > \lambda_c$, where λ_c is the critical value of λ for which the bifurcation occurs. This type of bifurcation is a saddle-node bifurcation.

Note that when dealing with PDEs, the equivalent of a fixed point is a time-independent solution that can either be homogenous (i.e. constant in space) or inhomogeneous (i.e. non-constant profile in space). One way of studying the stability of equilibrium solutions of PDEs (and their bifurcations to an extent) is using the so-called LSA. An example of method will be presented in section 2.5.2. Such an analysis can predict the growth or decay of perturbations near an equilibrium solution, thus providing information about the stability of the equilibrium solution. By these means, one can derive when one steady state is preferred from another and determine the bifurcation point from the onset of instability. A famous example is the formation of Bénard (convection) cells in the Rayleigh–Bénard problem (Figure 2.19).

In this problem, there are various types of steady states and corresponding bifurcations. The critical parameter that governs these instabilities is called the Rayleigh number and for low values, where conduction prevails, a linear temperature profile is predicted. Upon reaching a critical value, a bifurcation occurs and convection becomes the dominant mechanism. In this case, the flow appears to be steady in time but periodic in space (Figure 2.19(b)). For more information, the reader should refer to Chelidze (2016) where the results of linear stability analyses for various boundary value problems of this type are presented.

In the following, we will present the study of a problem that is more relevant to geomechanics applications, the localization of shear and volumetric strain.

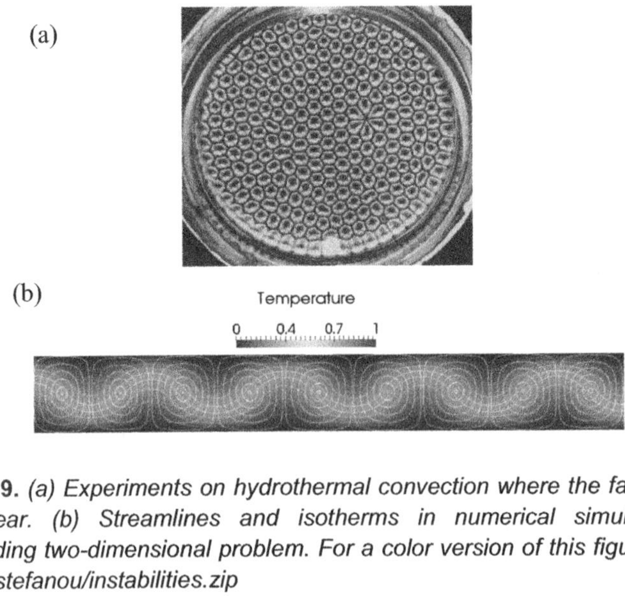

Figure 2.19. *(a) Experiments on hydrothermal convection where the famous Bénard cells appear. (b) Streamlines and isotherms in numerical simulation of the corresponding two-dimensional problem. For a color version of this figure, see www.iste.co.uk/stefanou/instabilities.zip*

2.5. From ODEs to PDEs

The above sections were dedicated to the notion of stability and bifurcation focusing on ODEs. In this section, we try to extend the above concepts to PDEs that cover a variety of systems including mechanics of solids, in general, and geomechanics. A classical problem of bifurcation and instability is strain localization in materials. Strain localization is frequently manifested as thin bands where deformation is localized. Depending on the kinematics of strain localization, three main types of deformation bands are distinguished. These are dilation (or extension) bands, shear bands and compaction bands (Bernard *et al.* 2002). Whatever their type is, deformation bands appear at the moment that the homogeneous deformation of a system becomes an unstable equilibrium solution. In other words, the system bifurcates to a non-homogeneous solution where the strain is localized. The classical approach for determining when this localization takes place is based on calculating the determinant of the acoustic tensor (Rudnicki and Rice 1975).

2.5.1. *Deformation bands and the acoustic tensor*

Consider a homogeneous, homogeneously deformed solid subjected to quasi-static increments of deformation. Let us assume that after an increment, a deformation band is formed, which breaks the aforementioned homogeneity of the deformation field (and consequently of the stress field), as shown in Figure 2.20. The displacement field remains continuous across the boundaries of the band, but its gradient does not (different strains inside the band):

$$[\![\Delta u_i]\!] = 0 \quad \text{and} \quad [\![\Delta u_{i,j}]\!] = g_i n_j , \qquad [2.34]$$

where [[.]] denotes discontinuity across the deformation band boundary (e.g. $[[a]] = a^+ - a^-$), n_i is the orientation vector of the deformation band with $i = 1,2,3$ the three-dimensional space, u_i the displacement field and Δ denotes the increment of a field. $(\ )_{,i}$ denotes derivation in terms of x_i.

The vector g_i describes the direction of the discontinuity and its inner product with the orientation of the band, n_i, determines the type of the deformation band (see Figure 2.20). In particular, if $n_i g_i = 0$, the deformation band is a pure shear band, if $n_i g_i = -1$ a pure compaction band and if $n_i g_i = +1$ a pure dilation (extension) band. The intermediate states, $0 < n_i g_i < 1$ and $-1 < n_i g_i < 0$, correspond, respectively, to dilatant and contracting shear bands.

In quasi-static conditions, the stress vector has to be continuous across the deformation band boundary:

$$[\![\Delta t_i]\!] = [\![\Delta \sigma_{ij}]\!] n_j = 0 \qquad [2.35]$$

Consider the class of materials that for a small increment Δ, the constitutive law can be written (linearized) as follows:

$$\Delta \sigma_{ij} = L_{ijkl} \Delta u_{k,l} \qquad [2.36]$$

The tensor L_{ijkl} can be continuous across the boundary of the band ($[\![C_{ijkl}]\!] = 0$) or discontinuous in the sense that elastic unloading can occur outside the band, while continued inelastic loading continues within the band. In the first case, we say that we have *continuous bifurcation*, whereas in the second case, we say that we have *discontinuous bifurcation*. It is shown that continuous bifurcation precedes discontinuous bifurcation (Rice and Rudnicki 1980).

By replacing equations [2.36] into [2.35] and using [2.34], we obtain:

$$[\![\Delta \sigma_{ij}]\!] n_j = L_{ijkl} [\![u_{k,l}]\!] n_j = n_j L_{ijkl} n_l \, g_k = 0 \qquad [2.37]$$

The tensor $\Gamma_{ik} = n_j L_{ijkl} n_l$ is called an *acoustic tensor*. If its determinant is not zero, then the g_k has to be zero, which means that the deformation is continuous along the assumed deformation band. In other words, no discontinuity of the gradient of the displacement field can appear across the boundary of the deformation band and the homogeneous solution prevails. Otherwise, if

$$|\Gamma_{ik}| = 0 \qquad [2.38]$$

the homogeneous solution seizes to be the only one and deformation bands are possible. For an orientation n_i, the type of the deformation band is given by the eigen-vector g_i.

The above condition for strain localization is independent of the material constitutive behavior as long as equation [2.36] can be written. For instance, for an elastoplastic material whose plastic behavior is a function of the first and second invariants of the stress tensor (Figure 2.21), Issen and Rudnicki (2000) (see also Bésuelle and Rudnicki 2000) showed that under axisymmetric compression conditions of loading strain localization is possible when the hardening modulus becomes smaller than a critical value h_{cr} for given values of μ and β (see Figure 2.22).

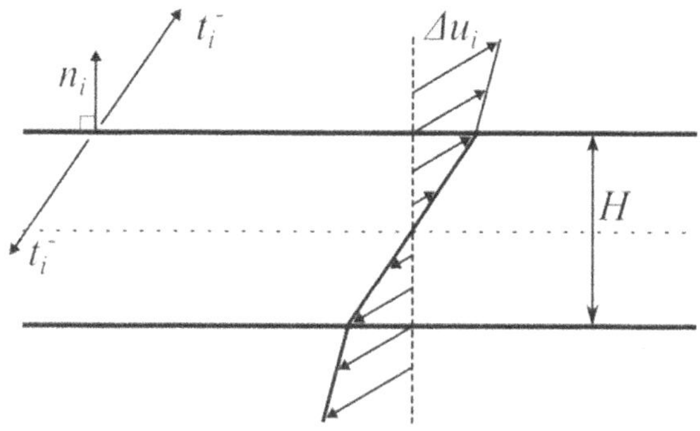

Figure 2.20. *Schematic representation of a deformation band and of the discontinuity of the displacement field*

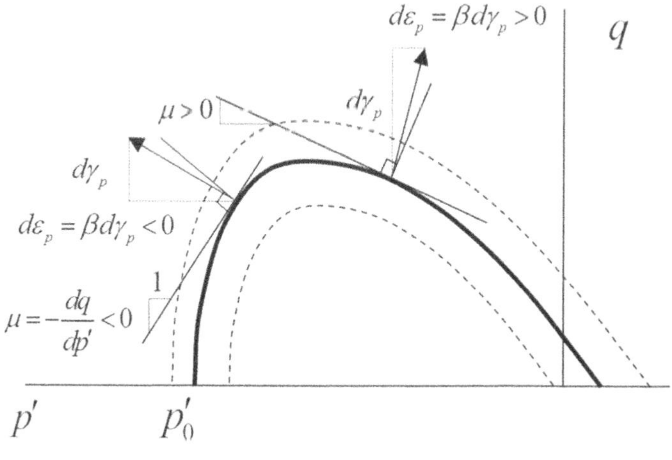

Figure 2.21. *Elastoplastic yield envelope with hardening/softening (dotted lines). Compression is considered negative*

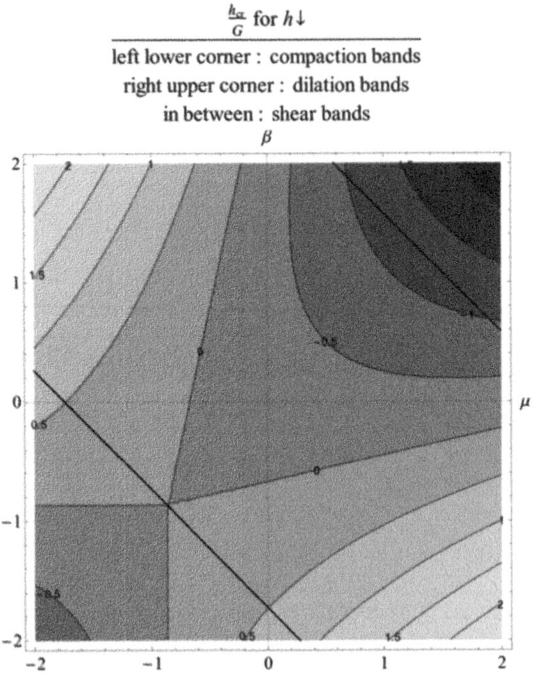

Figure 2.22. *Critical hardening values as a function of the β and μ for strain localization (Issen and Rudnicki 2000). Note that for non-associate plastic flow rule, localization can occur even with hardening (h_{cr} > 0). The above figure was derived by using the acoustic tensor criterion for localization (equation [2.38]). For a color version of this figure, see www.iste.co.uk/stefanou/instabilities.zip*

For the above derivations, we considered quasi-static conditions. If we remove this restriction, the jump of the shear stresses at the boundary of the shear band is not necessarily zero due to acceleration (not in equilibrium). From the linear momentum balance, we obtain:

$$[\![\Delta t_i]\!] = [\![\Delta\sigma_{ij}]\!] n_j = -\rho c [\![\gamma_i]\!], \qquad [2.39]$$

where c is the velocity of a propagating discontinuity in direction n_i such that $[[\gamma_i]] = [[\Delta v_i]] = -c g_i$ (see Hadamard conditions on propagating discontinuities; Hadamard 1903; Lemaitre *et al.* 2009). Inserting equation [2.36] into [2.39] and using [2.34], we get:

$$\left(n_j L_{ijkl} n_l - \rho c^2 \delta_{ik} \right) g_k = 0 \qquad [2.40]$$

This equation shows that if there are accelerating waves ($[[\gamma_i]] = -cg_i \neq 0$), the eigenvalues of the tensor Γ_{ij} are equal to the square root of their wave velocity c^2. This is why Γ_{ij} is called an *acoustic tensor*. The condition of localization derived in quasi-static conditions (equation [2.38]) corresponds to $c=0$ or, in other words, to the existence of stationary acceleration waves.

2.5.2. *Deformation bands as an instability problem*

The same condition with equation [2.38] can be derived by studying the stability of the homogeneous solution of the continuous system. In this case, we do not have a system of ODEs anymore for which we saw how to investigate the stability of an equilibrium point, but a PDE. Nevertheless, the bifurcation analysis approach is similar. Stability is defined in section 2.2.1.

The general PDEs of the problem are:

$$\sigma_{ij,j} = \rho \ddot{u}_i, \qquad [2.41]$$

where ρ is the density of the material and the double dot represents the second time derivative (acceleration). Suppose again a homogeneous, homogeneously deformed solid that it is in equilibrium:

$$\sigma^*_{ij,j} = 0 \qquad [2.42]$$

Considering the same class of materials that can be linearized around the above equilibrium point such as equation [2.36] to hold, we get:

$$\sigma_{ij} = \sigma^*_{ij} + \Delta\sigma_{ij} = \sigma^*_{ij} + L_{ijkl}\Delta u_{k,l} \qquad [2.43]$$

where, as in section 2.5.1, Δu_i is the increment in displacements, i.e. $\Delta u_i = u_i - u^*_i$. Δu_i can also be seen as a perturbation of the reference, homogeneous solution u^*_i (see section 2.2.1). By injecting the above equation in [2.41] and using [2.42], we obtain:

$$L_{ijkl}\Delta u_{k,lj} = \rho \Delta \ddot{u}_i \qquad [2.44]$$

Note that L_{ijkl} is calculated at u^*_i and therefore it is independent of Δu_i (Taylor expansion of σ_{ij} around ε^*_{ij}). Therefore, equation [2.44] is a linear PDE that can be solved by the separation of variables (or Fourier transform). The linearization of the stress tensor around the equilibrium point (equation [2.36]) is central in strain localization analysis. Consequently, the conditions derived either in section 2.5.1 or in the current one are valid as far as this linearization is possible.

Using the method of separation of variables, $\Delta u_i = X(x_k)U_i(t)$. Replacing in [2.44], we obtain:

$$L_{ijkl}X_{,lj}\;U_k(t) = \rho X\,\ddot{U}_i \qquad [2.45]$$

This equation has sinusoidal solutions in terms of X. Moreover, we are looking for deformation bands, which are planar as shown in Figure 2.20. Therefore, the solution in terms of X takes the form $X(x_i) = e^{ikn_ix_i}$, where n_i is the orientation vector of the deformation band as in the previous section and k is the wave number (of the perturbation). Note that λ is the wavelength corresponding to the wave number k ($k = 2\pi/\lambda$) in order to satisfy the boundary conditions at the boundary of the deformation band $\lambda = H/N$, where N is a integer. Inserting X in equation [2.45] and by setting $\dot{U}_i = V_i$, we obtain the following system of ordinary differential equations:

$$\begin{aligned}\dot{V}_i &= -\frac{1}{\rho}\left(\frac{2\pi}{\lambda}\right)^2 n_j L_{ijkl} n_l\, U_k \\ \dot{U}_i &= V_i\end{aligned} \qquad [2.46]$$

In this way, we transformed the PDEs of the problem to a system of ODEs, which we can study in the same way as in the previous sections. As shown in section 2.2, the above equations take solutions of the form $U_k(t) = g_k e^{st}$. After some algebraic manipulations, the eigenvalue problem becomes:

$$\left[n_j L_{ijkl} n_l + \rho\left(\frac{\lambda s}{2\pi}\right)^2 \delta_{ik}\right] g_k = 0\ , \qquad [2.47]$$

which is identical to [2.40] by setting $c = i\dfrac{\lambda s}{2\pi}$. Note that $\Delta u_i = U_i e^{i\frac{2\pi}{\lambda}n_px_p + st}$ $= e^{i\frac{2\pi}{\lambda}\left(n_px_p - i\frac{s\lambda}{2\pi}t\right)} = e^{i\frac{2\pi}{\lambda}(n_px_p - ct)}$, which describes a wave traveling with speed c. If the real part of s is positive, then the homogeneous solution u_i^* is unstable and the system bifurcates to a non-uniform solution, a band, with direction n_i. As before, the type of the deformation band (compaction, shear, dilation band) is determined by the product $n_i g_i$. It is worth emphasizing that the above condition is independent of the specific constitutive law, provided that it is rate-independent. For rate-dependent materials, a similar approach can be followed. The methodology is quite general and can be applied in many problems, including problems with multiphysical couplings, such as thermo-poro-chemo-mechanical couplings (e.g. Stefanou and Sulem 2014; Sulem and Stefanou 2015). Moreover, even though a Cauchy (Boltzmann) continuum was considered here, the same approach can be applied in Cosserat or even higher order continua (e.g. Mühlhaus and Vardoulakis 1988; Sulem *et al.* 2011).

If the eigenvalues of Γ_{ij} do not depend on the (perturbation) wavelength λ and s, then the acceleration wave velocity c is constant (does not depend on λ). If in addition they have a positive real part, the perturbation that propagates faster has zero wave length because for $\lambda \to 0$, $s \to \infty$. In other words, the minor imperfection in size will propagate faster and dominate the other imperfections of larger wavelength. This is why in the classical Cauchy continuum, which has no internal lengths, the deformation band thickness is zero (the localization takes place on a mathematical plane). The fact that the smallest perturbation propagates faster also justifies the mesh dependency on finite element calculations. For instance, in the frame of classical simulations in elastoplasticity of Cauchy rate-independent continua with softening behavior (or even in perfect plasticity), the numerically predicted shear band thickness depends on the finite element discretization and on the element size (Figure 2.23).

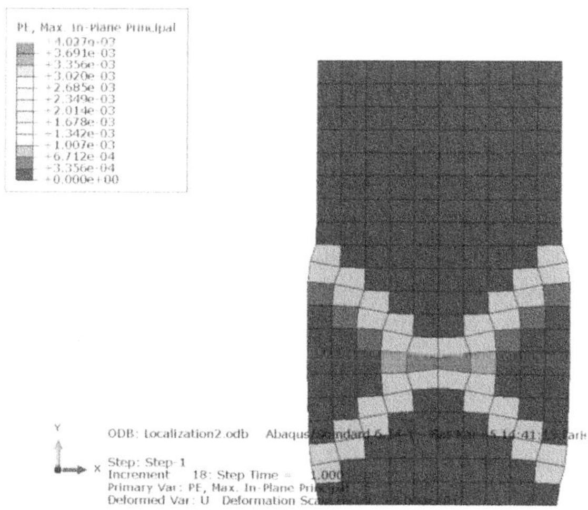

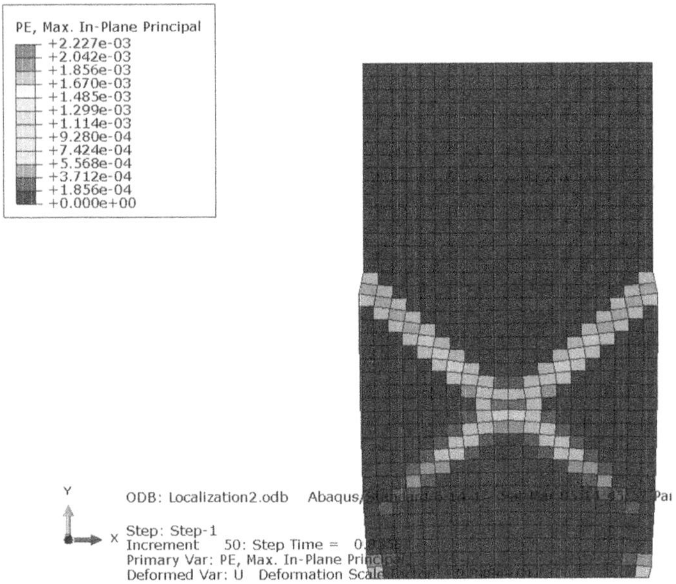

Figure 2.23. *Shear band formation and mesh dependency for a rate-independent elastoplastic, von Mises, Cauchy medium with strain softening. The shear band thickness is always one to two elements thick and therefore mesh dependent. The plastic strains and the global energy dissipation are also mesh dependent. Abaqus v6.14 was used for the simulations. For a color version of this figure, see www.iste.co.uk/stefanou/instabilities.zip*

2.6. Summary

The aim of this chapter was to give the basic ideas and tools of bifurcation theory and stability analysis. The definition of (Lyapunov) stability was given, as well as the fundamental theorems that allow us to study the stability of linear and nonlinear systems of ODEs. The notion of bifurcation was explained and illustrated through examples and a classification of the most common bifurcations and instabilities was presented. The focus was given on ODEs as their behavior is central for understanding bifurcation and stability. The study of PDEs is an extension of the ideas presented for ODEs and was presented in the last section. The strain localization conditions of homogeneously deformed solids were derived as an example (acoustic tensor). After studying this chapter, the reader will be able to distinguish the basic notions of stability and bifurcation and apply the different concepts in more complicated systems in geomechanics that are characterized by advanced constitutive law and multiphysical couplings.

2.7. Appendix

Let $\dot{y}=f(y,\mu)$ be a dynamical system with a bifurcation point at $y=y_0$ for $\mu=\mu_c$. A Taylor expansion of the equation yields:

$$\dot{x}=f(y_0,\mu_c)+(y-y_0)\left.\frac{\partial f}{\partial y}\right|_{(y_0,\mu_c)}+ \\ (\mu-\mu_c)\left.\frac{\partial f}{\partial \mu}\right|_{(x_0,\mu_c)}+\frac{1}{2}(y-y_0)^2\left.\frac{\partial^2 f}{\partial y^2}\right|_{(y_0,\mu_c)}+\ldots \quad [2.48]$$

In the case of saddle-node bifurcations, we have that the term $f(y_0,\mu_c)=0$ since y_0 is a fixed point and $\partial f/\partial y|_{(y_0,\mu_c)}=0$ by definition for this specific bifurcation. Therefore,

$$\dot{y}=a(\mu-\mu_c)+b(y-y_0)^2+\ldots, \quad [2.49]$$

where $a=\partial f/\partial \mu|_{(y_0,\mu_c)}$ and $b=1/2\partial^2 f/\partial y^2|_{(y_0,\mu_c)}$. Thus, for (y,μ) sufficiently close to (y_0,μ_c) along with $a,b\neq 0$, we can neglect the higher order terms resulting to the normal forms: $\dot{y}=\mu-y^2$ or $\dot{y}=\mu+y^2$.

2.8. References

Benallal, A., and Comi, C. (2000). On localization in saturated porous continua. *Comptes Rendus de l'Académie Des Sciences*, 328(12), 847–853. Available: https://doi.org/10.1016/S1620-7742(00)01277-0.

Benallal, A., and Comi, C. (2003). Perturbation growth and localization in fluid-saturated inelastic porous media under quasi-static loadings. *Journal of the Mechanics and Physics of Solids*, 51(5), 851–899. Available: https://doi.org/ 10.1016/ S0022-5096(02)00143-6.

Bernard, X.D., Eichhubl, P., Aydin, A. (2002). Dilation bands: A new form of localized failure in granular media. *Geophys. Res. Lett.*, 29(24), 2176.

Bésuelle, P. (2001). Compacting and dilating shear bands in porous rock: Theoretical and experimental conditions. *J. Geophys. Res.*, 106(B7), 13435–13442.

Bésuelle, P. and Rudnicki, J.W. (2000). Localization: Shear bands and compaction bands. *Int. Geophys.*, 89, 219–321.

Bigoni, D. and Hueckel, T. (1991a). Uniqueness and localization I: Associative and non-associative elastoplasticity. *Int. J. Solids Struct.*, 28(2), 197–213.

Bigoni, D. and Hueckel, T. (1991b). Uniqueness and localization II: Coupled elastoplasticity. *Int. J. Solids Struct.*, 28(2), 215–224.

Bratu, G. (1914). Sur les équations intégrales non linéaires. *Bull. la Société Mathématique Fr.*, 42, 113–142.

Brauer, F. and Nohel, J.A. (1969). *The Qualitative Theory of Ordinary Differential Equations: An Introduction*. Dover Publications, New York.

Chambon, R., Caillerie, D., Viggiani, G. (2004). Loss of uniqueness and bifurcation vs instability: Some remarks. *Rev. Française Génie Civ.*, 8(5–6), 517–535.

Chandrasekhar, S. (1961). *Hydrodynamic and Hydromagnetic Stability – International Series of Monographs on Physics*. Clarendon, Oxford.

Chelidze, D. (2016). *Fixed Point Eigenvalues and Phase Portrait*. The Wolfram Demonstrations Project.

Chen, H.T., Douglas, A.S., Malek-Madani, R. (1989). An asymptotic stability criterion for inhomogenous simple shear. *Q. Appl. Math.*, XLVII, 247–262.

Crowford, D.J. (1991). Introduction to bifurcation theory. *Rev. Mod. Phys.*, 63(4), 991–1037.

Gelfand, I.M. (1963). Some problems in the theory of quasi-linear equations. *Trans. Am. Math. Soc.*, 29, 295–381.

Hadamard, J. (1903). *Leçons sur la propagation des ondes et les équations de l'hydrodynamique*. Librairie Scientifique A. Hermann, Paris.

Hale, J.K. and Kocak, H. (1991). *Dynamics and Bifurcations*. Springer, New York, 1991.

Issen, K.A. and Rudnicki, J.W. (2000). Conditions for compaction bands in porous rock. *J. Geophys. Res.*, 105(B9), 21529.

Lemaitre, J., Chaboche, J.-L., Benallal, A., Desmorat, R. (2009). *Mécanique des matériaux solides*. Dunod, Paris.

Lyapunov, A.M. (1892). *The General Problem of the Stability of Motion*. University of Kharkov, Ukraine.

Lyapunov, A.M. (1966). *Stability of Motion*. Academic Press, New York.

Lyapunov, A.M. (1992). The general problem of the stability of motion. *Int. J. Control*, 55, 531–773.

Mühlhaus, H.B. and Vardoulakis, I. (1988). The thickness of shear hands in granular materials. *Géotechnique*, 38(3), 331–331.

Perrin, G and Leblond, J.B. (1993). Rudnicki and Rice's analysis of strain localization revisited. *J. Appl. Mech.*, 93, 842–846.

Rice, J.R. (1976). The localization of plastic deformation. In *Theoretical and Applied Mechanics (Proceedings of the 14th International Congress on Theoretical and Applied Mechanics)*, Koiter, W.T. (ed.). NorthHolland Publishing Co., Delft, 207–220.

Rice, J.R., Rudnicki, J.W. (1980). A note on some features of the theory of localization of deformation. *Int. J. Solids Struct.*, 16, 597–605.

Rudnicki, J.W., Rice, J.R. (1975). Conditions for the localization of deformation in pressure-sensitive dilatant materials. *J. Mech. Phys. Solids*, 23(6), 371–394.

Sprott, J.C. (2004). Dynamical models of love. *Nonlinear Dynamics Psychol. Life Sci.*, 8(3), 303–314.

Stefanou, I. and Sulem, J. (2014). Chemically induced compaction bands: Triggering conditions and band thickness. *J. Geophys. Res. Solid Earth*, 119(2), 880–899.

Strogatz, S.-H. (1988). Love affairs and differential equations. *Math. Mag.*, 35–35 61(1).

Strogatz, S.-H. (1994). *Nonlinear Dynamics and Chaos*. Perseus Books, Reading, Massachusetts.

Sulem, J., & Stefanou, I. (2016). Thermal and chemical effects in shear and compaction bands. *Geomechanics for Energy and the Environment*, 6, 4–21. Available at: https://doi.org/10.1016/j.gete.2015.12.004.

Sulem, J., Stefanou, I., Veveakis, E. (2011). Stability analysis of undrained adiabatic shearing of a rock layer with Cosserat microstructure. *Granul. Matter*, 13(3), 261–268.

Vardoulakis, I., Sulem, J. (1995). *Bifurcation Analysis in Geomechanics*. Blackie, Glascow.

Veveakis, E., Sulem, J., Stefanou, I. (2012). Modeling of fault gouges with Cosserat continuum mechanics: Influence of thermal pressurization and chemical decomposition as coseismic weakening mechanisms. *J. Struct. Geol.*, 38, 254–264.

Veveakis, E., Stefanou, I., Sulem, J. (2013). Failure in shear bands for granular materials: Thermo-hydro-chemo-mechanical effects. *Géotechnique Lett.*, 3, 31–36.

3

Material Instability and Strain Localization Analysis

Jean SULEM

Navier, Ecole des Ponts ParisTech, Université Paris-Est, Marne-la-Vallée, France

In this chapter, the principles of strain localization analysis applied to geomaterials are presented. Conditions for the formation of different types of deformation bands are given. Extension of the analysis to fluid-saturated porous media is also presented.

3.1. Introduction

Failure of many engineering structures is characterized by the formation and propagation of a failure plane. Laboratory experiments as well as field observations indicate that the orientation of the failure plane (or fault surface) is controlled by the directions of the principal stresses. In the well-established Mohr–Coulomb theory, the inclination θ of the failure surface with respect to the direction of the minimum (in absolute value) principal stress is given, where ϕ is the friction angle of the material. For typical values of Coulomb friction angle ϕ, 30°–50°, values of θ range from 60° to 70°, which is comparable with the range of observed failure plane inclinations. Mohr–Coulomb's theory is commonly used in geomechanics as the dominant feature in the behavior of geomaterials is its frictional character. The orientation of a failure surface can be deduced from the knowledge of the orientation of the principal stresses (not their magnitude) and one material property (the internal friction angle). Inversely, the orientation of the principal stresses can be simply

Instabilities Modeling in Geomechanics,
coordinated by Ioannis STEFANOU and Jean SULEM. © ISTE Ltd 2020.

deduced, using Mohr–Coulomb's theory, from the orientation of the failure plane and the knowledge of the friction coefficient of the material.

Although the simplicity of this approach has made it very useful, the predictions of Mohr–Coulomb's theory have been criticized because it is commonly observed that the complete inelastic response of a material influences the conditions of incipient failure and not only one material parameter such as the internal friction angle.

Moreover, for studying the response of a structure in the postfailure regime when the strength is suddenly dropping, one has to understand what failure physically means. The first observation was that what appears as a failure plane is in fact a zone of localized shear deformation. Thus, the study and the modeling of strain localization phenomena have proven to be very useful in the understanding of failure mechanisms. Based on the theoretical studies of material stability, as developed by Hadamard (1903) for elastic materials and later extended by Thomas (1961), Hill (1962) and Mandel (1966) for inelastic materials, the localization process is seen as an instability that can be predicted from the prefailure constitutive behavior of the material. The conditions for the onset of localization are thus established by seeking the possible critical conditions for which the constitutive equations of the material (in the prelocalized stage) may allow the existence of a bifurcation point for which the deformation mode will localize into a planar band (Rice 1976; Vardoulakis 1976). In this approach, the initiation of failure in the form of the incipient of a shear band is modeled as a constitutive instability and consequently a great number of studies have been dealing with the development of appropriate constitutive relationships, which can predict satisfactorily the onset of shear banding. Geomaterials are characterized by a non-associated and non-coaxial plastic behavior and this has important effects on the localization process (Rudnicki and Rice 1975). Numerous experimental studies on soils and rocks have been performed to understand the physical processes, which control strain localization and validate the theoretical and constitutive concepts (e.g. Vardoulakis 1980; Sulem *et al.* 1999; Bésuelle *et al.* 2000; Desrues and Viggiani 2004).

Shear band formation in the form of localized shear deformation is not the only possible localized deformation mode. Zones of localized deformation are sometimes observed in the direction normal to the maximum compressive stress as observed in highly porous rocks. These structures are interpreted as compactions bands (e.g. Baud *et al.* 2004). Deformation zones in the form of shear bands or compaction bands in geomaterials are observed on a very large range of scales from sub-millimetric (grain size) to kilometric scale (geological structure). The formation of deformation bands is associated with an induced heterogeneity of strength and of other material properties (e.g. porosity, grain size, pore size and permeability) in relation with the transformation at the microscale of the microstructure of the rock inside the band. In geological structures, shear zones also play a major role in the nucleation of earthquakes, landslides and slope failure. They are also of prime importance in the flow of water, gas and oil in the subsurface as they can serve as conduits or barriers for fluid and heat fluxes. On the other hand, compaction bands are usually characterized by a significant reduction of the pore space, which in most cases is accompanied by an important reduction in permeability (Olsson *et al.* 2002). Their presence can lead to largely anisotropic flow in fluid infiltrated porous rocks. Therefore, compaction bands are also of primary importance in reservoir mechanics for hydrocarbons production, CO_2 storage and mineral exploration.

The problem of modeling localized deformation in geomaterials is quite a challenging task due to the difficulties that are encountered while dealing with softening materials and moving elastoplastic boundaries. It is, however, one of the most interesting bifurcation problems. By asking the question of possible spontaneous change of the deformation mode for a given loading history and subsequent evolution of this secondary deformation mode, one may search for the conditions of uniqueness and stability of the corresponding boundary value problem. It turns out that the result of such analyses dealing with geomaterials depends primarily on the assumed physical nonlinearities, which are inherent to the underlying constitutive description and are in a lesser degree influenced by geometrical nonlinearities.

It is not possible in a single paper to review the important literature published over the past 40 years on the subject. We will thus choose to address some points in relation to the mathematical and physical background of bifurcation analysis as applied to geomechanics with some special emphasis on the effects of fluid and temperature.

3.2. Shear band model

Extensive presentation of shear band analysis in geomaterials can be found in Vardoulakis and Sulem (1995) (for a review, see also Bésuelle and Rudnicki 2004; Desrues and Chambon 2002). The strain localization analysis consists of searching the incipient of a shear band in a solid as a mathematical bifurcation condition for the deformation field. Considering an infinitesimal neighborhood of a point in an elastic-plastic solid, which is homogeneous as for the constitutive law and stress state, the strain localization phenomenon is understood as the appearance of a discontinuity in strain rates that marks the onset of non-uniform response. Such a bifurcation of the velocity gradient along a loading path can be caused by material destabilizing effects such as softening and lack of plastic normality in the constitutive law, as well as geometrical destabilizing effects such as large deformation affecting equilibrium equations. This bifurcation condition is obtained from (1) the constitutive relationships of the material, (2) the conditions of mechanical equilibrium across the shear band boundary and (3) the kinematic compatibility conditions, which express that the velocity field is to be continuous (no material discontinuity). The latter condition implies that only the normal component of the velocity gradient across the shear band is discontinuous, whereas the tangential one is continuous (weak discontinuity). The above conditions describe the so-called "continuous" bifurcation modes. It has been shown that the critical state for continuous bifurcation precedes the one for "discontinuous" bifurcation, where a discontinuity of the velocity field itself (and not only its gradient) is considered (Simo *et al.* 1993). A non-trivial solution for the condition of continuous bifurcation is a necessary condition for the shear band existence and provides both the shear band orientation and the deformation jump across the shear band. Rudnicki and Rice (1975) and Rice (1976) have obtained solutions for realistic elastoplastic constitutive relationships for geomaterials.

3.2.1. *Strain localization criterion*

According to the definition proposed by Hill (1962), a shear band is viewed as a thin layer that is bounded by two parallel material discontinuity surfaces of the incremental displacement gradient (Figure 3.1). These material discontinuity surfaces $D^{(1)}$ and $D^{(2)}$ are called shear band boundaries and their distance, $2d_B$, is the thickness of the shear band. Within the frame of constitutive theories without material length, the shear band thickness d_B is undetermined.

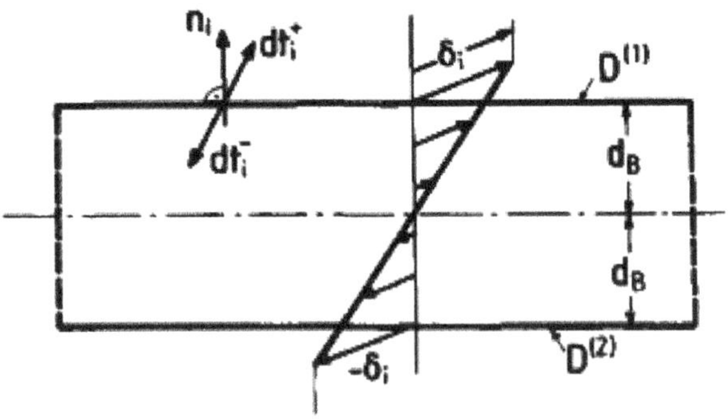

Figure 3.1. *Model of a deforming shear band with heat and fluid fluxes*

Let us denote by n the unit vector normal to the band. Inside the band the incremental displacement field $\Delta \boldsymbol{u}^1$ depends only on the position across the band; outside the band the rate of deformation is assumed to remain homogeneous. Assuming that the displacement field is continuous across the band, then according to Maxwell's theorem, only the normal derivative of this field may be discontinuous across the band. Accordingly, the following *kinematic compatibility conditions* hold:

$$\left[\Delta u_i\right] = 0 \text{ and } \left[\partial_j \Delta u_i\right] = g_i n_j , \qquad [3.1]$$

where [.] denotes the jump of the quantity across the shear band boundary. Note that, as emphasized by Bésuelle and Rudnicki (2004), the form of equation [3.1] requires that $\left[\partial_j \Delta u_i\right]$ has a vanishing intermediate eigenvalue and thus contains a plane of zero incremental displacement. Consequently, localization is favoured when the pre-bifurcation homogeneous field contains a plane of zero extension rates, as in plane strain, whereas highly destabilizing effects such as strong softening behavior is needed to generate shear band formation in axisymmetric deformation.

Incremental strains and stresses are linked through the constitutive relationships:

$$\Delta\sigma_{ij} = C_{ijkl}\partial_l \Delta u_k \qquad [3.2]$$

Equilibrium across the shear band boundary implies the following *static compatibility condition*:

$$\left[\Delta\sigma_{ij}\right] n_j = 0 \qquad [3.3]$$

We observe that there are two possibilities, namely that the constitutive behavior across the shear band boundaries is either continuous or discontinuous. Concerning discontinuous bifurcations, one has to examine the possibility that elastic unloading occurs outside the shear band, while continued elastic-plastic loading occurs within the band. If the elastoplastic constitutive law admits a single smooth yield surface and plastic potential, Rice and Rudnicki (1980) have shown that continuous bifurcation analyses provide the lower limit to the range of deformations for which discontinuous bifurcations can occur. Accordingly, we restrict ourselves here to the first possibility of continuous constitutive behavior, namely $\left[C_{ijkl}\right] = 0$.

Using the constitutive relationships [3.2], the static compatibility condition [3.3] and the kinematic compatibility conditions [3.1], we finally obtain:

$$\Gamma_{ik} g_k = 0\,, \tag{3.4}$$

where $\Gamma_{ik} = C_{ijkl} n_j n_l$ is the acoustic tensor. It follows that weak stationary discontinuities for the incremental displacement exist only if the acoustic tensor is singular:

$$\det \boldsymbol{\Gamma} = 0 \tag{3.5}$$

Equation [3.5] is the characteristic equation in terms of the direction cosines n_i of a statically, kinematically and materially admissible discontinuity surface. If the characteristic equation provides real solutions for the direction cosines n_i, discontinuity surfaces for the incremental displacement gradient exist and may also develop in due course of the deformation. Equation [3.5] is thus the *localization criterion*.

The vector dot product $\boldsymbol{n.g}$ determines the nature of the deformation band created at localization. The following classification describes the deformation band types (Figure 3.2):

$$\begin{cases} \boldsymbol{n.g} = 1: & \text{pure dilation band} \\ 0 < \boldsymbol{n.g} < 1: & \text{dilatant shear band band} \\ \boldsymbol{n.g} = 0: & \text{simple shear band} \\ -1 < \boldsymbol{n.g} < 0: & \text{compactive shear band} \\ \boldsymbol{n.g} = -1: & \text{pure compaction band} \end{cases} \tag{3.6}$$

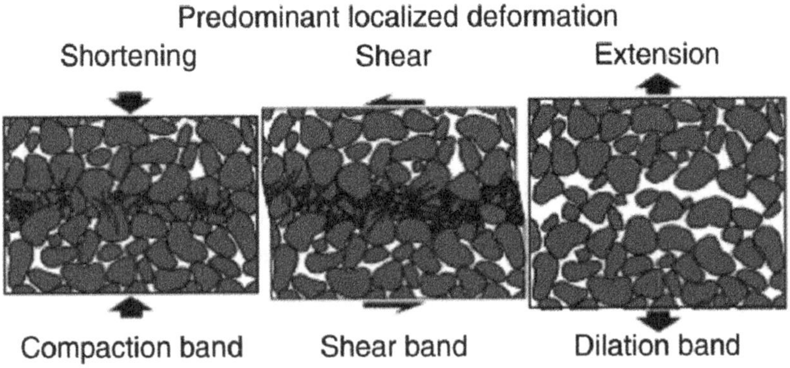

Figure 3.2. *Localized deformation with shear, compaction and dilation modes (Du Bernard* et al. *2002)*

In a simple shear band, the instantaneous relative movement is tangent to the band. In a dilatant (respectively, compactive) shear band, the angle between the unit vectors $\boldsymbol{n}$ and $\boldsymbol{g}$ is acute (respectively, obtuse) and thus the band exhibits some form of instantaneous expansion (respectively, contraction). Pure dilation and pure compaction bands exhibit little or no shear offset (Borja and Aydin 2004).

3.2.2. *Strain localization, loss of ellipticity and vanishing speed of acceleration waves*

The governing equations of an incremental boundary problem are obtained from the equations of equilibrium $\partial_j \Delta\sigma_{ij} = 0$ (for simplicity, we omit here the body forces), the constitutive relationships of the form given by equation [3.2], and the boundary conditions (prescribed tractions or displacements at the boundary of the considered body). Incorporating the constitutive equations [3.2] into the equilibrium equations and assuming piecewise linear incremental constitutive equations yield the following second-order differential system:

$$C_{ijkl}\partial_j \partial_l \Delta u_l = 0 \qquad [3.7]$$

The *ellipticity condition* of the above differential system is expressed as follows:

$$\forall \boldsymbol{n}, C_{ijkl} n_j n_l \text{ is strictly definite positive.} \qquad [3.8]$$

Consequently, the localization criterion [3.5] corresponds to the state of loss of ellipticity of the governing equations. They change type and turn from elliptic to hyperbolic. Shear bands are thus identified with the characteristic lines of the governing hyperbolic partial differential equations.

Let us consider now the propagation of acceleration waves in a solid body along the direction $\boldsymbol{n}$. Acceleration waves are weak discontinuities of the various mechanical fields across wavefronts, which propagate with speed c. One can show that the propagation speed c is the solution of an eigenvalue problem (see Vardoulakis and Sulem 1995) and that c is obtained from the following equation:

$$\det\left(\Gamma_{ik} - \rho c^2 \delta_{ik}\right), \qquad [3.9]$$

where ρ is the density of the material. Thus, if the acoustic tensor is strictly definite positive, all the velocities of acceleration waves are real. When all waves are able to propagate with real velocity, the material is stable in a dynamic sense. This is called

Hadamard's stability criterion (Hadamard 1903), first established for hyperelastic materials. Consequently, the localization criterion [3.5] corresponds to a state for which the velocity of wave propagation in the direction normal to the band is null (stationary wave).

3.3. Shear band formation in element tests on rocks

3.3.1. *Drucker–Prager model*

As seen above, the localization criterion depends on the constitutive relation. The Drucker–Prager plasticity model with non-associate flow rule is commonly used for porous rocks. The yield surface and the plastic potential are expressed as

$$F = \overline{\tau} - \mu(q - \sigma); Q = \overline{\tau} + \beta\sigma, \qquad [3.10]$$

where $\sigma = \sigma_{kk}/3$ is the mean stress (negative in compression), $\overline{\tau} = \sqrt{s_{ij}s_{ij}/2}$ is the Mises equivalent stress (with $s_{ij} = \sigma_{ij} - \sigma\delta_{ij}$), μ is the friction coefficient and β is the dilatancy coefficient. For low-porosity rock, inelastic response is dilatant and $\beta > 0$; however, compressed high-porosity rock typically experiences initial compaction, followed by either dilation or further compaction, depending on the stress state. Negative values for the dilatancy coefficient and negative values for the friction coefficient at high mean stress can thus be observed as for cap yield surface. The critical value of the hardening modulus for which the localization is satisfied (equation [3.5]) is given by Rudnicki and Rice (1975):

$$\frac{H_c}{G} = \frac{1+\nu}{9(1-\nu)}(\beta - \mu)^2 - \frac{1+\nu}{2}\left(N + \frac{\beta + \mu}{3}\right)^2, \qquad [3.11]$$

where G is the elastic shear modulus, ν is the Poisson's ratio, μ is the friction coefficient and β is the dilatancy coefficient. N is the normalized intermediate principal deviatoric stress ($N_{ij} = s_{ij}/\overline{\tau}$) and varies from $-1/\sqrt{3}$ for axisymmetric extension $(\sigma_1 = \sigma_2 > \sigma_3)$ to $1/\sqrt{3}$ for axisymmetric compression $(\sigma_1 > \sigma_2 = \sigma_3)$. The value $N = 0$ corresponds to pure shear $(\sigma_1 = -\sigma_3, \sigma_2 = 0)$. The shear band is in a plane parallel to the intermediate principal stress and its normal is inclined with an angle θ_B with respect to the σ_3-direction (most compressive direction) expressed as (Rudnicki and Olsson 1998):

$$\theta_B = \frac{\pi}{4} + \frac{1}{2}\arcsin\alpha, \text{ with } \alpha = \frac{(2/3)(1+\nu)(\beta+\mu) - N(1-2\nu)}{\sqrt{4-3N^2}} \quad [3.12]$$

The above equations give the largest critical value of the hardening modulus and the shear band orientation as established by Rudnicki and Rice (1975). As mentioned by Perrin and Leblond (1993), these solutions are valid only if $|\alpha| \leq 1$, thus when the magnitude of $\mu+\beta$ is small enough

$$\frac{N(1-2\nu) - \sqrt{4-3N^2}}{2(1+\nu)/3} \leq \beta + \mu \leq \frac{N(1-2\nu) + \sqrt{4-3N^2}}{2(1+\nu)/3} \quad [3.13]$$

The case for which $\beta + \mu > \frac{N(1-2\nu) + \sqrt{4-3N^2}}{2(1+\nu)/3}$ corresponds to the formation of dilation bands in the direction normal to the least compressive principal stress, $\theta_B = \pi/2$, whereas the case for which $\beta + \mu < \frac{N(1-2\nu) - \sqrt{4-3N^2}}{2(1+\nu)/3}$ corresponds to the formation of compaction bands in the direction parallel to the least compressive principal stress, $\theta_B = 0$ (Issen and Rudnicki 2000). The corresponding critical hardening modulus has the form

$$\frac{H_c^k}{G} = \frac{1+\nu}{9(1-\nu)}(\beta-\mu)^2 - \frac{1+\nu}{1-\nu}\left(\frac{1}{2}N_k - \frac{\beta+\mu}{3}\right)^2 - \left(1 - \frac{3}{4}N_k^2\right) \quad [3.14]$$

where $k = 1$ for the dilation bands and $k = 3$ for the compaction bands and N_1 and N_3 are the least and most compressive principal values of the normalized deviatoric stress tensor. These solutions for the critical hardening modulus and the orientation of the band are continuous over the limits of equation [3.13].

For plane strain deformation, an approximate solution for the shear band inclination (equation [3.12]) has been given by Vardoulakis (1980):

$$\theta_B = \frac{\pi}{4} + \frac{\phi_p}{4} + \frac{\psi_p}{4}, \quad [3.15]$$

where ϕ_p and ψ_p are, respectively, the friction angle and the dilatancy angle at peak values. This expression was first proposed by Arthur *et al.* (1977) on the basis of experimental observations.

3.3.2. *Non-coaxial plasticity*

In classical flow theory of plasticity, the direction of the plastic deformation is fixed with respect to the normal to the plastic potential. Consequently, the direction is fixed by the current state of stress and does not depend on the direction of the stress increment. The plastic deformation rate possesses the same principal axes as the stress tensor, which means that it is coaxial to the stress tensor. This is a strong assumption and its consequences on the prediction of shear localization have been discussed in many papers (e.g. Rudnicki and Rice 1975; Vardoulakis 1980; Papamichos and Vardoulakis 1995). In order to arrive to better predictions for shear band formation, one has to abandon the concept of the classical coaxial plasticity flow rule and resort to *hypoplasticity* flow rules, which consider one way or the other the effect of stress rate. Rudnicki and Rice (1975) have considered the contribution of a non-coaxial term motivated by a yield vertex plasticity model. Other examples of non-coaxial constitutive models have been proposed in the form of a deformation theory of plasticity (e.g. Vermeer and Schotman 1986; Sulem and Vardoulakis 1990). More generally, the incrementally nonlinear laws of Darve (1985) and Chambon and Desrues (1989) developed for granular soils and rocks are non-coaxial.

These modifications of the plasticity flow rule result in significant changes in material response for non-proportional loading paths (Desrues and Chambon 2002; Bésuelle and Rudnicki 2004).

3.3.3. *Cataclastic shear banding*

At high confinement, suppressed dilatancy may lead to grain crushing or *cataclasis* inside the shear band (El Bied *et al.* 2002) as shown in Figure 3.3, which in turn leads to substantial porosity and permeability reduction (Sulem and Ouffroukh 2006). At large scale, similar phenomena are observed in faulted zones when sheared. Usually in fault zones, two main domains can be identified: a fault core of small thickness constituted of highly comminuted ultra-cataclasites is surrounded with a damage zone, which consists of fractured host rock (e.g. Chester and Chester 1998). The ultracataclastic structure is the result of numerous earthquake ruptures. As shown, for example, in a recent study of Sulem *et al.* (2004) dedicated to the characterization of Aigion fault material in the Gulf of Corinth (Greece), the permeability of the fault core is very low, so this zone acts as an impervious barrier to transverse fluid flow, whereas the highly fractured damage zone around acts mostly as conduits for nearly along-strike flow. Similar observations can be found in the work of Wibberley and Shimamoto (2002).

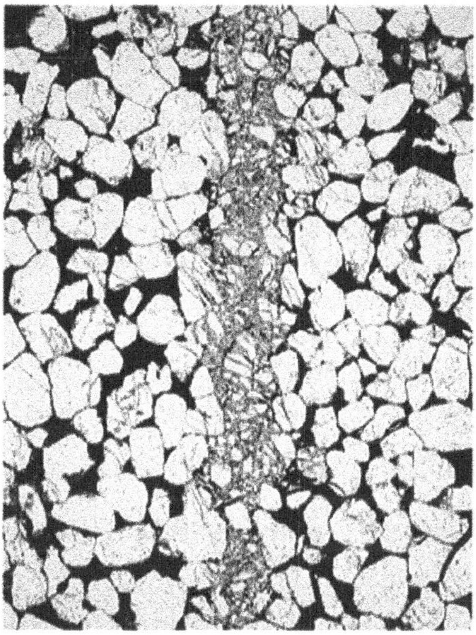

Figure 3.3. *Cataclastic shear banding in Fontainebleau sandstone (El Bied* et al. *2002)*

3.3.4. *Postlocalization behavior*

The various drawbacks and shortcomings of the classical continuum theory in connection with strain localization are related to the fact that in the postbifurcation regime, we deal in general with mathematically ill-posed governing equations. The origin of this undesirable situation can be traced back to the fact that conventional constitutive models do not contain material parameters with a dimension of length, so the shear band thickness (i.e. the extent of the plastically softening region) is undetermined. We can say that localization of deformation leads to a change in scale of the problem so phenomena occurring at the scale of the grain cannot be ignored anymore in the modeling of the macroscopic behavior of the material. Then, it appears necessary to resort to continuum models with microstructure to correctly describe localization phenomena. These generalized continua usually contain additional kinematical degrees of freedom (Cosserat continuum) and/or higher deformation gradients (higher grade continuum). Cosserat continua and higher grade continua belong to a general class of constitutive models, which account for the materials' microstructure. The contemporary formulation of these models is based on the work of Mindlin (1964) and Germain (1973a, 1973b). Rotation gradients and higher velocity gradients introduce a material length scale into the problem, which, as mentioned earlier, is necessary for the correct modeling of localization phenomena. This idea was widely publicized by the paper of Mühlhaus and Vardoulakis (1987). In this case, the underlying mathematical problem describing localization phenomena is "regularized" and the governing equations remain elliptic. Moreover, this technique allows robust computations to follow the evolution of the considered system in the postbifurcation regime and to extract additional information such as the shear band thickness or to assess the effect of scale. In the last two decades, large-scale numerical simulations that account for higher order continuum effects have been developed (Papanastasiou and Vardoulakis 1992; Zervos *et al.* 2001; Matsushima *et al.* 2002; Collin *et al.* 2006, 2016).

3.4. Strain localization in fluid-saturated porous media

3.4.1. *Strain localization criterion in fluid-saturated porous media*

The rate constitutive equations for saturated porous elastic-plastic media can be written in the following two alternative forms (Benallal and Comi 2003; Coussy 2004):

$$\begin{aligned} \dot{\boldsymbol{\sigma}} &= \mathbf{C}^d \dot{\boldsymbol{\varepsilon}} - \mathbf{K}\dot{p} \\ \dot{\boldsymbol{\sigma}} &= \mathbf{C}^u \dot{\boldsymbol{\varepsilon}} - \frac{1}{D}\mathbf{K}\dot{\zeta}, \end{aligned} \tag{3.16}$$

where $\boldsymbol{C}^d$ is the drained tangent modulus related to the stress rate under drained conditions (constant pore pressure p) and $\boldsymbol{C}^u$ is the undrained tangent modulus related to the stress rate under undrained conditions (constant fluid content ζ):

$$\begin{aligned} \mathbf{C}^d &= \mathbf{C}_e^d - \frac{1}{H^d}\left(\mathbf{C}_e^d : \frac{\partial F}{\partial \boldsymbol{\sigma}} \otimes \frac{\partial Q}{\partial \boldsymbol{\sigma}} : \mathbf{C}_e^d\right) \\ \mathbf{C}^u &= \mathbf{C}_e^u - \frac{1}{H^u}\left(\mathbf{C}_e^u : \left(\frac{\partial F}{\partial \boldsymbol{\sigma}} - \frac{bM}{3K^u}\frac{\partial F}{\partial p}\mathbf{1}\right) \otimes \left(\frac{\partial Q}{\partial \boldsymbol{\sigma}} - \frac{bM}{3K^u}\frac{\partial Q}{\partial p}\mathbf{1}\right) : \mathbf{C}_e^u\right) \end{aligned} \tag{3.17}$$

In the above equations, $\mathbf{C}_e^d$ (respectively, $\mathbf{C}_e^u$) is the drained (respectively, undrained) elastic tensor, F is the yield function, Q is the plastic potential, b is the Biot elastic coefficient, M is the Biot modulus, K^u is the undrained elastic bulk modulus, with

$$H^d = \frac{\partial Q}{\partial \boldsymbol{\sigma}} : \mathbf{C}_e^d : \frac{\partial F}{\partial \boldsymbol{\sigma}} + h^d$$

$$H^u = H^d + \frac{1}{N}(\mathbf{K} \otimes \mathbf{L})$$

$$\mathbf{K} = b\mathbf{1} + \frac{1}{H^d}\left(\frac{\partial Q}{\partial p} - b\frac{\partial Q}{\partial \boldsymbol{\sigma}} : \mathbf{1}\right)\mathbf{C}_e^d : \frac{\partial F}{\partial \boldsymbol{\sigma}} \quad [3.18]$$

$$\mathbf{L} = b\mathbf{1} + \frac{1}{H^d}\left(\frac{\partial F}{\partial p} - b\frac{\partial F}{\partial \boldsymbol{\sigma}} : \mathbf{1}\right)\frac{\partial f}{\partial \boldsymbol{\sigma}} : \mathbf{C}_e^d$$

$$D = \frac{1}{M} + \frac{1}{H^d}\left(\frac{\partial F}{\partial p} - b\frac{\partial F}{\partial \boldsymbol{\sigma}} : \mathbf{1}\right)\left(\frac{\partial Q}{\partial p} - b\frac{\partial Q}{\partial \boldsymbol{\sigma}} : \mathbf{1}\right)$$

and h^d is the drained plastic hardening modulus.

The stability of homogeneous deformation is investigated by considering an infinitesimal perturbation $\delta\mathbf{X}$ to the set $\mathbf{X}$ of all fields describing the response of the medium (displacements, stresses, pore pressure, etc.). The perturbation field is written in the form $\delta\mathbf{X} = \tilde{\mathbf{X}}\exp(i\xi\mathbf{n}.\mathbf{x} + st)$, where $\mathbf{n}$ is a unit vector, ξ is a wavenumber and s is the growth rate of the perturbation. Instability in the form of unbounded growth of the perturbation occurs when the following condition is met (Benallal and Comi 2003):

$$sD\det\left[\mathbf{n}.\mathbf{H}^u.\mathbf{n}\right] + k\xi^2\det\left[\mathbf{n}.\mathbf{H}^d.\mathbf{n}\right] = 0, \quad [3.19]$$

where k is the permeability of the medium. Thus, the rate of growth s is given by:

$$s = -\frac{k\xi^2}{D}\frac{\det\left[\mathbf{n}.\mathbf{H}^d.\mathbf{n}\right]}{\det\left[\mathbf{n}.\mathbf{H}^u.\mathbf{n}\right]} = 0 \quad [3.20]$$

From the above equation, it is seen that unbounded growth of the perturbation occurs whenever for some directions $\mathbf{n}$ we have:

$$\det\left[\mathbf{n}.\mathbf{H}^d.\mathbf{n}\right].\det\left[\mathbf{n}.\mathbf{H}^u.\mathbf{n}\right] < 0 \quad [3.21]$$

Therefore, strain localization occurs when the drained or the undrained acoustic tensor becomes singular:

$$\det\left[\mathbf{n}.\mathbf{H}^{d}.\mathbf{n}\right]=0$$

or $\quad$ [3.22]

$$\det\left[\mathbf{n}.\mathbf{H}^{u}.\mathbf{n}\right]=0$$

Thus, the two conditions must be checked on the real deformation path (which is not necessarily drained or undrained) to infer which one is met first. For associative behavior, it is shown that the singularity of the drained acoustic tensor occurs before the singularity of the undrained acoustic tensor, which means that instability occurs when the condition of localization of the underlying drained deformation is met (Benallal and Comi 2000). However, for non-associative behavior, instability is controlled either by the drained or the undrained properties, depending on the constitutive equations and the loading path.

3.4.2. *Stability analysis of undrained shear on a saturated layer*

The problem of the stability of a saturated rock layer that is sheared plane strain (Figure 3.4) has been addressed by Rice (1975), Vardoulakis (1985, 1986), Garagash and Rudnicki (2002) and Garagash (2005). We assume that the layer has a thickness L and is long enough so that we can consider that all the fields only vary with z (normal to the layer axis) and time t.

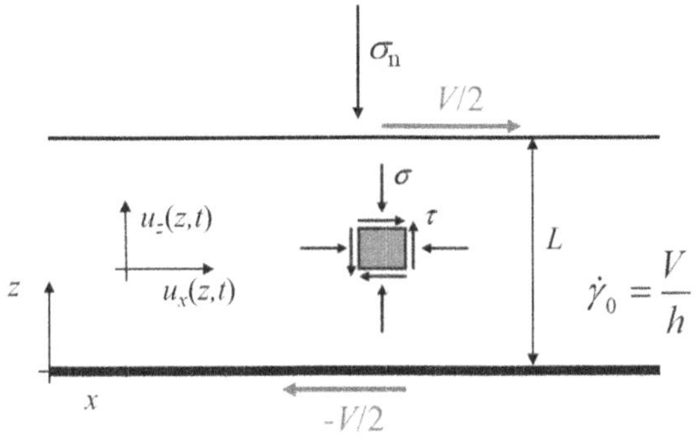

Figure 3.4. *Saturated rock layer under constant normal stress and sheared at constant velocity*

The shear strain γ and the volumetric strain ε are given by:

$$\gamma=\frac{\partial u_x}{\partial z} \quad \varepsilon=\frac{\partial u_z}{\partial z} \qquad [3.23]$$

For the one-dimensional spatial variation, the equilibrium equations reduce to:

$$\frac{\partial \tau}{\partial z} = 0 \quad \frac{\partial \sigma}{\partial z} = 0 \tag{3.24}$$

So the stresses are spatially uniform and vary only with time. For incompressible fluid and by using Darcy's law, the equation of fluid mass balance gives the following diffusion equation of the pore fluid p:

$$k \frac{\partial^2 p}{\partial z^2} = \frac{\partial \varepsilon}{\partial t}, \tag{3.25}$$

where k is the permeability of the rock.

Boundary conditions permit us to distinguish the globally drained case (imposed pore pressure on the boundaries) and the globally undrained case (zero fluid flux at the boundaries):

$$\begin{aligned} &\text{Drained case: } p(z = \pm L/2, t) = p_0 \\ &\text{Undrained case: } \frac{\partial p}{\partial z}(z = \pm L/2, t) = 0 \end{aligned} \tag{3.26}$$

To simplify the presentation, it is assumed that the solid grains are incompressible so that the constitutive equations can be written in terms of Terzaghi effective stress $\sigma - p$. One can easily extend the analysis to compressible grains and pore fluid by using the framework of Biot theory of porous media (e.g. Rice 1975). By separating the elastic and plastic parts of the deformation, the elastoplastic constitutive equations are written as:

$$\begin{aligned} &d\gamma = \frac{1}{G} d\tau + d\gamma^p \; ; d\varepsilon = \frac{1}{M} d(\sigma - p) + d\varepsilon^p \\ &d\gamma^p = \frac{1}{H}\left(d\tau - \mu d(\sigma - p)\right); d\varepsilon^p = \beta d\gamma^p \end{aligned} \tag{3.27}$$

where G is the elastic shear modulus, M is the elastic bulk modulus under uniaxial strain, H is the plastic hardening modulus, μ is the friction coefficient and β is the dilatancy coefficient.

The homogeneous drained response is described by the following incremental equations:

$$dp = 0, \quad d\tau = h d\gamma \text{ with } h = \frac{H}{1 + H / G} \qquad [3.28]$$

and the undrained homogeneous response by:

$$d\varepsilon = 0, \quad d\tau = h_u d\gamma \text{ with } h_u = \frac{H + \beta\mu M}{1 + (H + \beta\mu M) / G} \qquad [3.29]$$

The stability of homogeneous deformation under the globally undrained condition is studied by considering small perturbations of the various fields and by analyzing their evolution in time:

$$\gamma = \gamma_0 + \tilde{\gamma}\,; \varepsilon = \varepsilon_0 + \tilde{\varepsilon}\,; \sigma = \sigma_0 + \tilde{\sigma}\,; \tau = \tau_0 + \tilde{\tau}\,; p = p_0 + \tilde{p} \qquad [3.30]$$

Using the equilibrium and the constitutive equations, the diffusion equation for the perturbation of the pore pressure [3.29] is written as:

$$\frac{\partial \tilde{p}}{\partial t} = c \frac{\partial^2 \tilde{p}}{\partial z^2} \text{ with } c = \kappa M \frac{H}{H + \mu\beta M} \qquad [3.31]$$

This perturbation can be decomposed in Fourier modes with wavelength λ:

$$\tilde{p} = P(0) e^{st} \cos\left(\frac{2\pi z}{\lambda}\right), \qquad [3.32]$$

where s is the growth coefficient in time of the perturbation (Lyapunov coefficient). The zero fluid flux boundary condition is satisfied for $\lambda = L / n$, where n is an integer.

From equations [3.35] and [3.36], we deduce the expression of s:

$$s = -c\left(\frac{2\pi}{\lambda}\right)^2 \qquad [3.33]$$

The stability condition is given by the sign of s. If $c < 0$, then $s > 0$, which corresponds to a perturbation that grows exponentially in time (instability). Perturbations with the smallest wavelength grow faster. Note that if $c < 0$, equation [3.31] represents a diffusion equation with a negative diffusivity coefficient or in other words a diffusion equation for which time runs backwards.

For a dilatant material ($\beta > 0$), the instability occurs when the hardening modulus H becomes equal to zero, i.e. at the peak of the drained stress–strain curve ($h = 0$) and when the growth coefficient of the perturbation with infinitesimal wavelength grows the fastest. For a contractant material ($\beta < 0$), the instability occurs at the peak of the undrained stress–strain curve ($h_u = 0$) and the growth coefficient of the perturbation is infinite for all wavelengths.

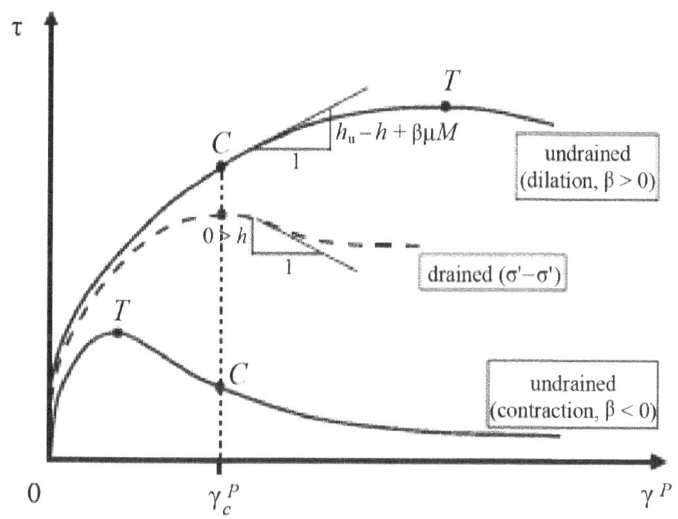

Figure 3.5. *Shear stress–plastic shear strain curves in drained and undrained conditions for a dilatant (β > 0) or contractant (β < 0) material (adapted from Garagash and Rudnicki 2002)*

The above conclusions indicate that the above mathematical problem is ill-posed and needs to be regularized. This can be done by introducing rate dependency in the constitutive equations (Vardoulakis 1986; Garagash 2005). For that, strain rate dependency is introduced in the plastic yield criterion f by considering that the friction coefficient μ depends on the plastic shear strain and its rate. Accordingly, two hardening moduli $h = \dfrac{\partial f}{\partial \gamma^p}$ and $\delta = \dfrac{\partial f}{\partial \ln \dot{\gamma}^p}$ are introduced.

$$\tau = f\left(\sigma - p, \gamma^p, \dot{\gamma}^p\right)$$
$$\mu = \frac{\partial f}{\partial(\sigma - p)}, \; h = \frac{\partial f}{\partial \gamma^p}, \; \delta = \frac{\partial f}{\partial \ln \dot{\gamma}^p} \qquad [3.34]$$

The same stability analysis as the one presented above can be developed and the following quadratic equation for the growth coefficient of the perturbation s is obtained:

$$\frac{\delta}{\dot{\gamma}_0^p}s^2+\left(h+M\beta\mu+Mk\frac{\delta}{\dot{\gamma}_0^p}\left(\frac{2\pi}{\lambda}\right)^2\right)s+Mkh\left(\frac{2\pi}{\lambda}\right)^2=0 \qquad [3.35]$$

The instability condition ($s>0$) is given by:

$$h+M\beta\mu+Mk\frac{\delta}{\dot{\gamma}_0^p}\left(\frac{2\pi}{\lambda}\right)^2<0 \qquad [3.36]$$

This condition occurs first for the largest wavelength (i.e. $\lambda=L$). In other words, for a contractant material under undrained shear, diffuse instability occurs first. One can make the following remarks: (1) strain rate sensitivity is amplified by pore fluid diffusion to delay the instability, (2) higher shearing rate leads to earlier instability, (3) if the layer thickness is small enough, diffuse instability may be prevented (scale effect) and localized instability will occur first.

3.5. Conclusion

Bifurcation analysis and localization theory constitute the basis of contemporary continuum theory of failure as a natural extension of classical theory of strength of materials. The basic notions in relation with the shear band model are presented. The bifurcation condition corresponds to the singularity of the acoustic tensor. It is shown that the localization criterion corresponds to a state for which the velocity of wave propagation in the direction normal to the band is null. Extension to fluid-saturated media is presented. For associative behavior, it is shown that instability occurs when the condition of localization of the underlying drained deformation is met. However, for non-associative behavior, instability is controlled either by the drained or the undrained properties, depending on the constitutive equations and loading path. The inelastic volume changes play a key role in the diffuse or localized instability.

3.6. References

Arthur, J.R.F., Dunstan, T., Chua, K.S. (1977). Induced anisotropy in a sand. *Géotechnique*, 27(1), 13–30.

Baud, P., Klein, E., Wong, T.F. (2004). Compaction localization in porous sandstones: Spatial evolution of damage and acoustic emission activity. *J. Struct. Geol.*, 26(4), 603–624.

Benallal, A. and Comi, C. (2000). On localization in saturated porous continua. *Comptes Rendus de l'Académie des Sciences*, 328, 847–853.

Benallal, A. and Comi, C. (2003). Perturbation growth and localization in fluid-saturated inelastic porous media under quasi-static loadings. *J. Mech. Phys. Solids*, 51, 851–899.

Du Bernard, X., Eichhubl, P., Aydin, A. (2002). Dilation bands: A new form of localized failure in granular media. *Geophys. Res. Lett.*, 29(24), 2176.

Bésuelle, P. and Rudnicki, J.W. (2004). Localization: Shear bands and compaction bands. *In International Geophysics: Mechanics of Fluid-Saturated Rocks*, Y. Guéguen and M. Bouteca (eds), Elsevier.

Bésuelle, P., Desrues, J., Raynaud, S. (2000). Experimental characterisation of the localisation phenomenon inside a Vosges sandstone in a triaxial cell. *Int. J. Rock Mech. Min. Sci.*, 37(8), 1223–1237.

El Bied, A., Sulem, J., Martineau, F. (2002). Microstructure of shear zones in Fontainebleau sandstone. *Int. J. Rock Mech. Min. Sci.*, 39(7), 917–932.

Borja, R.I. and Aydin, A. (2004). Computational modeling of deformation bands in granular media I: Geological and mathematical framework. *Comput. Methods Appl. Mech. Eng.*, 193(27–29), 2667–2698.

Chester, F.M. and Chester, J.S. (1998). Ultracataclasite structure and friction processes of the Punchbowl fault, San Andreas system, California. *Tectonophysics*, 295(1–2), 199–221.

Collin, F., Chambon, R., Charlier, R. (2006). A finite element method for poro mechanical modelling of geotechnical problems using local second gradient models. *Int. J. Numer. Methods Eng.*, 65(11), 1749–1772.

Collin, F., Kotronis, P., Pardoen, B. (2016). Numerical modelling of Multiphysics couplings and strain localization. In *ALERT Doctoral School 2016: Modelling of Instabilities and Bifurcation in Geomechanics*, J. Sulem, I. Stefanou, E. Papamichos, M. Veveakis (eds). Available: http://alertgeomaterials.eu/data/school/2016/2016_ALERT_schoolbook.pdf.

Coussy, O. (2004). *Poromechanics*. John Wiley and Sons, Chichester, England.

Darve, F. (1985). *Les lois incrementales non-linéaires, Manuel de rhéologie des géomatériaux*. Presse des Ponts et Chaussées, Paris, France.

Desrues, J. and Chambon, R. (1989). Shear band analysis for granular materials: The question of incremental non-linearity. *Ingenieur-Archiv*, 59(3), 187–196.

Desrues, J. and Chambon, R. (2002). Shear band analysis and shear moduli calibration. *Int. J. Solids Struct.*, 39(13–14), 3757–3776.

Desrues, J. and Viggiani, G. (2004). Strain localization in sand: An overview of the experimental results obtained in Grenoble using stereophotogrammetry. *Int. J. Numer. Anal. Methods Geomech.*, 28(4), 279–321.

Garagash, D.I. (2005). Diffuse vs. localized instability in compacting geomaterials under undrained conditions. In *Geomechanics: Testing, Modeling and Simulation. Proceedings of the First Japan-US Workshop on Testing, Modelling and Simulation*, Yamamuro, J.A., Koseki, J. (eds). ASCE, Geotechnical Special Publication, USA.

Garagash, D.L. and Rudnicki, J.W. (2002). Stability of undrained deformation of fluid-saturated geomaterials. *In Plasticity, Damage and Fracture at Macro, Micro and Nano Scales: Proceedings of Plasticity '02: the Ninth International Symposium on Plasticity and Its Current Applications*, A.S. Khan, O. Lopez-Pamies (eds), 492–494, Neat Press, Fulton, USA.

Germain, P. (1973). The method of virtual power in continuum mechanics. Part 2: Microstructure. *SIAM J. Appl. Math.*, 25(3), 556–575.

Germain, P. and Lee, E.H. (1973). On shock waves in elastic-plastic solids. *J. Mech. Phys. Solids*, 21(6), 359–382.

Hadamard, J. (1903). *Leçons sur la propagation des ondes et les équations de l'hydrodynamique*. Librairie Scientifique A. Hermann, Paris.

Hill, R. (1962). Acceleration waves in solids. *J. Mech. Phys. Solids*, 10(1961), 1–16.

Issen, K.A., Rudnicki, J.W. (2000). Conditions for compaction bands in porous rock. *J. Geophys. Res.*, 105(B9), 21529.

Mandel, J. (1966). Conditions de stabilité et postulat de Drucker. In *Rheology and Soil Mechanics/Rhéologie et Mécanique des Sols*, Kravtchenko, P.M. Sirieys (eds). International Union of Theoretical and Applied Mechanics, Springer, Berlin.

Matsushima, T., Chambon, R., Caillerie, D. (2002). Large strain finite element analysis of a local second gradient model: Application to localization. *Int. J. Numer. Methods Eng.*, 54(4), 499–521.

Mindlin, R.D. (1964). Micro-structure in linear elasticity. *Arch. Ration. Mech. Anal.*, 16(1), 51–78.

Mühlhaus, H.B. and Vardoulakis, I. (1987). The thickness of shear bands in granular materials. *Géotechnique*, 37(3), 271–283.

Olsson, W.A., Holcomb, D., Rudnicki, J.W. (2002). Compaction localization in porous sandstone: Implications for reservoir mechanics. *Oil Gas Sci. Technol.*, 57(5), 591–599.

Papanastasiou, P.C. and Vardoulakis, I.G. (1992). Numerical treatment of progressive localization in relation to borehole stability. *Int. J. Numer. Anal. Methods Geomech.*, 16(6), 389–424.

Papamichos, E. and Vardoulakis, I. (1995). Shear band formation in sand according to non-coaxial plasticity model. *Géotechnique*, 45(4), 649–661.

Perrin, G. and Leblond, J.B. (1993). Rudnicki and Rice's analysis of strain localization revisited. *J. Appl. Mech.*, (93), 842–846.

Rattez, H., Stefanou, I., Sulem, J. (2018a). The importance of thermo-hydro-mechanical couplings and microstructure to strain localization in 3D continua with application to seismic faults. Part I: Theory and linear stability analysis. *Journal of the Mechanics and Physics of Solids*, 115, 54–76.

Rattez, H., Stefanou, I., Sulem, J., Veveakis, M., Poulet, T. (2018b). The importance of thermo-hydro-mechanical couplings and microstructure to strain localization in 3D continua with application to seismic faults. Part II: Numerical implementation and post-bifurcation analysis. *Journal of the Mechanics and Physics of Solids*, 115, 1–29.

Rice, J.R. (1975). On the stability of dilatant hardening for saturated rock masses. *J. Geophys. Res.*, 80(11), 1531–1536.

Rice, J.R. (1976). The localization of plastic deformation. In *Theoretical and Applied Mechanics (Proceedings of the 14th International Congress on Theoretical and Applied Mechanics)*, Koiter, W.T. (ed.). NorthHolland Publishing Co., Delft, The Netherlands, 207–220.

Rice, J.R. and Rudnicki, J.W. (1980). A note on some features of the theory of localization of deformation. *Int. J. Solids Struct.*, 16, 597–605.

Rudnicki, J.W. and Rice, J.R. (1975). Conditions for the localization of deformation in pressure-sensitive dilatant materials. *J. Mech. Phys. Solids*, 23(6), 371–394.

Rudnicki, J.W. and Olsson, W.A. (1998). Reexamination of fault angles predicted by shear localization theory. *Int. J. Rock Mech. Min. Sci.*, 35(4–5), 512–513.

Simo, J.C., Oliver, J., Armero, F. (1993). An analysis of strong discontinuities induced by strain-softening in rate-independent inelastic solids. *Comput. Mech.*, 12(5), 277–296.

Sulem, J. and Ouffroukh, H. (2006). Shear banding in drained and undrained triaxial tests on a saturated sandstone: Porosity and permeability evolution. *Int. J. Rock Mech. Min. Sci.*, 43(2), 292–310.

Sulem, J. and Vardoulakis, I. (1990). Bifurcation analysis of the triaxial test on rock specimens. A theoretical model for shape and size effect. *Acta Mech.*, 83(3–4), 195–212.

Sulem, J., Vardoulakis, I., Papamichos, E., Oulahna, A., Tronvoll, J. (1999). Elasto-plastic modelling of Red Wildmoor sandstone. *Mech. Cohesive-frictional Mater.*, 4(3), 215–245.

Sulem, J., Vardoulakis, I., Ouffroukh, H., Boulon, M., Hans, J. (2004). Experimental characterization of the thermo-poro-mechanical properties of the Aegion Fault gouge. *Comptes Rendus – Geosci.*, 336, 455–466.

Thomas, T.Y. (1961). *Plastic Flow and Fracture in Solids*. Academic Press, New York, USA.

Vardoulakis, I. (1976). Equilibrium theory of the shear bands in plastic bodies. *Mech. Res. Commun.*, 3, 209–214.

Vardoulakis, I. (1980). Shear band inclination and shear modulus of sand in biaxial tests. *Int. J. Numer. Anal. Methods Geomech.*, 4, 103–119.

Vardoulakis, I. (1985). Stability and bifurcation of undrained plane rectilinear deformations on water-saturated granular soils. *International Journal for Numerical and Analytical Methods in Geomechanics*, 9, 399–414.

Vardoulakis, I. (1986). Dynamic stability analysis of undrained simple shear on water-saturated granular soils. *International Journal for Numerical and Analytical Methods in Geomechanics*, 10, 177–190.

Vardoulakis, I. and Sulem, J. (1995). *Bifurcation Analysis in Geomechanics*. Blackie, Glasgow, Scotland.

Vermeer, P.A. and Schotman, G.J. (1986). An extension to the deformation theory of plasticity. In *2nd International Symposium on Numerical Models in Geomechanics*. Balkema, Rotterdam, The Netherlands, 33–41.

Wibberley, C.A.J. and Shimamoto, T. (2002). Internal structure and permeability of major strike-slip fault zones: The median tectonic line in Mie Prefecture, Southwest Japan. *J. Struct. Geol.*, 25, 59–78.

Zervos, A., Papanastasiou, P., Vardoulakis, I. (2001). Modelling of localisation and scale effect in thick-walled cylinders with gradient elastoplasticity. *Int. J. Solids Struct.*, 38(30–31), 5081–5095.

4

Experimental Investigation of the Emergence of Strain Localization in Geomaterials

Pierre BÉSUELLE

3SR, Université Grenoble Alpes and CNRS, France

Failures of geomaterials, including soils, rocks and concretes, are generally associated with a localized deformation. For about 40 years, strain localization phenomena have been investigated experimentally in different materials. Most of these studies have been motivated by the theoretical background of shear band analysis using bifurcation theory. Strain field measurements have been developed to characterize strain localization, especially for sand specimens. Full-field methods, imaging tools and experimental loading apparatus have evolved considerably over past 15 years. This chapter looks at on the contributions of recent developments on the characterization of the strain localization process. The emergence of strain localization involves the progressive evolution from diffuse to localized deformation. The text introduces the methods used and then shows some selected experimental results obtained from some sands and porous rocks.

4.1. Introduction

Localized deformation is a ubiquitous phenomenon in geomaterials (soils, rocks and concrete). It occurs over a vast range of size scales, from the microscale level of grains to faults extending over hundreds of kilometers. It occurs in a variety of forms, as a concentration or coalescence of cracks; a distinct, planar frictional surface; a gouge zone of finely comminuted material; or simply a region of higher shear strain or relative grain movements. In geomaterials, the severe shearing in regions of localized deformation may be accompanied by dilatancy (inelastic volume increase) and/or compaction (inelastic volume decrease) as well as by chemical alteration. If the material is fluid-saturated, as is frequently the case, inelastic volume changes can induce the flow of fluid or changes in pore pressure, which will affect the response. Localization occurs under a variety of conditions that depend on the material and the loading process (e.g. mean stress and loading rate). Although most frequently associated with the formation of faults under nominally brittle conditions or shear bands semi-brittle conditions, localization can also occur by cataclastic flow of rocks at higher mean stresses and by ductile shearing at temperatures and pressures typical of depths of 10–15 km into the Earth's crust. A large transition zone bridges the brittle and ductile regimes. This distinction between the different types of localization can also be a matter of observation scale.

Sometimes, the occurrence of localization essentially coincides with failure. For example, in the axisymmetric loading of porous rock or sand specimens, homogeneous deformation is more or less terminated by the sudden appearance of a sharp fault or shear band in the sample and by rapid loss of strength. However, the transition from homogeneous deformation to clearly localized deformation is still an open question as far as eventual precursors are concerned. In other instances, incipient localization may be only the first step in a continuing process: a shear band can be deactivated but still exist (non-reversible deformation) and can be potentially reactivated later. In other circumstances, a shear band can evolve as a fault, marking an inability of the material to support large strain or highly concentrated microdeformation mechanisms, and switch to a material discontinuity.

Obviously, the phenomenon of localization in geological materials is both rich and complex and its consequences can be important. Seismic activity on large faults is a threat to human life and is a concern for the siting of structures. Shear zones and fractures of a more modest scale can be problems for foundations and underground structures of all types. The formation of a localized slip surface is frequently a predecessor to slope failure. More generally, however, features of localized deformation control not only the mechanical behavior but also the transport behavior of the crust. For example, faults often form impermeable barriers that trap hydrocarbons or bound aquifers. The presence of localized zones of deformation may disrupt efforts to withdraw or inject fluids into the subsurface. Although faults and shear zones typically inhibit flow perpendicular to their strike, they can also act as high-permeability channels for flow along the strike, and carry fluid much further and more rapidly than predicted by assuming a homogeneous formation.

The importance of localization has motivated numerous experimental, theoretical and numerical works in the field of geomechanics, especially since the 1970s. However, open questions exist about how the occurrence of localization and its subsequent evolution are related to the stress state, deformation history, material properties and the coupling of deformation with the transport of fluid and heat and with chemical alteration.

In terms of theoretical approach, ever since the pioneering works of Hill (1962) and Rice (1976), bifurcation theory considers strain localization as a loss of a uniqueness problem. If one considers a homogeneous body subjected to boundary conditions such that one possible solution for the next increment is additional homogeneous deformation, the bifurcation approach seeks conditions for which an alternative solution is possible, corresponding to localized deformation in a planar band (see Vardoulakis and Sulem (1995) and Bésuelle and Rudnicki (2004) for comprehensive reviews).

Numerical modeling of strain localization concerns, in particular, the practical aspect of the band thickness. The absence of a length scale in the classical constitutive models causes numerical simulations to depend spuriously on the mesh size (in terms of band thickness and *de facto* energy dissipation). Recent developments have introduced a length scale in the continuum using micromorphic models (Chambon *et al.* 2001; Cosserat 1909; Fleck and Hutchinson 1997; Mühlhaus and Vardoulakis 1987) or non-local models (Bažant *et al.* 1987; Bažant and Pijaudier-Cabot 1988; Vardoulakis and Aifantis 1991), to cite a few studies.

Concerning experimental characterization, a large amount of work has also been produced, on soil and rock mechanics (see, for example, Vardoulakis and Sulem (1995); Desrues (1998); Paterson and Wong (2005) for extended reviews). Pioneering studies have characterized strain localization by *postmortem* or destructive analysis (especially in rock mechanics). At the end of mechanical tests, after the specimens have been removed from the loading device, the patterns of localization, shear band orientation, local porosity and grain scale damage are analyzed with regard to experimental and loading conditions (boundary conditions, mean stress level, stress rate, material initial state, fluid saturation, etc.).

This chapter attempts to review recent experimental observations about *the progressive evolution from diffuse to localized deformation.* Spatial descriptions of physical quantities and the time evolution of these fields are needed for this aim, in contrast to more conventional measurement techniques based on global measurement by transducers positioned outside of the loading device and/or at the specimen boundaries (no field measurement) or on *postmortem* characterization (no time evolution). Such techniques that are referred to as full-field measurements became more and more popular during the last two decades (Viggiani and Hall 2012). They are potentially well-adapted to detecting, during laboratory tests, the transition from the initial (quasi-) homogeneous regime to the localized regime.

Full-field methods represent a long list of different techniques. Kinematic (displacement) field measurements seem obvious for detecting strain localization.

Once the displacement field is measured, the strain field can be deduced quite easily by applying continuum mechanics. Many techniques exist, including among others, image correlation, the grid method, speckle interferometry and the moiré method (Grédiac 2004; Vautrin *et al.* 2002). Other methods for non-kinematic measurements exist based on imagery techniques of physical processes, such as ray absorption, wave propagation or attenuation, temperature and electrical resistivity. Some methods use tracer emission detection (e.g. gamma ray emission). Generally, these methods are not direct and need reconstruction algorithms to quantify the physical properties: X-ray tomography, neutron tomography, ultrasonic tomography, electrical resistivity tomography, magnetic resonance tomography, positron emission tomography, dynamic thermal tomography, etc. Acoustic emission monitoring is also interesting for geomaterials. The acoustic events (AEs) generated during deformation (due to microcracking) are detected by transducers on the specimen boundaries and are located by inverse analysis, giving a field of AEs resolved in time (e.g. Fortin *et al.* (2009)). A review of the different full-field measurement methods is beyond the scope of this chapter. We will simply present the digital image correlation (DIC) method because it is the most popular kinematic field method in geomechanics. The method can be used with classical images from a digital high resolution camera (2D measurement) or combined with more complex imaging techniques such as X-ray tomography (3D measurements). The success of DIC in geomechanics comes, in 2D, from the relative simplicity of the specimen preparation in generating a needed random speckle, the availability of commercial codes and the trend in decreasing costs of digital cameras. Moreover, in 3D (volume), the possibility to use the natural heterogeneity of the geomaterial as a random speckle is crucial.

To continue the chapter, some selected results will be presented to illustrate the transition from diffuse to localized deformation in sands and porous rocks. Note that we have chosen to present results coming, in large part, from our research group in Grenoble. We chose to select some results that we know well, and not necessarily the most advanced results from the literature each time. However, this selection should give quite a wide view of the subject. The first results concern the strain localization in a Hostun sand observed in plane strain compression. Then, the strain localization in sands will be presented at a smaller scale at the scale of the grains because of X-ray *computed tomography* (CT). Afterward, the localization in two porous rocks (a Vosges sandstone and a clayey rock) observed under plane strain compression will be discussed.

4.2. Methods

This section describes the methods and tools used to obtain the results presented in below section. DIC is first presented. The method became popular in geomechanics several years ago. It has superseded the *false relief stereophotogrammetry* (FRS), which was used in the 1970s–1990s (Butterfield *et al.*

1970; Desrues and Viggiani 2004). FRS is based on photographs taken from a fixed viewpoint at different times of the specimen's deformation. An essential feature of FRS is that the deformation is directly perceived as a fictitious relief (hence the name of the method) by using the well-known stereoscopic effect on successive pairs of photographs. The relief observed is due to the (plane) deformation taking place in the time interval between the two photographs, and not the real 3D geometry of the object. When two photographs are viewed in stereo, displaced regions appear elevated, with the elevation being proportional to the magnitude of the displacement. In the presence of a shear band, the deforming specimen appears as two planes of different elevation connected by a slope.

X-ray CT is then described, which allows imaging of the specimens in 3D (volume). X-ray CT can directly detect the shear bands in specimens undergoing dilatancy or compaction (mass density change). However, the combination of X-ray CT with DIC is a much more powerful tool for detecting incipient shear bands, because it measures the full strain tensor field (including shear strain) inside the specimen and not only the mass density change (volume strain field).

The adaptation of experimental loading devices for geomaterial to full-field measurements will be discussed. The devices need to be transparent to visible light (classical camera) or to X-rays for X-ray CT. This adaptation of the loading device can become complex, especially for rock mechanics due to the high strength and stiffness of the specimens.

4.2.1. *Digital image correlation*

DIC provides a displacement field measurement between two images of the specimen. One of the images is called the reference image and the image corresponding to a deformed state of the specimen is called the deformed image. The displacement field corresponds to the transformation that matches the reference image to the deformed image. The measure at a discrete set of positions (nodes) using a correlation algorithm is based on a principle of optical flow conservation. Then, the strain tensor field can be deduced from the displacement vector field using spatial derivation methods for a discrete field. Note that an extended description of the method can be found in Hall (2012) and a state of the art description of good practices can be found in Bornert *et al.* (2008).

The nodal displacements of the specimen are determined by correlating small subsets (centered on the nodes) of the two digital images of the sample. Originally, DIC was applied to photographs of a plane surface of the specimen (e.g. Peters and Ranson (1982); Sutton *et al.* (1983)), but has been extended to volume 3D pictures of the specimen (e.g. Bay *et al.* (1999); Verhulp *et al.* (2004)).

We call $\underline{X}$ the position in the reference picture and $\underline{x}$ the position in the deformed picture. The displacement is defined by $\underline{u}(\underline{x}) = \underline{x} - \underline{X}$. u, v and w are the displacement components of $\underline{u}$ in the space coordinates x, y and z of $\underline{X}$, respectively. The displacement gradient tensor is defined by

$$\underline{\underline{F}} = \underline{\underline{I}} + \frac{\partial \underline{u}}{\partial \underline{X}} = \begin{pmatrix} 1 + \frac{\partial u}{\partial x} & \frac{\partial u}{\partial y} & \frac{\partial u}{\partial z} \\ \frac{\partial v}{\partial x} & 1 + \frac{\partial v}{\partial y} & \frac{\partial v}{\partial z} \\ \frac{\partial w}{\partial x} & \frac{\partial w}{\partial y} & 1 + \frac{\partial w}{\partial z} \end{pmatrix}$$

The three main types of DIC can be described in the following:

– 2D-DIC needs one camera and two pictures of a flat surface of the specimen (defined as $z = z_0$). The components of the displacement vector $u(x, y, z_0)$ and $v(x, y, z_0)$ are measured, which allows the components F_{xx}, F_{xy}, F_{yx} and F_{yy} to be determined. The surface of the specimen can be prepared with a random speckle.

– 3D-surface DIC (stereovision plus stereocorrelation) needs two cameras (and requires special calibration of the two cameras' frame) and two times two pictures of the surface of the specimen. The surface is not necessarily flat. The components of the displacement vector $u(x, y, z_0)$, $v(x, y, z_0)$ and $w(x, y, z_0)$ are measured, which allow the components F_{zx} and F_{zy} to be determined, in addition to the previous components. The derivative with respect to z is not accessible, it is still restricted to surface analysis. It is applicable to the evaluation of non-planar objects and out-of-plane deformations.

– 3D-volumetric DIC (also known as digital volume correlation or DVC) needs a 3D imaging device (e.g. X-ray tomograph) and two 3D volume images. The three components of the displacement vector $u(x, y, z)$, $v(x, y, z)$ and $w(x, y, z)$ are measured, which allow all of the gradient tensor's component to be determined. It is applicable to analysis of 3D deformation including internal deformation, but requires the object to have an internal "character" that may be followed (correlated) between images.

The Green–Lagrange strain tensor is defined as

$$\underline{\underline{E}} = \frac{1}{2}\left(\underline{\underline{F}}^T . \underline{\underline{F}} - \underline{\underline{I}}\right)$$

and the linearized strain tensor (for small strain assumption) is defined by

$$\underline{\underline{\varepsilon}} = \frac{1}{2}\left(\underline{\underline{F}}^T + \underline{\underline{F}}\right)$$

The components of $E_{xx} = 0.5\,(F_{xx}^2 + F_{yx}^2 + F_{zx}^2 - 1)$, $E_{xy} = 0.5\,(F_{xx}.F_{xy} + F_{yx}.F_{yy} + F_{zx}.F_{zy})$ and $E_{yy} = 0.5\,(F_{xy}^2 + F_{yy}^2 + F_{zy}^2 - 1)$

can be computed exactly with 3D-surface DIC, while they are approximated with 2D-DIC under the assumption that the out-of-plane displacement is quite homogeneous with respect to x and y, which is correct in plane strain deformation (no out-of-plane displacement). The approximation is also correct in the case of small stretching and small rotation, when $\underline{\underline{E}}$ tends toward $\underline{\underline{\varepsilon}}$.

The displacement field is estimated for a set of nodes that are. 2D or 3D Digital images correspond to some gray-level pixelated pictures. Each node is characterized by the gray-level distribution in its vicinity, called a subset, which is composed of pixels (2D) or voxels (3D). The subset from the reference image is associated with a subset in the deformed image that optimizes a similarity with the reference subset. The similarity is measured with a correlation coefficient. The difference in the positions of the two subsets corresponds to the displacement of the node. The process is repeated for each node selected on the reference image to obtain the displacement field.

The deformation of the subset, centered on the position $\underline{X}_0$, from the reference image to the deformed image can be approximated at the first order:

$$\phi(\underline{X}) = \underline{X} + \underline{u}(\underline{X}) \approx \underline{X} + \underline{u}(\underline{X}_0) + \frac{\partial \underline{u}}{\partial \underline{X}}(\underline{X}_0).(\underline{X} - \underline{X}_0),$$

where $\underline{u}(\underline{X}_0)$ is the searched displacement of the subset. The main difference between the correlation methods comes from how to estimate the similarity between the reference and the deformed subset, including the choice of correlation function and the assumptions on the transformation of the subset. The assumed transformation can simply be a rigid translation ($\partial\underline{u}/\partial\underline{X}\,(\underline{X}_0) = 0$) or can take into account a rigid rotation and a stretching of the subset.

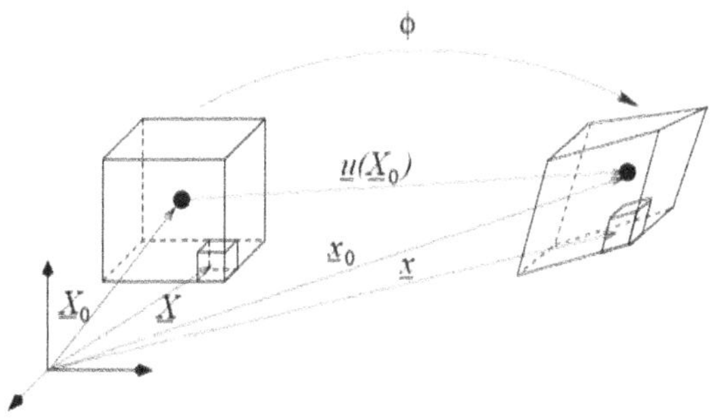

Figure 4.1. *Material deformation of the subset (from Bornert* et al. *(2004))*

The DIC method is decomposed into several steps, including an approximation of the displacement at an integer number of pixels and a subpixel refinement (Figure 4.2):

1) The set of nodes distributed on the reference image is defined. Generally, the nodes are regularly spaced, with a given number of pixels for the distance between nodes.

2) The subset around the node is determined, which is generally a square (2D) or a cube (3D) with a size of a few pixels (voxels).

3) The zone of research (zone of interest) is determined and the most similar subset in the deformed image is searched.

4) For all possible positions in the research area, a correlation coefficient is measured corresponding to a displacement of an integer number of pixels, assuming a rigid displacement (no deformation of the subset). The position that maximizes the similarity coefficient is guessed as the best approximation.

5) The previous approximation is refined by a subpixel algorithm, because the true displacement rarely corresponds to an integer numbers of pixels. Generally, the subset size in this step is smaller than in step 2. Moreover, the zone of research (step 3) is reduced to very few pixels (voxels).

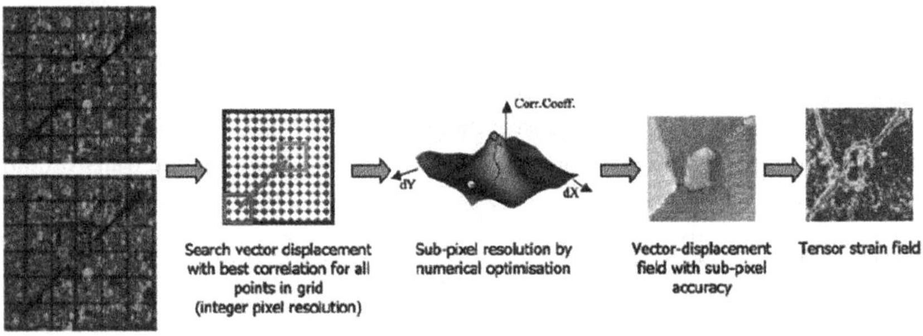

Figure 4.2. *Schematic of a 2D-surface DIC analysis approach (from Hall (2012)). For a color version of this figure, see www.iste.co.uk/stefanou/instabilities.zip*

There are different kinds of subpixel approximations (e.g. Pan *et al.* (2006)). The simplest is based on an interpolation of the correlation coefficient for positions around the approximation in 4. The coefficient is computed for rigid displacement of the subset with no stretching. The subpixel approximation corresponds to a maximum of the interpolation of the correlation coefficient. More complex methods account for the stretching of the subset, which generally corresponds to a linear approximation (which can also be a second-order function). Then, the parameters of the transformation are searched to optimize the correlation coefficient. The optimization can concern the displacement, rotation and stretching parameters. However, this process can be time consuming. To reduce the computation time, the stretching of the subset is sometimes determined out of the optimization loop (Bornert *et al.* 2004) and can be updated during an iterative algorithm. The zone of research during the subpixel correlation needs to be restrained to maintain a reasonable computation time. Several optimization algorithms exist, that are more or less robust and efficient (Newton–Raphson, Levenberg–Marquart, etc.). The subpixel correlation needs an interpolation of one of the two images to work on a continuous gray-level picture (not pixelated). The selection of the interpolation scheme is a key factor and directly affects the calculation's accuracy and convergence efficiency. A first- and second-order continuity is preferable to assure the smoothness of the intensity function and its derivatives.

The gray-level distribution in the subset can be interpreted as the signature of the central node. This implies the need of a "character" of the image to be able to distinguish the subset in a unique manner. With some geomaterials, the natural heterogeneity of their material structure is sufficient to serve as speckle pattern. This is especially important for volume-3D DIC. With the surface DIC methods, a synthetic pattern can be deposited onto the specimen surface. The speckle pattern acts as a fixed reference that follows the movements of the material, meaning that its morphology is an important parameter that directly affects the accuracy of the measured displacements. A synthetic pattern can be obtained using by a layer of white paint and an ultra-fine layer of black paint droplets such as using an airbrush. The speckle density, gray-level contrast and characteristic size affect the quality of the measure. This implies that each speckle pattern must comply with a predetermined quality requirement to be considered appropriate for imaging. It should be adapted to the pixel(voxel) size of the digital images (see Bornert *et al.* (2008)).

DIC can be adapted as per material's specificities. For example, for a granular material studied using X-ray CT, DIC can be used to measure a "continuous strain field", using the standard procedure described above. DIC can also be used to measure the kinematic displacement (and rotation) of each grain; this is the *discrete* volumetric DIC (Hall *et al.* 2010). Before the DIC procedure, the grains have to be individually identified through image segmentation. Then, the DIC subset has to be mapped onto each grain, using an *ad hoc* mask for each grain. Then, the (quasi)standard DIC procedure can be used to search the rigid displacement and rotation that optimize the correlation coefficient of each grain between the two images.

4.2.2. *X-ray computed tomography*

X-rays are a high-frequency form of electromagnetic radiation discovered at the end of the 19th Century by W. C. Röntgen. They can penetrate matter, depending on the wavelength spectrum. Many interactions exist between X-ray photons and matter, including photoelectric absorption. This absorption depends on the X-ray's characteristics and the atomic number of the matter. X-rays are thus a type of radiation that is able to be transmitted through matter, and is sensitive to the atomic number of the material that it is radiating through. For geomaterials, this absorption depends on the mass density and the mineralogy.

An *X-ray radiograph* is a two-dimensional measurement of the amount of X-ray photons arriving at each point onto a detector during a given exposure time, which is an integration of the X-ray attenuation of the matter being traversed along the path of the X-rays.

X-ray computed tomography is a method of reconstructing a 3D field of an X-ray attenuation coefficient inside an object, starting from a series of different projections (radiographs) of the object. The different projections are generally acquired by a relative rotation of the object and an image acquisition system. The source and detector rotate in medical applications. In industrial/laboratory applications such as this one, the specimen is usually rotated – with a vertical axis of rotation – by a rotation stage. Note that this technique is fully detailed in a number of textbooks on tomography, such as Hsieh (2009).

The kind of X-ray source distinguishes medical/industrial/laboratory tomographs from synchrotron facilities. The firsts are X-ray vacuum tubes that emit a cone-beam in the direction of the object and detector. Synchrotron light sources are large particle accelerator facilities. The main differences with respect to the previous small sources, for our applications, are the considerably higher brightness (photon flux) and capability for quasi-parallel beams (which can be advantageous in term of imaging), which allow fast scanning, (quasi)monochromatic X-ray light, phase contrast reconstruction, etc. The cone-beam of a laboratory tomograph is also interesting because the field of view changes, depending on the relative position of the object with respect to the detector and the source and the specimen can be enlarged on the detector due to geometric magnification (see Ketcham and Carlson (2001) for an introduction to X-ray CT for geomaterials). The experimental results on geomaterials that will be described in the next section come from the X-ray scanner in Laboratoire 3SR (Grenoble). A detailed description of the scanner can be found in Andò (2015).

4.2.3. *Experimental devices for* in situ *full-field measurements*

4.2.3.1. *A plane strain apparatus for soils*

Desrues and co-workers originally developed a plane strain compression apparatus adapted for full-field measurement using one classical camera. The design of this device (the biaxial apparatus, hereafter) shares its underlying concept with those developed by Vardoulakis and Goldscheider (1981) and Drescher *et al.* (1990), in which the biaxial apparatus was specifically conceived to allow free shear band formation in a soil specimen.

Within this device, a 35 mm thick prismatic sand specimen, surrounded by a latex membrane, is mounted between two rigid walls that induce plane strain conditions (see Figure 4.3). The initial nominal height and width of the specimen (in the plane of deformation) can vary in the range of 75–350 mm and 90–175 mm, respectively. The side walls are 50 mm thick glass plates that allow photographs to be taken of a specimen's in-plane deformation during the test. All surfaces in contact with the specimen are enlarged and lubricated with silicone grease to minimize friction. The lower and upper loading platens house porous stones connected to drainage lines. The top platen is free to rotate and slide horizontally in the plane of deformation, although it can be locked. This allows free lateral displacement of the upper part of the specimen once a shear band forms due to deviatoric loading. A large cell, filled with silicone oil, surrounds the specimen. The cell can sustain up to 2 MPa and has two opposite pairs of large Plexiglas windows on its lateral surface. Strain-controlled axial loading is applied through a screw jack that rests atop the device.

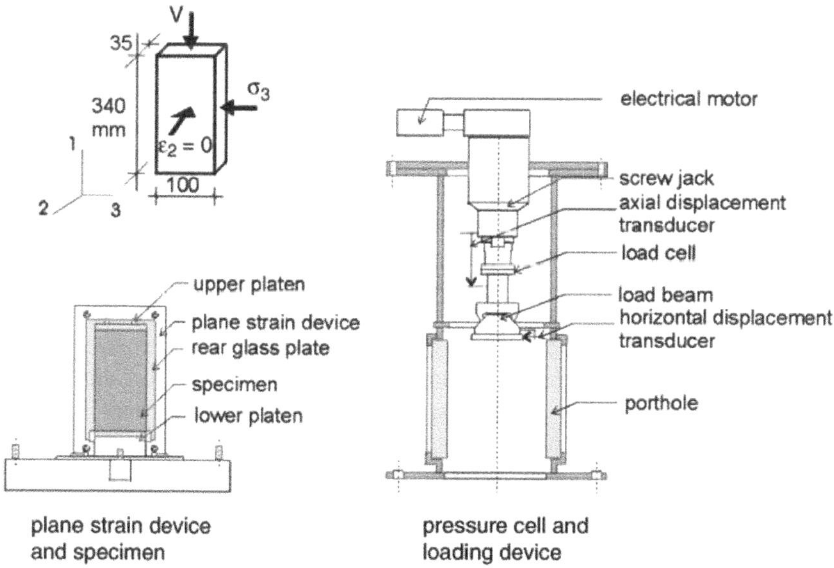

Figure 4.3. *Schematic diagram of the plane strain apparatus for soils (from Desrues and Viggiani (2004))*

The jacket around the specimen can be observed through the glass plate. The jacket follows the deformation of the specimen. Because the full-field measurement used was FRS, a grid was drawn on the jacket; however, a random speckle pattern for DIC analysis can be deposited as well.

4.2.3.2. *A true triaxial cell for rocks*

A true triaxial apparatus was developed at Laboratoire 3SR in Grenoble (France), with the aim of characterizing the onset of localization and the postlocalization regime in rocks (see Bésuelle and Lanatà (2016) for more details). The device can apply three independent stresses along the three space directions of a prismatic rock specimen, and allows its visualization under loading. A simplified schematic of the apparatus is shown in Figure 4.4. The specimen's surfaces, perpendicular to the principal stress (direction 1) and intermediate compressive stress (direction 2), are in contact with rigid platens, which are moved by two perpendicular pistons, as in some triaxial setups developed previously (Haimson and Chang 2000; Mogi 1971). The two surfaces of the specimen, perpendicular to the minor stress (direction 3), are free to deform because the stress is applied by a confining fluid (by means of a soft silicon membrane). Since deformation bands and cracks are generally parallel to the intermediate stress (at least for an isotropic material, or an anisotropic material with specific orientation of the symmetry directions), the specimen is free to deform and fail with no kinematic constraints imposed on the formation of the failure zone. Moreover, one of the two surfaces perpendicular to the intermediate stress (direction 2) is in contact with a hard window, in order to observe the specimen under loading. The two pistons can be controlled according to either a stress or displacement mode. The intermediate stress can be regulated to ensure that there is no deformation along this direction, thus allowing the application of plane strain loading, which implies that the kinematic of the surface in contact with the window is representative of the kinematic of the whole specimen, up to strain localization and beyond that level of stress. The surface of the specimen opposed to the horizontal piston is in contact with a thick, transparent, sapphire window (7), whose surfaces are optically polished. This surface of the specimen can be observed and photographed. For contact symmetry, the surface on the side of the horizontal piston is in contact with a thin sapphire platen, in order to ensure the same boundary conditions. The minor stress in direction 3 is applied to the two lateral surfaces of the specimen by a confining fluid. The specimen remains hydraulically isolated from this fluid by means of a silicone membrane (Figure 4.5(a)). The membrane contains the specimen and wraps around the four loading caps (two caps along direction 1 and two along direction 2). This arrangement ensures that there is direct contact between the specimen and loading platens, particularly the sapphire window. It is worth noting that along the axial direction (direction 1), a special arrangement of platens with a wedge shape (8), placed between the specimen and the top and bottom loading caps, ensures that the loading caps have the same thickness of the specimen along the intermediate stress (direction 2) (see Figure 4.5(b)), and larger than the specimen along direction 3. If a compression or extension of the specimen occurs along direction 2, the set of wedges automatically adapt to this variation, thus avoiding an extrusion of the membrane during the deformation of the specimen. In both the axial and horizontal directions, a loading cap (Figure 4.4, elements 9–10) is in contact with a piston by means of a spherical head, which allows the compensation, during the setup, of small parallelism offsets of the two opposite surfaces of the prismatic specimen. However, due to friction between the surfaces of the spherical head, once the contact is firmly established, they do not move anymore, and as such, this component does not show a degree of freedom (rotation) during the loading.

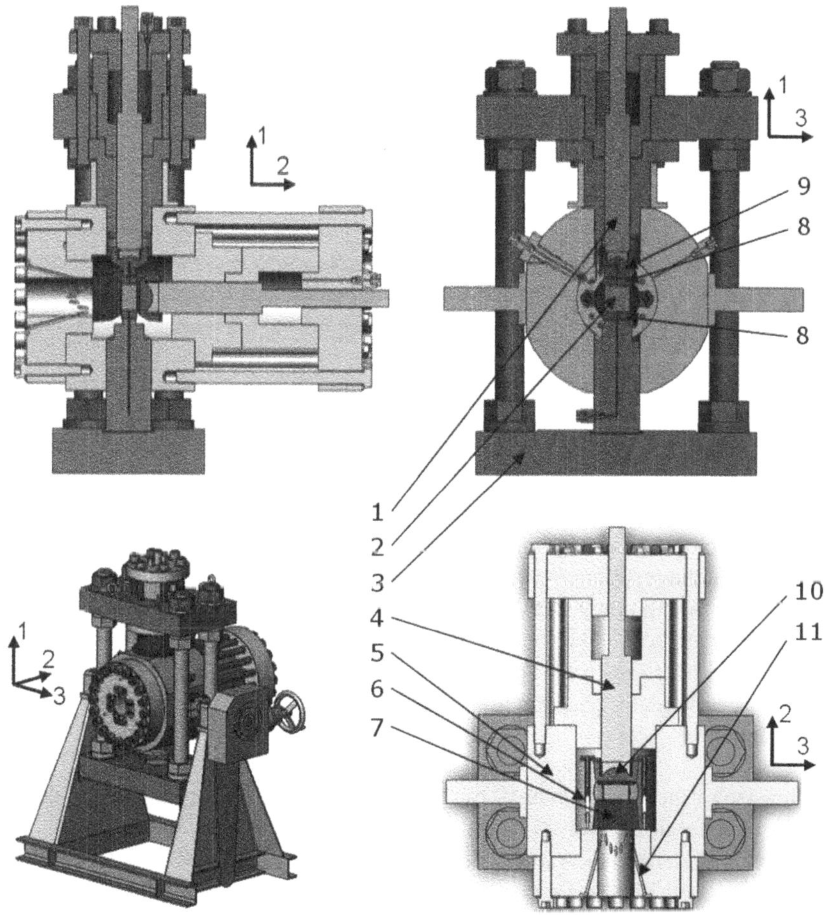

Figure 4.4. *Schematic diagram of the plane strain apparatus for rocks (from Bésuelle and Lanatà (2016)). For a color version of this figure, see www.iste.co.uk/stefanou/instabilities.zip*

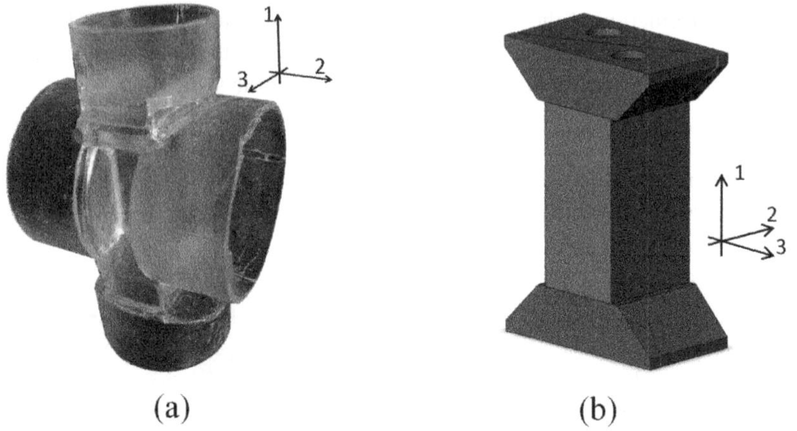

Figure 4.5. *(a) Photograph of the silicon membrane that separates the specimen from the confining fluid and (b) scheme of the adaptable wedges at the top and bottom of the specimen (from Bésuelle and Lanatà (2016))*

The surface of the specimen receives light through the sapphire window from a set of 20 LEDs, focused onto the sample surface using 20 optical fibers, which pass through orifices (11) made on the frame. These provide even and homogenous illumination for the taking of photographs, since the light power of each LED can be separately adjusted. Photographs of the visible surface of the specimen were taken throughout the loading with a high resolution camera. DIC analysis can thus be carried out on the resulting images to compute the displacement and strain fields. It is worth noting that to apply DIC, the observed specimen must have a random pattern, which is here artificial. The four surfaces of the specimen, which are in contact with the three rigid loading caps and the window, are lubricated. A thin layer of this lubricant is sufficiently transparent to allow good visualization of the pattern through the window.

The size of the specimen is 50 mm in the axial direction, 30 mm in the direction of intermediate stress and 25 or 50 mm in the direction of minor stress, which corresponds to slenderness ratios (the ratio between the height and width) of two and one, respectively. The loading cell can apply a maximum confining pressure of 100 MPa, while the axial and the horizontal pistons can apply forces of 500 and 700 kN, respectively. These values correspond to a differential stress, with respect to the confining pressures of 670 and 530 MPa, respectively, for a specimen with a slenderness ratio of two, and half of these values for a 50 mm width specimen.

4.2.3.3. *An X-ray transparent triaxial cell*

The system allows microtomography to be performed on a specimen under load (*in situ*), using a specifically built load frame that can be placed in the X-ray beam, either from a laboratory scanner or synchrotron microtomography beamline (Lenoir *et al.* 2007). Figures 4.6 and 4.7 show the experimental setup used at ESRF. The apparatus includes a small triaxial cell and a loading device designed specifically for microtomography on small soil or rock specimens under load (Lenoir 2006). The triaxial apparatus is practically the same as a conventional triaxial testing system, except for its much smaller size and the shape of its confining cell, which was designed to be as transparent as possible to the X-rays. A 10 mm thick confining cell made of polycarbonate was used, that is both transparent to the X-rays and capable of sustaining the 10 MPa confinement pressure and the tensile reaction force. The axial load and hence the deviator stress are applied in a displacement controlled manner using a motor-driven screw actuator. The loading system, which can be placed in the X-ray beamline without interfering with the tomographic scans, is quite compact and light, which is important because it sits directly on the translation and rotation stage during the experiment. The system has a maximum loading capacity of 7.5 kN. The same device can be used in a laboratory scan (Andò 2015).

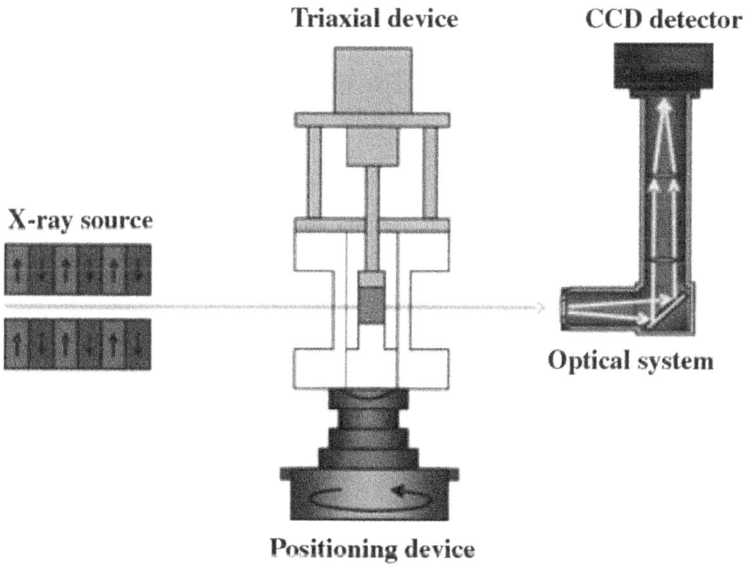

Figure 4.6. *Schematic of the X-ray microtomography setup for triaxial testing (from Lenoir* et al. *(2007))*

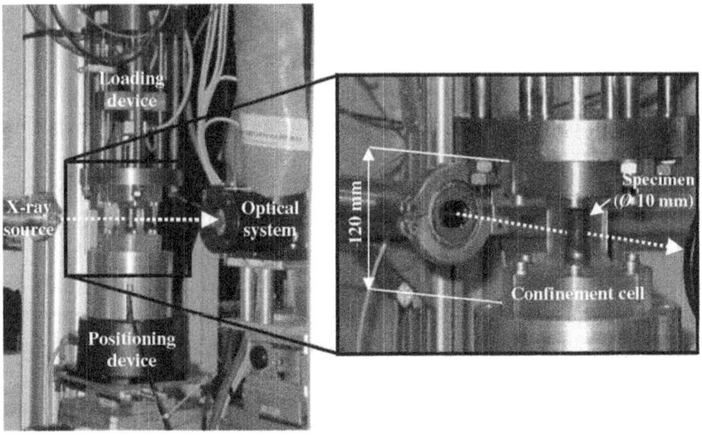

Figure 4.7. *Tomography setup for triaxial testing at the beamline ID15A at ESRF. Complete setup on the beamline (left) and zoom on the rock specimen inside the triaxial cell (right) (from Lenoir* et al. *(2007))*

Higher capacity in terms of confining pressure and axial loading are needed to experiment on stiffer rocks. It can be necessary to work with stiffer (and with low X-ray absorbance) cell materials such as PEEK, aluminum, beryllium and carbon fiber (e.g. Takemura and Oda (2004)).

4.3. Selected materials

4.3.1. *Hostun sand*

Hostun sand is produced in a quarry close to the commune of Hostun in the Department of Drôme, in the Rhône-Alpes region of France. It is approximately 60 km from Grenoble. The material is excavated as clumps made of a range of different silica grain sizes and lightly cemented by kaolin. The silica particles making up these clumps have been crushed *in situ* from larger rocks, and consequently have not been transported very far. Since the particles were generated by crushing and have not been transported very far, they are very angular. After quarrying, the clumps are washed to remove the kaolin, and are then sorted by grain size. Depending on the year of extraction (and the version of sieves), the D_{50} of the Hostun sand called S28 and then HN31 is about 0.34 mm. A uniform gradation of the natural sand was used, with 100% of the material passing a 0:63 mm sieve and retained on a 0:16 mm sieve. The conventional minimum and maximum volumetric weights are 13.24 and 15.99 kN/m^3, respectively, and the specific gravity G_s is 2.65 (see Flavigny *et al.* (1990); Desrues and Viggiani (2004) and Andò (2015) for more details).

4.3.2. *Caicos ooids sand*

Caicos ooids are made of $CaCO_3$ and more than 96% of the material is Aragonite, the rest being calcite and "high-magnesium calcite". The D_{50} of this material is around 0.42 mm. This sand comes from the Caicos platform in the British West Indies. Ooids grow from small seeds such as shell fragments, small quartz or calcite grains, and progressively become larger. They grow in marine environments, where the waves drive the accretion of material around the seed. Two principal mechanisms exist for an ooid to grow: by physical attachment of material as the ooid rolls (like a snowball) or by precipitation over the surface of the ooid. Both of these mechanisms tend to generate rounded grains – from whence the name comes – although the images acquired show some grains that are prolate spheroids (see Andò (2015) for more details).

4.3.3. *Vosges sandstone*

The tested rock is a natural sandstone coming from the Woustviller quarry in the Vosges mountains, Eastern France (Bésuelle *et al.* 2000). It is a pink quartz sandstone (quartz = 93%), a few percent of which is feldspar and white mica. The sandstone is poorly cemented, its cohesion is due to the interpenetration between the grains. Its porosity is about 22%. The dimension of the grains fluctuates between 0.15 and 0.45 mm with a mean value of about 0.30 mm. Its uniaxial compression strength is about 35 MPa.

4.3.4. *Callovo–Oxfordian clayey rock*

The material tested is Callovo–Oxfordian argillite, which has been considered as a potential host rock for radioactive waste disposal facilities, from the ANDRA Underground Research Laboratory (URL) located in Bure (Meuse/Haute-Marne, Eastern France) at approximately 500 m below the ground surface in the clay unit (Armand *et al.* 2017). It is a sedimentary rock composed of a clay matrix (40 to 60%), with grain inclusions of carbonates and tectosilicates (Robinet *et al.* 2012). The structure presents a microscopic scale described by a microcrystalline and porous clay matrix and a mesoscopic scale related to the organization of the non-porous carbonate and tectosilicate grains (with grain size in the range of 0.1–100 μm), which are embedded into the clay matrix. Centimetric or multi-centimetric pipe shapes filled by heavy minerals (pyrite) are the markers of bioturbation (David *et al.* 2007). At the investigated depth, the material has an extremely low permeability (5.10^{-20} to 5.10^{-21} m^2), a porosity of 17% (Armand *et al.* 2017) and a water content of about 7% (Conil *et al.* 2018). Its uniaxial compressive strength is about 20 MPa.

4.4. Strain localization in sands

The progressive strain localization in sands is analyzed below. First, some results based on FRS of plane strain compression tests on Hostun sand are presented from the work of Desrues and co-workers (Desrues and Viggiani 2004; Desrues 2004). In this application of FRS, the photographed image is the side of a sand specimen deforming under load, and the source of the differences between the successive images is the deformation process undergone by the specimen. FRS is now obsolete because of the relative low space resolution of the displacement field (with respect to DIC); however, experimental studies are still important. Then, a work that combines X-ray CT and DIC is presented. The analysis was in 3D, with a high space resolution and was extended for grain-scale analysis (Andò 2015; Andò *et al.* 2012b)). A former work, based on *in situ* X-ray CT during triaxial tests is also presented to discuss the volume strain inside shear bands (Desrues *et al.* 1996).

4.4.1. *Plane strain compression by FRS*

Three tests from Desrues and Viggiani (2004) were selected to discuss the emergence of strain localization in Hostun sand. Tests were performed in the plane strain apparatus described in section 4.2.3.1, with a lateral pressure of 80–90 kPa and a slenderness ratio (H/W) of 2–2.2, and the two loading platens in contact with the specimen were locked (no translation, no rotation), excepted in test shf06 for which rotation was allowed. The initial densities of the specimens corresponded to a dense state for shf06, a loose state for shf03 and an intermediate state for shf00.

Figure 4.8 shows the corresponding stress strain response obtained for each selected test, in terms of the effective stress ratio t/s' versus the global axial strain. Figures 4.9–4.11 show the stereophotogrammetry-based shear strain intensity ($\varepsilon_s = (\varepsilon_1 - \varepsilon_3)/2$) and volumetric strain ($\varepsilon_s = (\varepsilon_1 + \varepsilon_3)/2$) for each photographic increment. The size of the symbols is proportional to the value of the relevant quantity (note that the symbol scale is different for each increment). As far as volumetric strain is concerned, square symbols are for dilatancy and hexagons are for contractancy. The photograph numbers are noted on the relevant curve of the effective stress ratio t/s' versus the global axial strain (see Figure 4.8).

The evolution of the shear strain field of test shf00 (Figure 4.9) showed that the shear band pattern was partially in place at (or even just before) the stress peak (increment 4–5), as compared to the postpeak pattern (increment 5–6), which showed a shear band reflected on the top platen. The shear band was not yet reflected in increment 4–5. Two other bands were visible but less active; one was conjugated to the main shear band, and the second one was subparallel and initiated more or less from the top-left specimen corner. These two bands were deactivated in the next increment. The strain field was not homogeneous in the previous increments 3–4 and

1–2; however, the spatial resolution of FRS does not unambiguously show whether the deformation is localized or not. Non-homogeneity can be a consequence of an initial heterogeneity in the specimen (void ratio heterogeneity) or early strain localization (or both). Volumetric strains were localized in the same band as the shear strains (increments 4–5 and 5–6) and are mostly dilative inside the shear band.

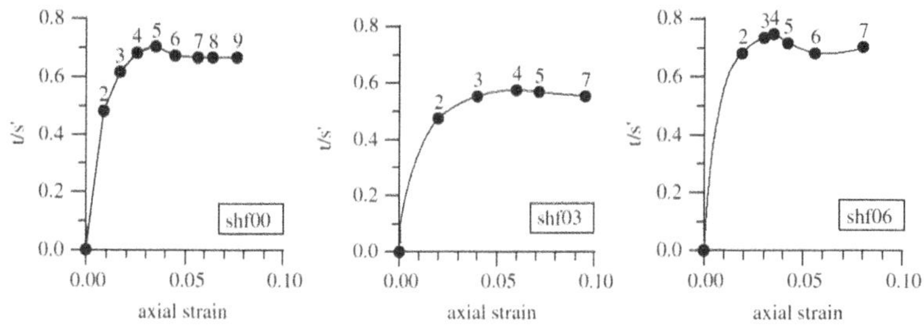

Figure 4.8. *Stress strain responses from various tests; the numbers noted on each curve are the photographs numbers (from Desrues and Viggiani (2004))*

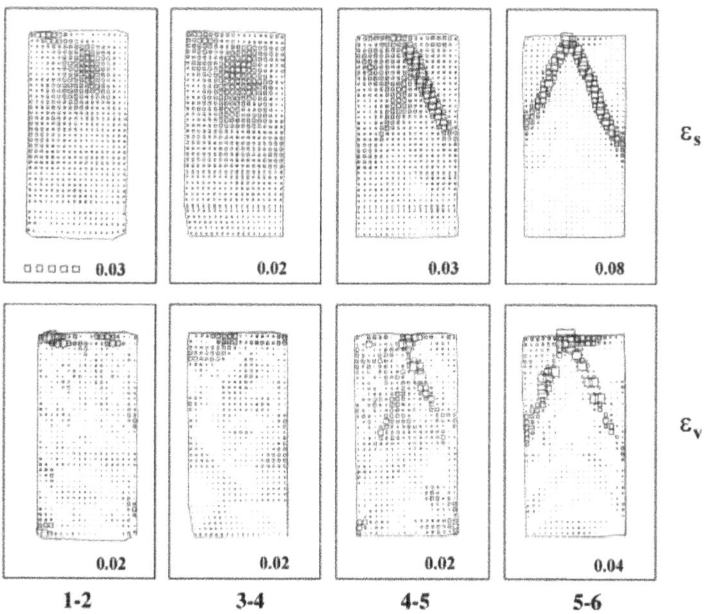

Figure 4.9. *Test shf00: stereophotogrammetry-based incremental fields of shear strain intensity (top row) and volumetric strain (bottom row) (from Desrues and Viggiani (2004))*

In test shf03 (Figure 4.10), the incremental shear strain fields (top row) indicated that a localized region first formed during increment 2–3 (before stress peak), showing two (or even three) parallel bands. Then, the pattern switched in the subsequent increment 3–4 (at the peak), resulting in a conjugate deformation pattern. A shear band could be observed in the last available photographic increment (4–5), which corresponded to the main band in increment 2–3 that was reactivated. In the first increment (1–2), the deformation was slightly non-homogeneous; however, the spatial resolution of FRS does not allow the deformation regime to state on. The final deformation pattern was characterized by volumetric strains that were localized in the same band as the shear strains. Both contractive and dilative incremental behaviors were exhibited inside the band.

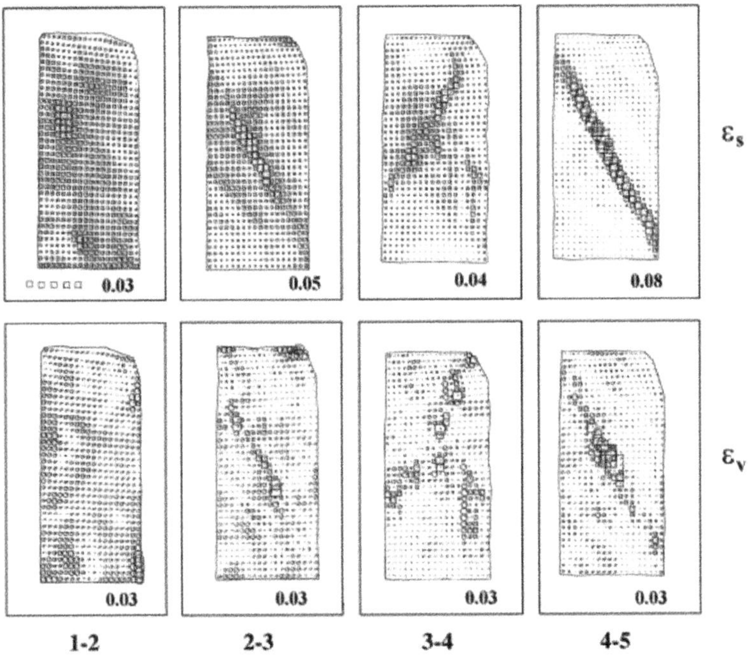

Figure 4.10. *Test shf03: stereophotogrammetry-based incremental fields of shear strain intensity (top row) and volumetric strain (bottom row) (from Desrues and Viggiani (2004))*

In test shf06 (Figure 4.11), the shear strain fields (top row) indicated that two parallel zones of strain localization formed in the middle portion of the specimen in increment 3–4 (shortly prior to the stress peak). Note that the shear strains were always larger in the left shear band, which was the only one maintained throughout the test and finally intersected the entire specimen, whereas the other shear band eventually disappeared. A non-homogeneous deformation already existed during increment 2–3 if not, to a lesser extent, from the beginning of the test (increment 1–2). This heterogeneous shear strain field, could not clearly be qualified as localized (due to the FRS space resolution); however, it evoked a crossed shear bands pattern. The conjugated "bands" seemed very large; however that these large bands were an apparent combination of thinner parallel bands cannot be ignored. The authors of the paper called the apparent transformation of the deformation pattern a phenomenon of the progressive "condensation" of a centrally located heterogeneity of the deformation. However, from our experience and observations on thinner cohesive materials (presented later), this apparent condensation could be interpreted as a phenomenon involving the selection of numerous bands in which some of them stay active and the others bands are deactivated. Volumetric strain fields (bottom row) essentially confirm such a picture. Note that the soil is always dilating in the localized regions.

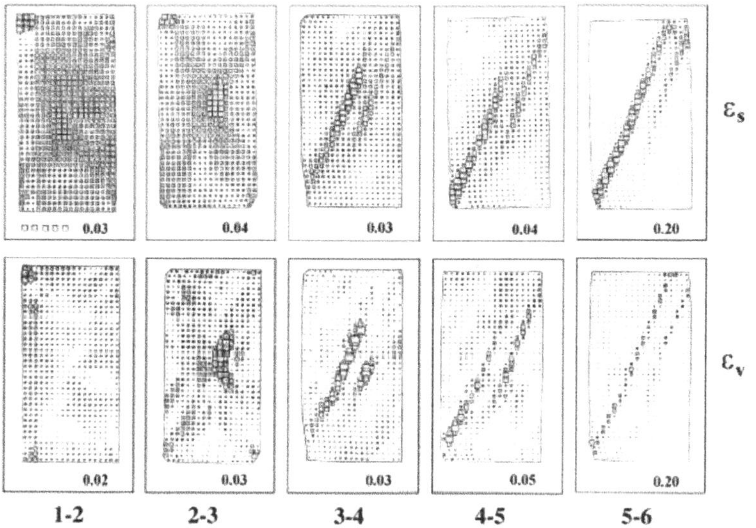

Figure 4.11. *Test shf06: stereophotogrammetry-based incremental fields of shear strain intensity (top row) and volumetric strain (bottom row) (from Desrues and Viggiani (2004))*

The understanding of the strain localization process in sands given by these results is limited by the space resolution of the FRS. However, it can be concluded that the strain localization is not a sudden appearance of shear bands in an initially homogeneous strain field. The evolution is quite progressive and heterogeneity starts well before the stress peak, probably by early strain localization. We can now consider more recent results obtained with more effective full-field methods.

4.4.2. *Triaxial compression by X-ray CT and DIC*

Here, we analyzed the COEA01 test on the Caicos ooid sand (Andò 2015; Andò *et al.* 2012a,b). This triaxial compression test was performed with a confining pressure of 100 kPa on a small dense specimen (10 mm diameter and 20 mm height). The test was performed *in situ* in the X-ray CT apparatus of Laboratoire 3SR (Grenoble) with a voxel size of 15 μm, using a transparent triaxial cell (see section 4.2.3.3). The axial loading was applied from the bottom of the specimen. The kinematic field was measured using both continuous volumetric DIC (described in section 4.2.1) and a grain-tracking algorithm (based on grain identification and not on image correlation, see Andò (2015)).

Figure 4.12 shows two vertical slices taken from the 3D image of the specimen before the axial loading. Figure 4.13 shows the stress–strain response and global volume curve measured for this test. There was a stress peak followed by a strain softening. There was a small contraction at the beginning of the axial loading; thereafter, the specimen dilated.

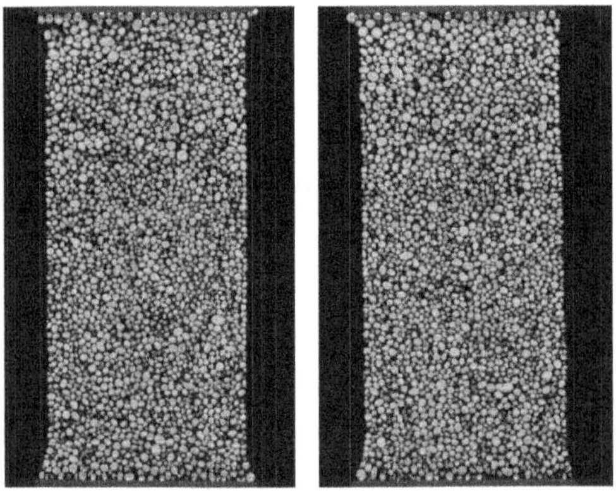

Figure 4.12. *Test COEA01: Slices of the X-ray CT scan of specimen COEA01 in its initial configuration. These slices are taken are* 90° *from each other (from Andò (2015))*

The kinematic field was measured at the grain scale using a grain-tracking algorithm. The displacement vector and the Euler rotation angles were determined for each grain (Figure 4.14). The grain displacement was higher at the bottom than at the top due to the loading conditions. The displacement field showed an evolution from a diffuse deformation regime to a localized regime, clearly exhibiting an inclined shear band. The grain rotation field was more instructive. The rotation field was quite homogeneous at the beginning (increments 01-02 and 02-03), but was localized with a shear band at the end (from increment 09-10). The deformation pattern was more complex just before and after the stress peak (increments 05-06 and 06-07, respectively). Several bands could be identified; while some were parallel to the final main band, others were conjugated with respect to this final band. This pattern could also be identified in increment 04-05.

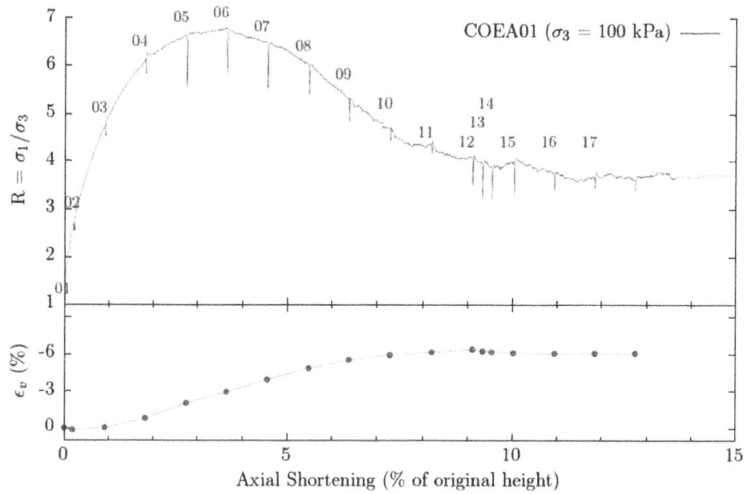

Figure 4.13. *Test COEA01: Stress strain response; the numbers noted on curve are the X-ray CT scan numbers (from Andò (2015))*

The equivalent continuum strain field was determined by two methods: (i) using the grain displacement measured by the grain tracking and (ii) by a continuous volumetric DIC (Figure 4.15). Note that the comparison and consistency of the two measures exclude the assumption of artifacts of the measure. The above observations of the rotation fields were confirmed. The transition from an initially homogeneous strain field (increment (01-02) and a localized regime with one active shear band at the end (from increment 09-10) was complex. The intermediate deformation pattern showed several shear bands that were parallel and conjugated, before and just after the stress peak. The number of active bands decreased during the subsequent loading step and was progressively reduced to a low number of shear bands generally one band (for a small specimen size).

The same full-field analysis was extended to Ottawa sand (D_{50} = 0.25 mm) and Hostun sand (D_{50} = 0.34 mm) (Andò 2015). A comparison of the tests on the three sands is presented in Figure 4.16, which shows the deviatoric strain calculated in five key increments for these three tests:

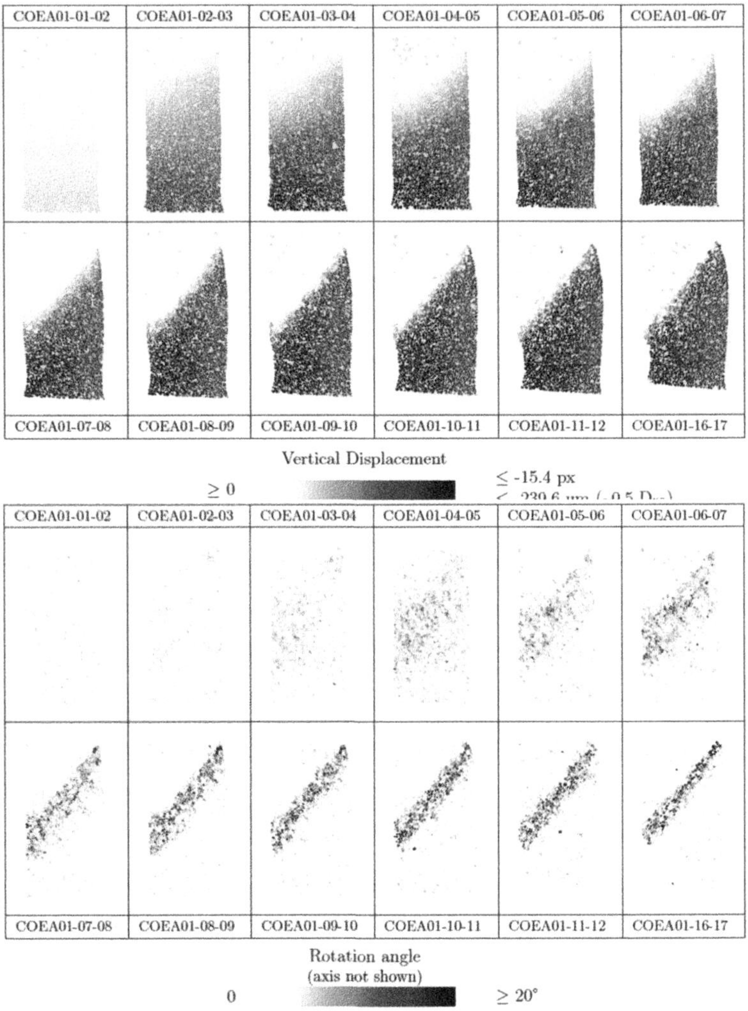

Figure 4.14. *Test COEA01: (top) slice of the grain displacement field during the test (numbers corresponds the scan numbers). (bottom) slice of the grain rotation (from Andò (2015)). For a color version of this figure, see www.iste.co.uk/stefanou/instabilities.zip*

– A: the first increment, close to the maximum slope of increase in porosity;

– B and C: two increments at the maximum slope of local porosity increased, one at the beginning and one at the end;

– D: one increment when the gradient of porosity reduction started to change; and

– E: the last increment at the beginning of the plateau when the local porosity increased.

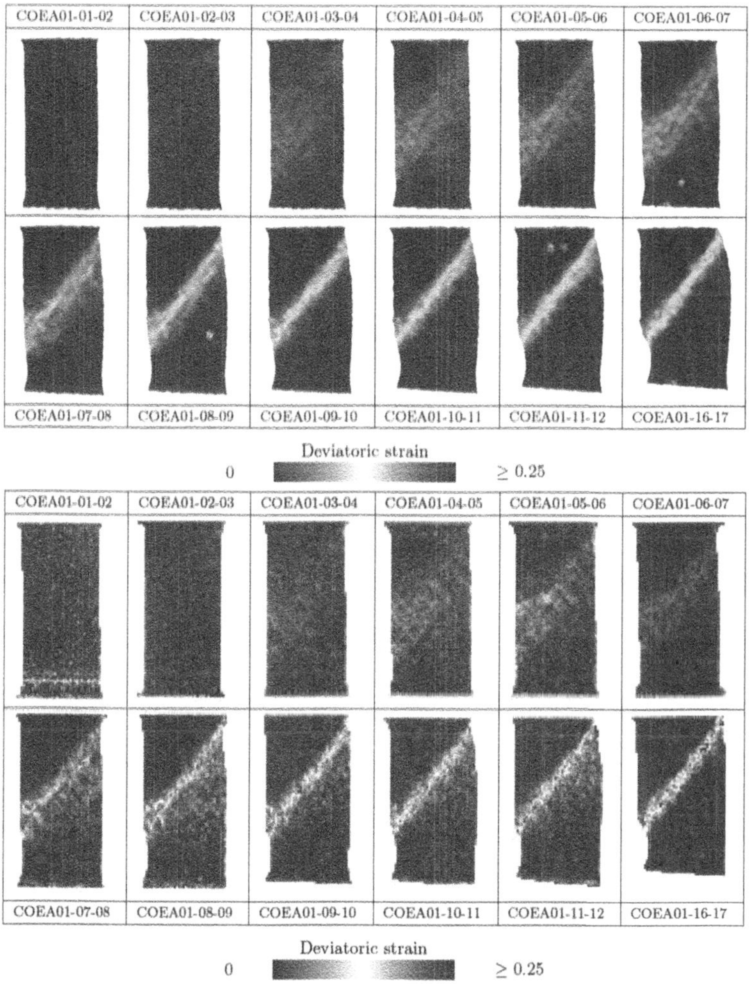

Figure 4.15. *Test COEA01: (top) slice of the strain field during the test, measured from the grain displacement field by grain tracking algorithm (numbers corresponds the scan numbers). (bottom) slice of the strain field, measured by a continuous volumetric DIC (from Andò (2015)). For a color version of this figure, see www.iste.co.uk/stefanou/instabilities.zip*

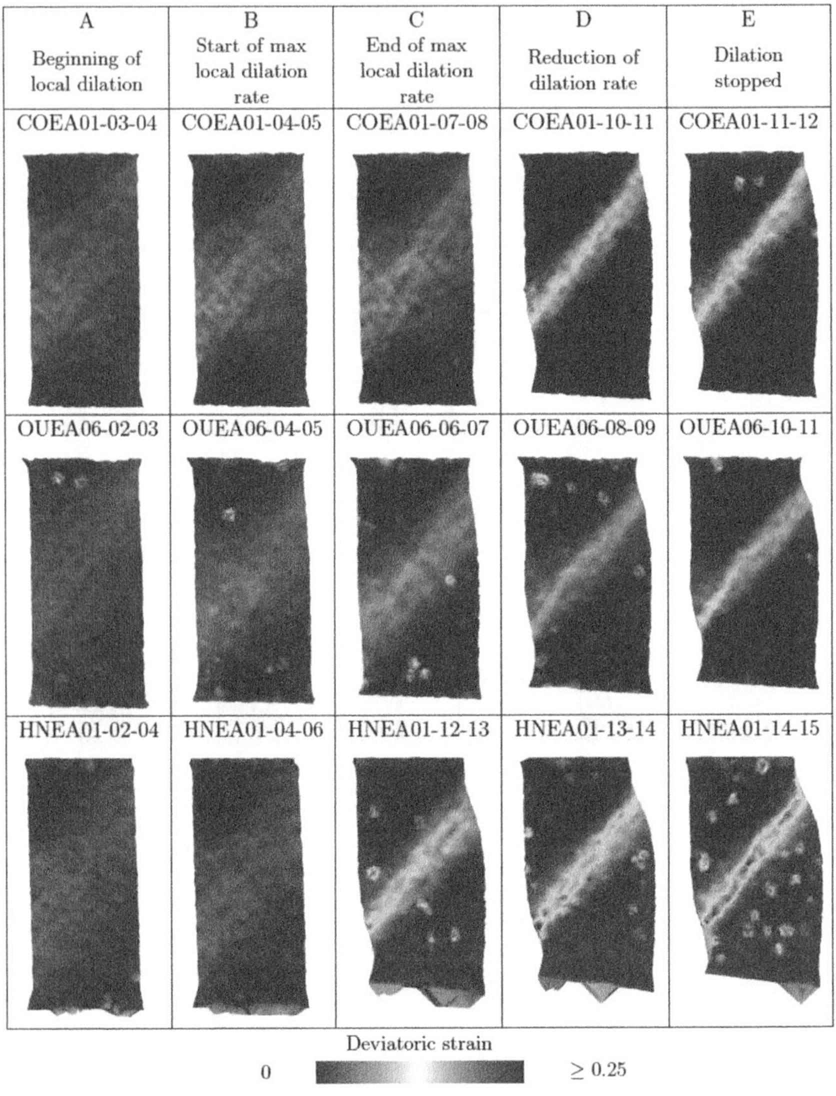

Figure 4.16. *Comparison of the strain field during tests on three sands: (top) test COEA01 on Caicos ooids sand, (middle) test OUEA06 on Ottawa sand and (bottom) test HNEA01 on Hostun sand (from Andò (2015)). For a color version of this figure, see www.iste.co.uk/stefanou/instabilities.zip*

This figure shows that the beginning of the local change in porosity had a clear meaning: some very slight, diffuse features of localized strains could be seen for all specimens in increment A (Andò 2015). Increments B and C for every specimen lie on either side of the peak. For COEA01, some considerable conjugate structures are visible in states B and C that were visible in specimen HNEA01 at state B. These were less visible in specimen OUEA06. In increment C, HNEA01 showed two parallel main bands. What was noticeable in all specimens for increments B and C was that the diffuse zone of localization visible in increment A was concentrated in space. There was a progressive process of shear band deactivation a kind of selection in which very few bands stay to be active at the end. In increment D, the shear band was in a much more recognizable form compared to each specimen's residual state. By increment D, all of the specimens had started the "second phase" of localized strain in which the strain was concentrated in a very limited area. By increment E, the shear band was very close to fully mature: the second phase of strain concentration crossed all of the specimens, and the increments all corresponded to increments just at the end of strain softening.

The more recent results of Desrues *et al.* (2018) are also interesting to consider. The test presented by the authors was performed using the same X-ray tomograph as before, but on a larger sample (70 mm diameter and 140 mm height cylinder) of Hostun sand in a dense state, with a lower image resolution (voxel size in the order of 100 μm). The 3D image correlation by volume shows richer localization structures than previously (Figure 4.17), due to a higher sample size to grain size ratio. The analysis and 3D visualization shows that the localization of the deformation begins in a rather diffuse way (many thin bands) well before the stress peak (before image 10 or even 06).

To conclude, the results confirm that strain localization is not an abrupt process. Quite the contrary, there is a progressive evolution from a quite homogeneous deformation toward a relatively diffuse shear band pattern before the stress peak, and the number of active bands progressively decreases to converge to very few shear bands.

However, the observation of a pattern with several bands does not mean that all bands work at the same time. They can be active successively. This is suggested by Le Bouil *et al.* (2014), which used spatially resolved diffusing wave spectroscopy to measure a strain quantity field on very short loading increments. The authors worked on a glass bead specimen with a plane strain compression apparatus. They showed the intermittency of the early bands, although the behavior of glass beads sometimes differs from that of natural sand. In the same vein, on a 2D granular material (rolls) solicited in the press $1\gamma2\varepsilon$ (Laboratoire 3SR, Grenoble) constituted by a deformable parallelogram, Viallon-Galinier *et al.* (2018) show that the localization of the deformation results from intermittent events organized spatially in time.

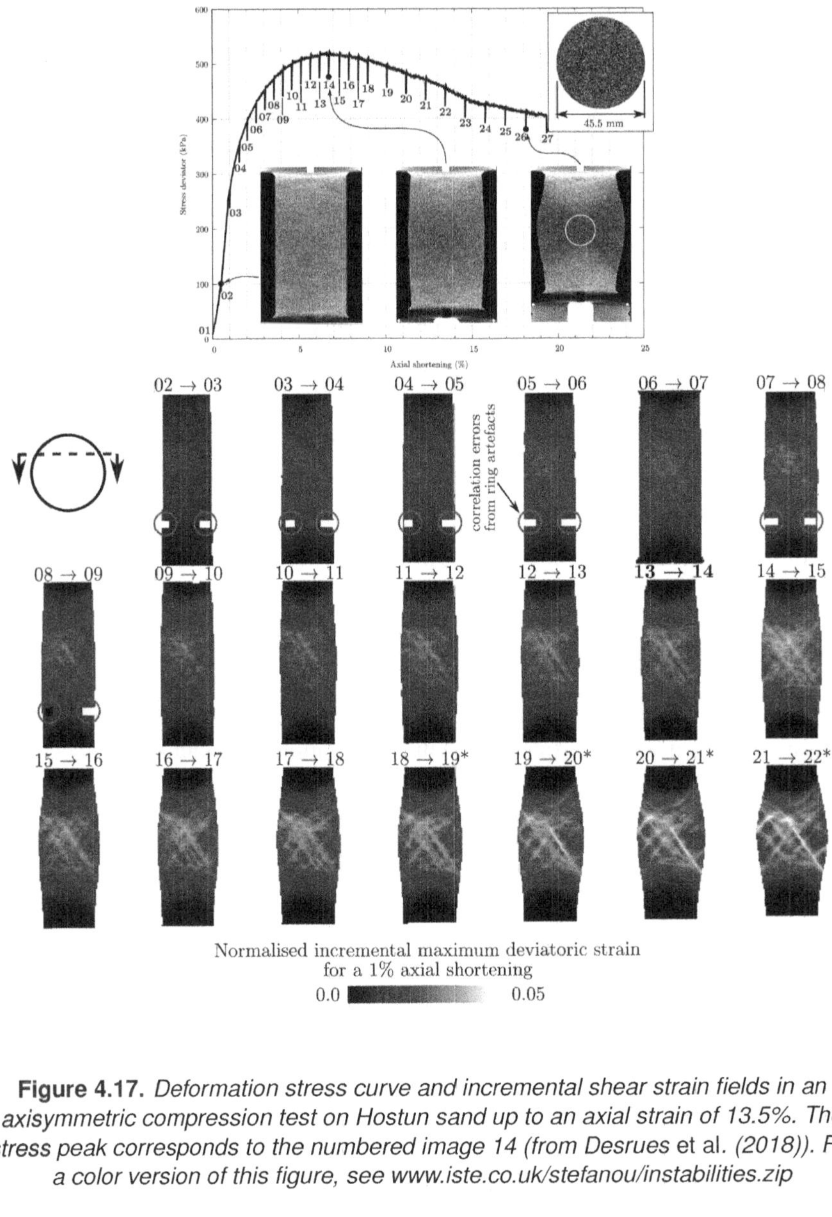

Figure 4.17. *Deformation stress curve and incremental shear strain fields in an axisymmetric compression test on Hostun sand up to an axial strain of 13.5%. The stress peak corresponds to the numbered image 14 (from Desrues* et al. *(2018)). For a color version of this figure, see www.iste.co.uk/stefanou/instabilities.zip*

4.4.3. *Triaxial compression by X-ray CT, the critical void ratio*

We report here some observations obtained by X-ray CT on the volume strain inside shear bands (Desrues *et al.* 1996). The study used a medical tomography apparatus that had poor spatial resolution with respect to the actual standard: the voxel size was $0.7 \times 0.7 \times 4\ \mathrm{mm}^3$. However, the results were interesting concerning the evolution of the void ratio in shear bands. The results introduced the notion of a *critical void ratio* inside shear bands. Several tests were performed on Hostun sand at the same confining pressure (60 kPa). Several initial specimen void ratio were obtained by sand pluviation, with different pluviation heights depending on the expected void ratio. The specimen size had a diameter and height of 100 mm. The local void ratio was determined from X-ray scanning, using a linear relation between the void ratio and X-ray absorption (gray-level in 3D pictures). The authors measured the void ratio before strain localization and then both inside and outside the shear bands after strain localization. They observed that, independently of the initial specimen void ratio (either loose or dense specimens), the void ratio inside the bands tended to move toward a unique value during the specimen loading (Figure 4.18). For dense specimens, the material dilated in the band, while for loose specimens, the porosity decreased lightly in the band.

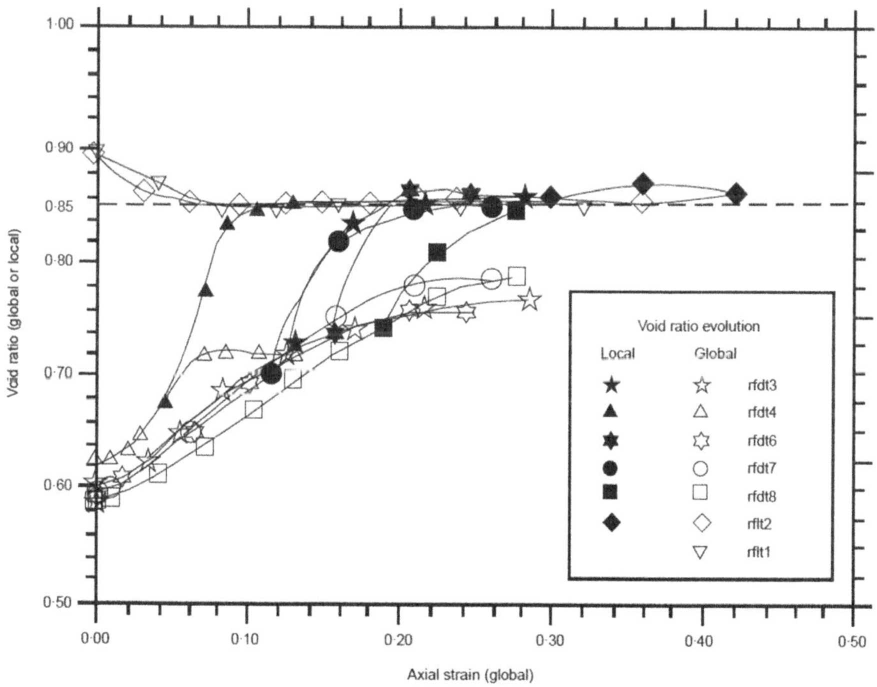

Figure 4.18. *Comparison of triaxial tests on loose and dense Hostun sand specimens: void ratio evolution inside and outside shear bands (from Desrues* et al. *(1996))*

To conclude, once the strain is localized in a specimen, the volume strain in the specimen is itself localized. The dilation (or compaction) of the specimen is mainly concentrated in the deformation bands. Here, both the space resolution and porosity change detection are too poor to clearly understand the transition from a homogeneous to a localized regime, compared to section 4.4.2. However, we have to keep in mind that the first observations of the diffuse shear band patterns in the previous section were associated with the specimen dilation, suggesting that this dilation originates from the early strain localization.

4.5. Strain localization in porous rocks

We consider here the emergence of strain localization in some porous rocks. Many studies exist that characterized strain localization using *post-mortem* analysis (see Paterson and Wong (2005) for a review), which are out of the scope of this chapter. Two experimental studies are considered here on two different rocks: Vosges sandstone and a clayey rock. Tests were performed on the true triaxial cell described in section 4.2.3.2, using 2D full-field measurement.

4.5.1. *Strain localization in Vosges sandstone*

The mechanical behavior and failure by strain localization in Woustviller red Vosges sandstone was extensively studied through axisymmetric compression and extension tests in Bésuelle *et al.* (2000); Bésuelle (1999). The recent work of Lanatà (2014) from which selected results are presented here has extended the analysis to plane strain compression tests. Minor stress was applied by a fluid and was constant, the axial loading was controlled in displacement and the loading was controlled in the third direction (intermediate principal stress direction) to satisfy zero displacement.

Two tests were selected: test BxR_GVR_06 was performed at a lateral stress of 20 MPa and test BxR_GVR_11 at 50 MPa. From Bésuelle *et al.* (2000), we know that the behavior of the rock at the lowest mean stress is quite brittle, the mean stress dependency on the deviatoric stress peak envelope is positive (cohesive-frictional behavior) and the shear bands are dilative. At the highest mean stress, the behavior was more ductile, the mean stress dependency on the deviatoric stress peak envelope was negative (cap surface), the shear bands were compactive and the inclination of the bands with respect to the axial loading direction was also higher than for a lower mean stress (Bésuelle 2001a,b).

Figure 4.19 shows the stress–strain response and global volumic curve measured for the test BxR_GVR_06. There was first a quasi linear part of the stress versus strain curve, followed by an incurvation up to a stress peak and a progressive and then sudden strain softening. The loss of control during the softening was due to the relative low cell stiffness compared to the specimen's stiffness. The loop of the curve after the fast softening was due to a delay in the axial piston pressure control. The stress level stabilized at the end of the test. The global volume strain compacted at the beginning and then dilated. The maximum compaction corresponds more or less to the loss of linearity of the upper curve. At the end of the test, the volume strain rate vanished.

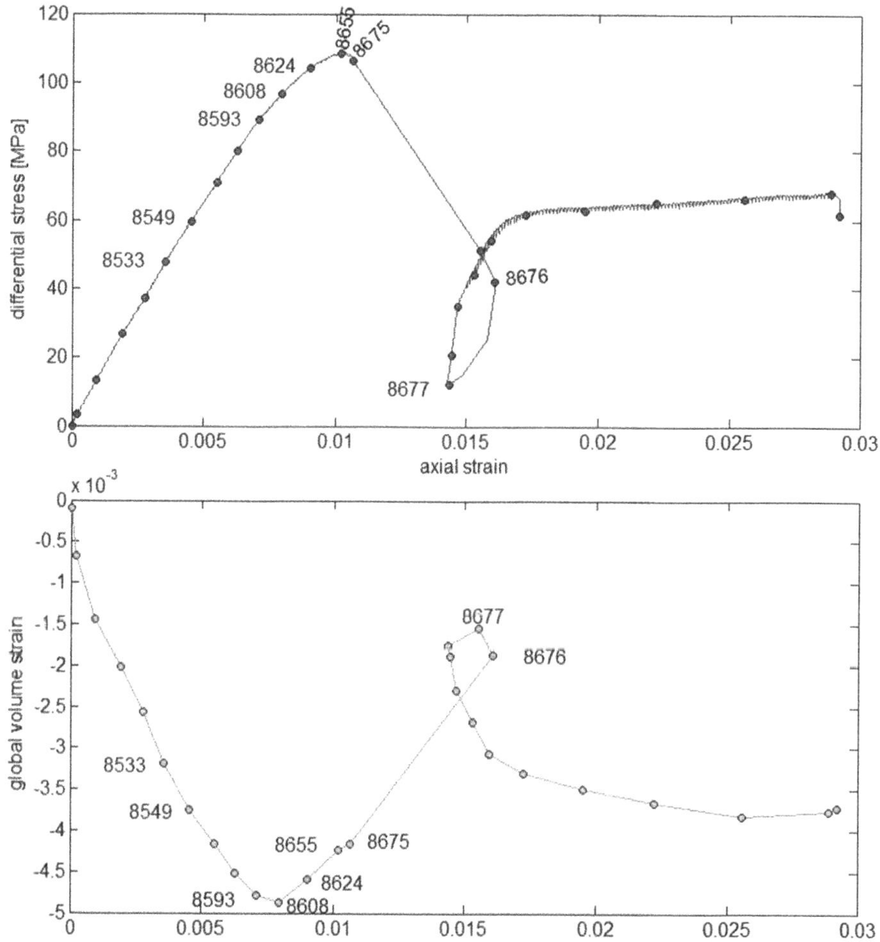

Figure 4.19. *Test BxR_GVR_06: (top) differential stress ($\sigma_1 - \sigma_3$) versus axial strain and (bottom) volume strain versus axial strain; the numbers noted on each curve are the selected photographs numbers (from Lanatà (2014)). For a color version of this figure, see www.iste.co.uk/stefanou/instabilities.zip*

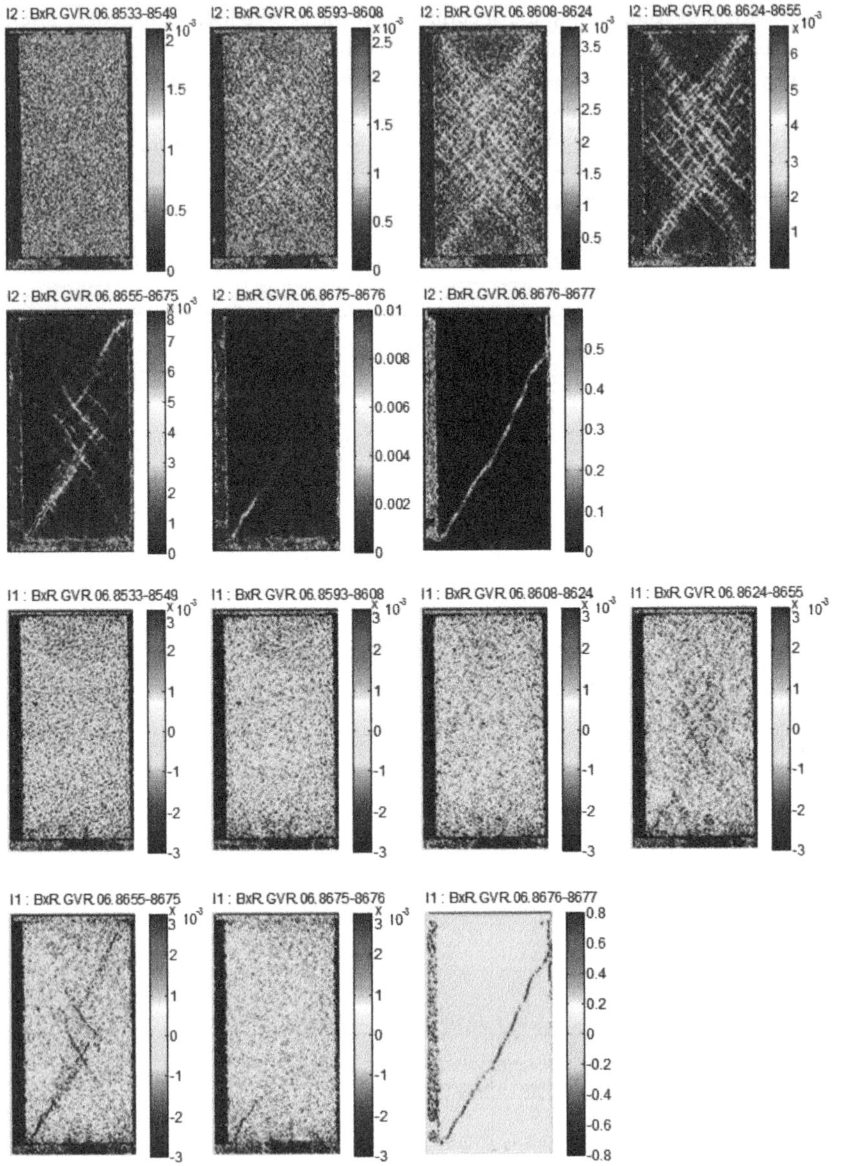

Figure 4.20. *Test BxR_GVR_06: (the two first lines) incremental second strain tensor invariant (shear strain) fields; the numbers at the top of pictures correspond to the photographs numbers (the two last lines), incremental first strain tensor invariant (volume strain) fields (from Lanatà (2014)). For a color version of this figure, see www.iste.co.uk/stefanou/instabilities.zip*

Some photographs were selected during the test (Figure 4.19): the incremental strain fields between these images are presented in Figure 4.20. The strain field was quite homogeneous during the linear part of the stress versus strain curve (increment 8533–8549). The small fluctuations in both shear and volume strains correspond to the noise of the DIC measure. Once the stress versus stain curve significantly left its quasi-linear regime, which corresponded to a significant reduction of the specimen volume compaction (increment 8593–8608), the shear strain field was no longer homogeneous. We distinguished a dense, diffuse strain localization band pattern: these bands were parallel and conjugated. This pattern was more visible in the next increment (8608–8624). A similar pattern was visible in the volume strain field, showing dilatancy in the bands. This step corresponded to dilatancy in the global volume. Before the stress peak (increment 8624–8655), the complex pattern was less dense and some shear bands still existed (plastic strain) but were deactivated. The active shear bands continued to dilate. After the peak (increment 8655–8675), the number of active bands significantly decreased and the shear and volume dilatancy became increasingly concentrated in a decreasing number of bands. During the abrupt softening (the position of photograph 8676 in Figure 4.19 is not well-defined due to a fast failure), one main shear band started to propagate from the bottom left corner and dominated the other bands. After the softening, the band was fully propagated through the specimen and was still dilating. The dilantancy decreased later (not shown). The inclination of the early bands before the stress peak with respect to the axial loading direction was about 45°, which decreased to around 30° at the end of the test.

Figure 4.21 shows the stress–strain response and global volume curve measured for the test BxR_GVR_11. The curve is similar to that of test BxR_GVR_06, but less brittle. There is first a quasi-linear part in the stress versus strain curve, followed by an curve up to a stress peak and a progressive and then a plateau on the curve. The global volume strain compacted from the beginning to the end. This kind of response was already described in Bésuelle *et al.* (2000). At the end of the test, the volume strain rate almost vanished.

Some photographs were selected during the test (Figure 4.21) and the incremental strain fields between these images are presented in Figure 4.22. The strain field was quite homogeneous during the linear part of the stress versus strain curve (increment 3138–3158). The small apparent strain concentration on the right side of the specimen was probably an artifact due to the optical effect of the thin lubricant layer deposited on the specimen's surface. Once the stress versus strain curve significantly left its quasilinear regime (increment 3196–3213), the shear strain field was no longer homogeneous. We distinguished a dense, diffuse strain localization band pattern: these bands were parallel and conjugated. This pattern was more visible in the next increment (3213–3230). A similar pattern was visible on the volume strain field, showing compaction in the bands. Before the stress peak (increment 3230–3242), the complex pattern was less dense and some shear bands still existed (plastic strain) but were deactivated. The active shear bands continued to be compactant. After the peak (increment 3242–3259), two main conjugated bands were still active and were around several short bands. The main bands were not very straight and they seemed to result from the coalescence of many smaller bands. In the following strain softening, only one band persisted, which crossed the top-left specimen corner. It also seems to result from the coalescence of many smaller bands. In the residual step (on the stress plateau), a second main band appeared, parallel to the previous one. Additionally, several conjugated small bands existed in the relay area between the main band. The bands persisted in the next increments (not shown). The volume strain was compactant in the bands. The inclination of the early bands before the stress peak, with respect to the axial loading direction, was about 45°, which stayed around 45° at the end of the test. This was also observed in Bésuelle *et al.* (2000) based on *postmortem* X-ray CT scans.

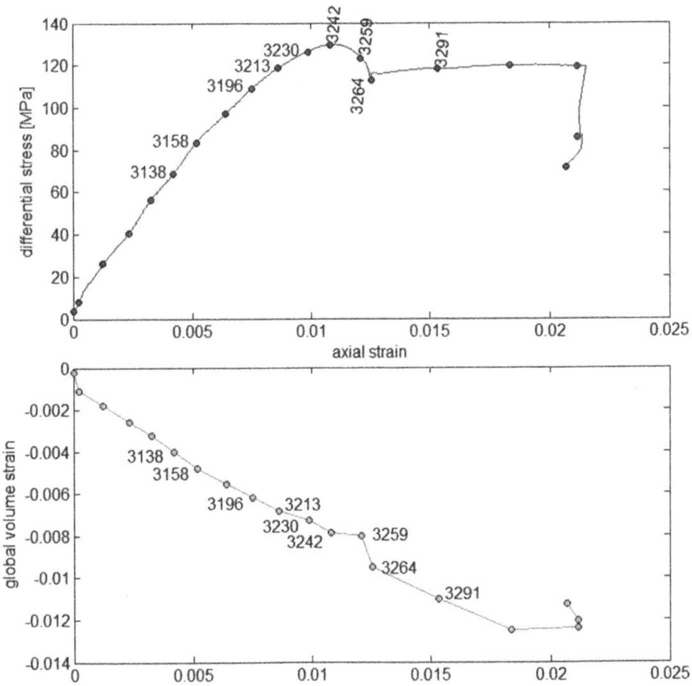

Figure 4.21. *Test BxR_GVR_11: (top) differential stress ($\sigma_1 - \sigma_3$) vs axial strain and (bottom) volume strain versus axial strain; the numbers noted on each curve are the selected photographs numbers (from Lanatà (2014)). For a color version of this figure, see www.iste.co.uk/stefanou/instabilities.zip*

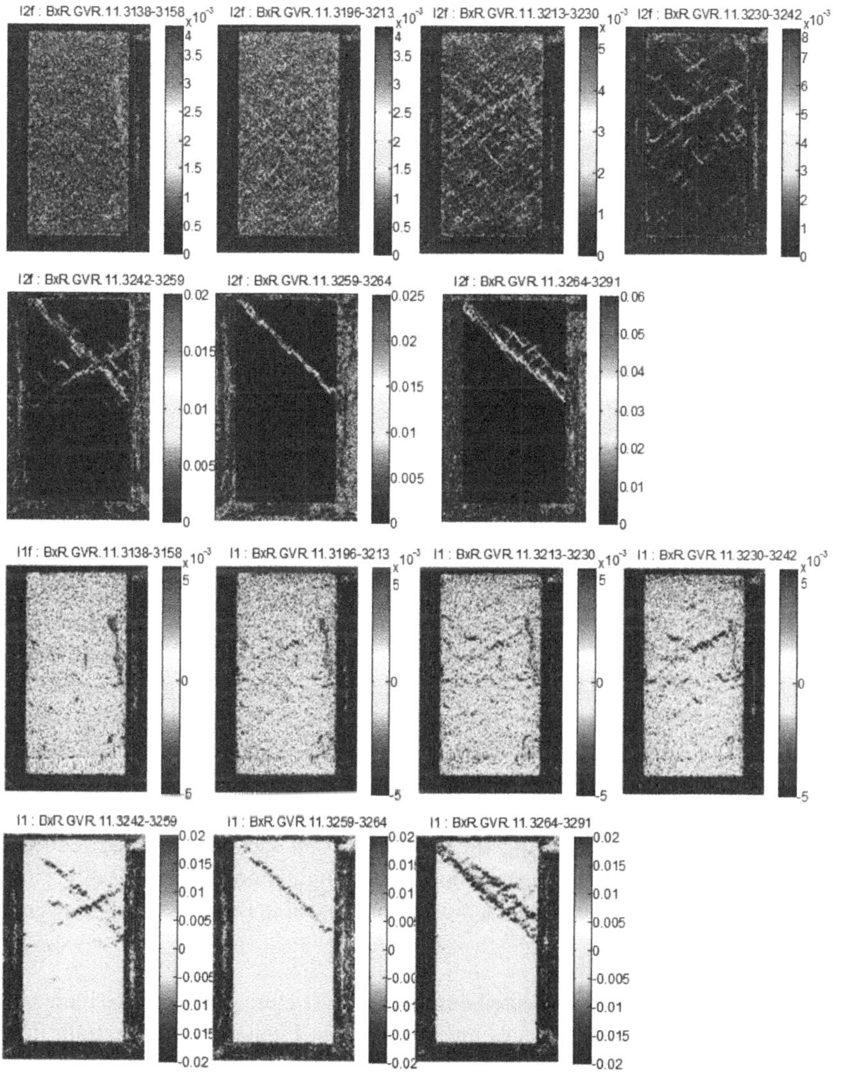

Figure 4.22. *Test BxR_GVR_11: (the two first lines) incremental second strain tensor invariant (shear strain) fields; the numbers at the top of pictures correspond to the photographs numbers (the two last lines), incremental first strain tensor invariant (volume strain) fields (from Lanatà (2014)). For a color version of this figure, see www.iste.co.uk/stefanou/instabilities.zip*

To conclude, these observations show the progressive nature of the strain localization process in porous sandstone. The strain localization is dependent on mean stress. A common observation between the two tests, one of which was performed at 20 MPa lateral stress and the other at 50 MPa, was the early diffuse strain localization, well before the stress peak. Once the stress versus strain curve lost its linearity, a complex pattern of parallel and conjugated bands appeared, oriented at about 45° with respect to the axial direction. At 20 MPa, the bands dilated and induced a global volume dilatancy, while they were compactant at 50 MPa and extended the global specimen compacting regime. The number of active bands decreased during the subsequent loading to a very small number after the peak. The inclination of the bands, after the peak, stayed constant at 50 MPa, while it decreased significantly at lower mean stress.

4.5.2. *Strain localization in a clayey rock*

We selected two tests performed on Callovo–Oxfordian clayey rock. Tests BxR_COx_06 and BxR_COx_13 (issued from cells EST32896 and EST2614, respectively) were performed at a lateral stress of 12 MPa, which corresponded to the *in situ* stress of the material.

Figure 4.23 shows the stress–strain response and global volume curve measured for the test BxR_COx_06. There was small curvature at first, then a quasi linear part in the stress versus strain curve, followed by an curvature up to a stress peak and a progressive and then sudden strain softening. The loss of control during the softening was due to the relative low cell stiffness compared to the specimen's stiffness. The loop of the curve after the fast softening was due to a delay in the axial piston pressure control. The stress level stabilized at the end of the test (plateau). The global volume strain compacted at the beginning and the compaction was progressively reduced up to a maximum, after which the specimen dilated slightly and, then the volume strain rate vanished. The maximum compaction corresponded to the stress peak of the upper curve.

Some photographs were selected during the test (Figure 4.23) and the incremental strain fields between these images are presented in Figure 4.24. The strain field is quite homogeneous during the linear part of the stress versus strain curve (increment 0025-0101). The small fluctuations in both shear and volume strains corresponded to the noise of the DIC measure. However, at the lower third of the specimen there was a shear stress concentration on the left side. This area corresponded to a large calcite inclusion very close the observed surface of the specimen, which was observed by X-ray CT after the test. The contrast in stiffness between the large inclusion and the rock matrix induced this strain concentration. Once the stress versus strain curve showed a small curvature (increment 0146-0236), the appearance of shear strain field changed. It was not noisy like the previous increment, but instead seemed structured.

A very dense pattern of parallel and conjugated shear bands seemed to emerge. This pattern was confirmed in the next increments 0169-0272, 0272-0312 and 0312-0344. During these increments, one first observes that the initial strain concentration detected in the first increment propagated progressively as a shear band. Moreover, during the prepeak steps, the number of active bands of the pattern decreased progressively. Only three bands were still active at the stress peak (increment 0344-0354), two of them propagated from the top-right and bottom-left corners. These two bands continued to propagate during the strain softening (increment 0371-0378). The term *fault propagation* should be more appropriate here; indeed, the strong strain intensity suggests a material discontinuity. Only one fault was active on the stress plateau: the fault initiated from the top-right corner. The propagation of this fault during the previous increment (0371-0378) showed that the process zone in front of the fault was itself composed of several parallel and conjugated shear bands.

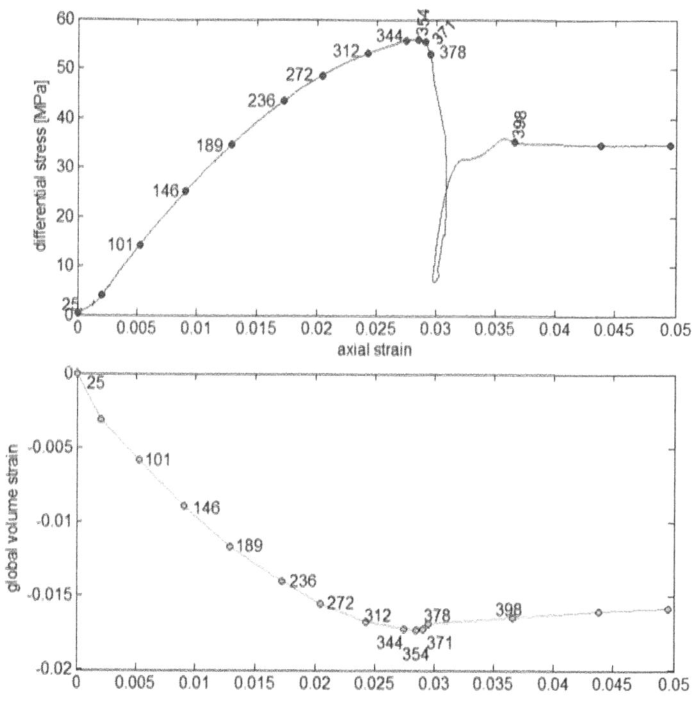

Figure 4.23. *Test BxR_COx_06: (top) differential stress ($\sigma_1 - \sigma_3$) versus axial strain and (bottom) volume strain versus axial strain; the numbers noted on each curve are the selected photographs numbers (from Bésuelle (2012)). For a color version of this figure, see www.iste.co.uk/stefanou/instabilities.zip*

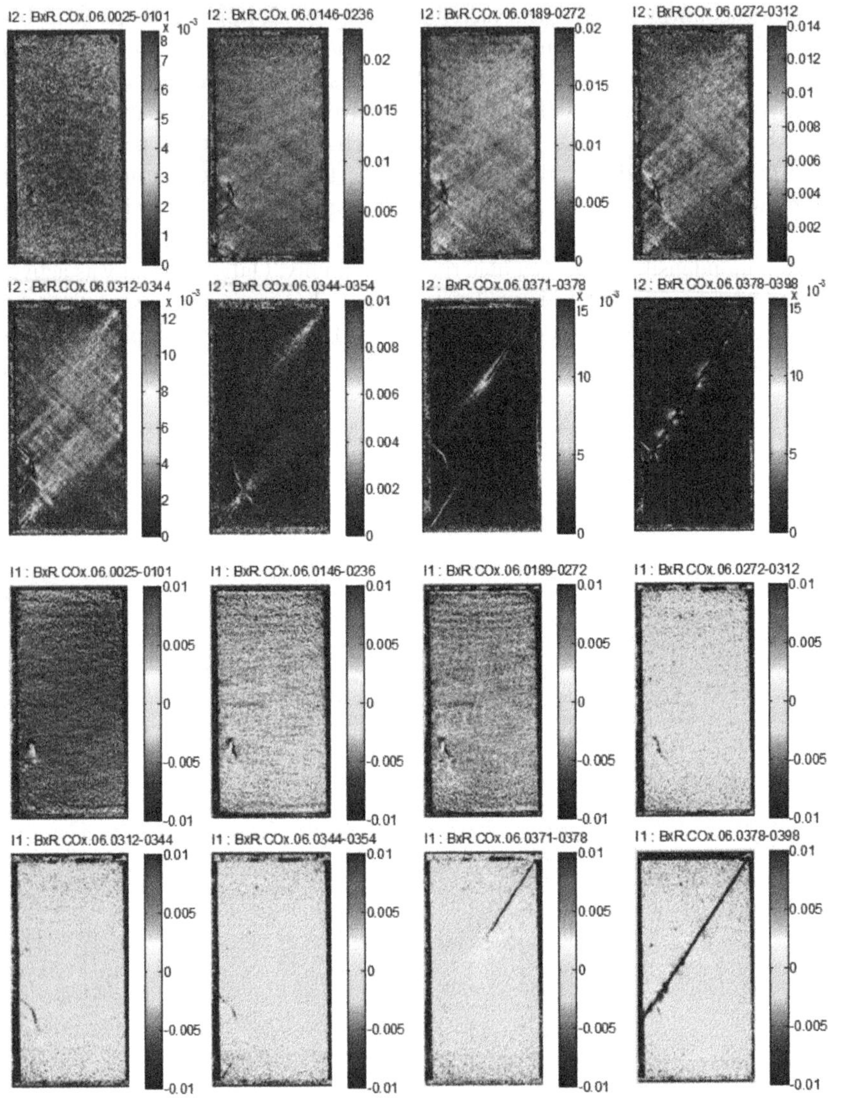

Figure 4.24. *Test BxR_COx_06: (the two first lines) incremental second strain tensor invariant (shear strain) fields; the numbers at the top of pictures correspond to the photographs numbers (the two last lines), incremental first strain tensor invariant (volume strain) fields (from Bésuelle (2012)). For a color version of this figure, see www.iste.co.uk/stefanou/instabilities.zip*

As far as the volume strain fields are concerned, some horizontal "bands" were observed, that corresponded to a well-known artifact of the DIC method, generally due to the subpixel step. They had no physical meaning. The first volume strain localization was observed at increment 0371-0378, where the fault dilated (propagation in a mixed mode I and II). This corresponded to the global dilatancy of the specimen. The volume strain in the last increment showed some area of compaction and dilatancy around the fault. This is probably evidence of roughness in the fault, where some areas resist sliding more. The inclination of early bands before the stress peak, with respect to the axial loading direction, was about 45°.

Figure 4.25 shows the stress–strain response and global volume curve measured for the test BxR_COx_13, performed in the same conditions as for BxR_COx_06. It differed a little from the previous test because of the strong initial curvature of the stress versus strain curve and its lower stress peak. This questions the low experimental reproducibility on such materials, mainly due to their natural heterogeneity.

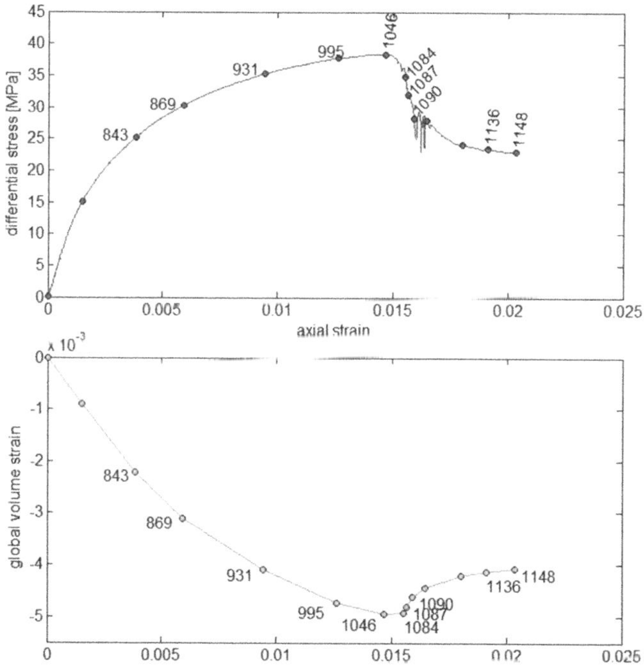

Figure 4.25. *Test BxR_COx_13: (top) differential stress ($\sigma_1 - \sigma_3$) versus axial strain and (bottom) volume strain versus axial strain; the numbers noted on each curve are the selected photographs numbers (from Bésuelle (2012)). For a color version of this figure, see www.iste.co.uk/stefanou/instabilities.zip*

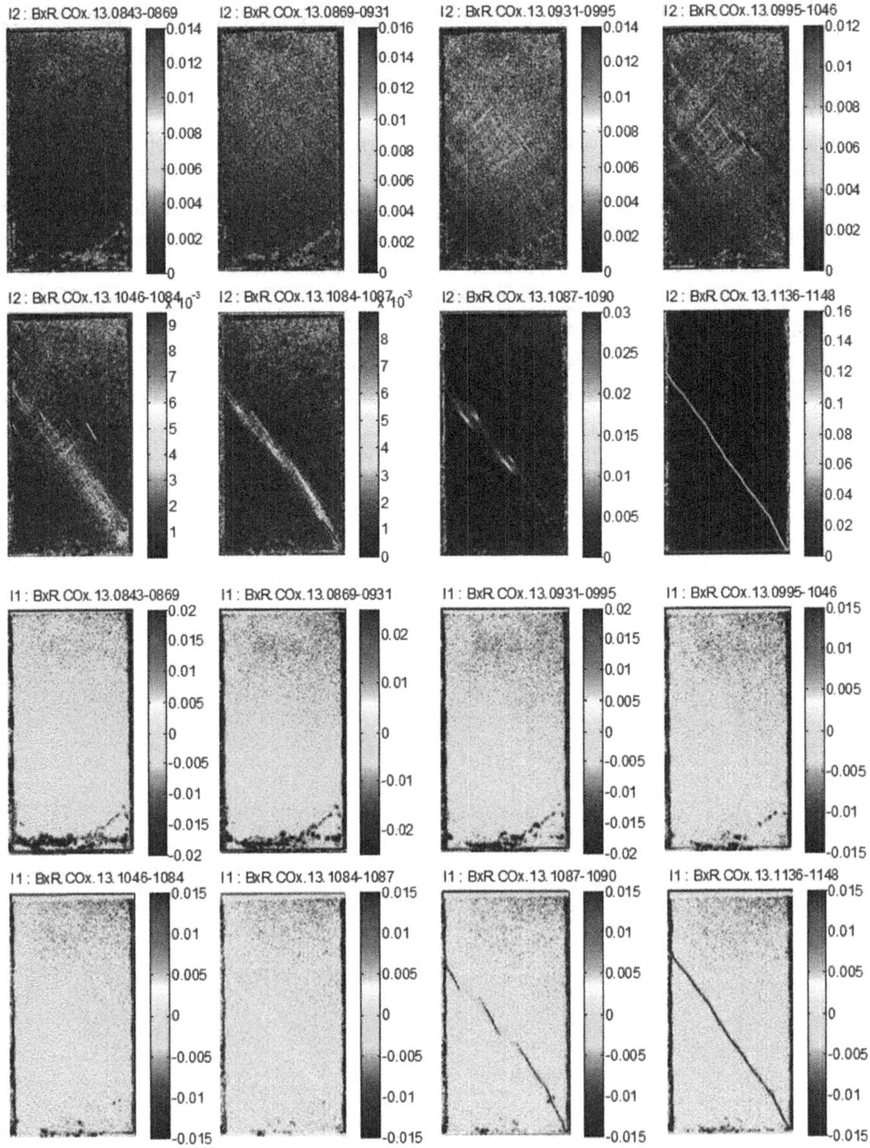

Figure 4.26. *Test BxR_COx_13: (the two first lines) incremental second strain tensor invariant (shear strain) fields; the numbers at the top of pictures correspond to the photographs numbers (the two last lines), incremental first strain tensor invariant (volume strain) fields (from Bésuelle (2012))*

The incremental strain fields are presented in Figure 4.26. The quality of the digital image is not as good as in the previous test. This can be due to the quality of the synthetic speckle or the camera's focus adjustment, which impacted the quality of the DIC results and the ability to detect small details. The results were not as fine as those for test BxR_COx_06. However, the usual dense pattern of parallel and conjugated shear bands was observed in increments 0931-0995 and 0995-1046. More interesting was the evolution after the stress peak. A large band composed of several parallel and conjugated bands was observed in increment 1046-1084. During the following increment, the width of the large band decreased progressively to converge to one band, which became a fault. The transition from strain localization to a fault here resulted from a condensation, during which the external bands were progressively deactivated, and the central band stayed active to evolve toward a fault.

Also the process of localization was progressive in a clayey rock and showed some early shear bands in a complex pattern, well before the stress peak. The early bands were very numerous, thin and close together. The thickness of the shear bands in a material with such a fine microstructure could not be measured with the digital camera used here. However, the band thickness was quite obvious directly linked with the microstructure internal's lengths. The volume strain inside the band showed a small dilatancy for this clay rock and the mean stress level used here. The two tests showed two examples of transition from shear bands to faults. One was by a fault propagation in a strain-localized pattern, with evidence of strain localization in the process zone. The other came from a phenomenon of condensation: a large band comprised several thinner bands became condensed by a deactivation of external thin bands. The few remaining bands became faults that coalesced to form a major fault.

4.6. Conclusion

Until now, the phenomenon of strain localization in geomaterials was understood as a propagation of shear bands or the birth of an abrupt shear band from a diffuse strain field associated with the stress peak of the specimen's response. The strain softening of the material in the bands induced a global softening of the specimen. Historically, this aspect came from observations of the specimens after the tests, in which relatively simple patterns of shear bands or faults were generally observed. This was also enforced by the theoretical approach of shear band analysis using bifurcation theory. This approach is generally quite consistent with experimental results at the stress peak.

The development of more efficient imaging tools and methods for full-field measurement has allowed better quantification of the deformation with better space and time resolutions and at a smaller scale. The recent results on a broad range of soil and porous rock suggest that the transition from a diffuse regime to a localized regime is more progressive than what is recalled above. A complex pattern of numerous parallel and conjugated shear bands appears well before the stress peak. These diffuse shear band patterns can generate volume strain, dilatancy or compaction, depending on the material and stress state. Some bands are progressively deactivated, and close to the stress peak, only a few shear bands stay active that concentrates the deformation. They are generally prone to strain softening, inducing global strain softening. In cohesive materials, these shear bands can evolve as faults or gouge zones of granular materials. This transition process of the deformation mode has been observed on loading paths of the compression type in plane strain or axisymmetric compression; it would be interesting to check if it is becoming generalized on other types of loading.

The recent trend of developing *in situ* microexperiments to quantify the micromechanisms of deformation at the scale of the microstructure should improve our understanding of early strain localization. In parallel, early band prediction appears to be a challenge for constitutive law development and numerical computation. They do not seem to fit into a bifurcation analysis because they appear early in the loading process. Their understanding will probably have to involve taking into account the variability of the microstructure of geomaterials; heterogeneity, although initially low, could have an important role.

4.7. References

Andò, E.C. (2015). Experimental investigation of microstructural changes in deforming granular media using x-ray tomography. PhD thesis, University of Grenoble, France.

Andò, E.C., Hall, S.A., Desrues, J., Viggiani, G., Bésuelle, P. (2012a). Experimental micromechanics: A triaxial test on sand observed at the grain scale. *Géotech. Lett.*, 2(3), 107–112.

Andò, E.C., Hall, S.A., Viggiani, G., Desrues, J., Bésuelle, P. (2012b). Grain-scale experimental investigation of localised deformation in sand: A discrete particle tracking approach. *Acta Geotechnica*, 7(1), 1–13.

Armand, G., Conil, N., Talandier, J., Seyedi, D.M. (2017). Fundamental aspects of the hydromechanical behaviour of Callovo-Oxfordian claystone: From experimental studies to model calibration and validation. *Comput. Geotech.*, 85, 277–286.

Bažant, Z. and Pijaudier-Cabot, G. (1988). Nonlocal continuum damage, localization instability and convergence. *J. Appl. Mech.*, 55(2), 287–293.

Bažant, Z., Lin, F.-B., Pijaudier-Cabot, G. (1987). Yield limit degradation: Nonlocal continuum model with local strain. In *Computational Plasticity. Models, Software and Applications*, Owen, D., Hinton, E., Oñate, E. (eds). Pineridge Press, Swansea, 1757–1780.

Bay, B.K., Smith, T.S., Fyhrie, D.P., Saad, M. (1999). Digital volume correlation: Three-dimensional strain mapping using x-ray tomography digital volume correlation: Three-dimensional strain mapping using x-ray tomography. *Experim. Mech.*, 39(3), 217–2264.

Bésuelle, P. (1999). Déformation et rupture dans les roches tendres et les sols indurés: comportement homogène et localisation. PhD thesis, University of Grenoble, France,

Bésuelle, P. (2001a). Compacting and dilating shear bands in porous rock: Theoretical and experimental conditions. *J. Geophys. Res.*, 106(B7), 13435–13442.

Bésuelle, P. (2001b), Evolution of strain localisation with stress in a sandstone: Brittle and semi-brittle regimes. *Phys. Chem. Earth, Part A*, 26(1-2), 101–106.

Bésuelle, P. (2012). Localisation des déformations dans l'argilite du Callovo-Oxfordien. Internal communication.

Bésuelle, P. and Lanatà, P. (2016). A new true triaxial cell for field measurements on rock specimens and its use in the characterization of strain localization on a Vosges sandstone during a plane strain compression test. *Geotech. Test. J.*, 39(5), 1–12.

Bésuelle, P. and Rudnicki, J. (2004). Localization: Shear bands and compaction bands. In *Mechanics of Fluid-Saturated Rocks*, Guéguen, Y., Boutéca, M. (eds). Academic Press, Elsevier, 219–321.

Bésuelle, P., Desrues, J., Raynaud, S. (2000). Experimental characterisation of the localisation phenomenon inside a Vosges sandstone in a triaxial cell. *Int. J. Rock Mech. Min. Sci.*, 37(8), 1223–1237.

Bornert, M., Chaix, J., Doumalin, P., Dupré, J.-C., Fournel, T., Jeulin, D., Maire, E., Moreaud, M., Moulinec, H. (2004). Mesure tridimensionnelle de champs cinématiques par imagerie volumique pour l'analyse des matériaux et des structures. *Instrum., Mes., Métrolog.*, 3–4, 43–88.

Bornert, M., Brémand, F., Doumalin, P., Dupré, J.-C., Fazzini, M., Grédiac, M., Hild, F., Mistou, S., Molimard, J., Orteu, J.-J., Robert, L., Surrel, Y., Vacher, P., Wattrisse, B. (2008). Assessment of digital image correlation measurement errors: Methodology and results. *Experim. Mech.*, 3, 353–370.

Butterfield, R., Harkness, R., Andrawes, K. (1970). A stereo-photogrammetric method for measuring displacements fields. *Géotechnique*, 20(3), 308–314.

Chambon, R., Caillerie, D., Matsushima, T. (2001). Plastic continuum with microstructure, local second gradient theories for geomaterials: Localization studies. *Int. J. Solids Struct.*, 38(46–47), 8503–8527.

Conil, N., Talandier, J., Djizanne, H., de La Vaissière, R., Righini-Waz, C., Auvray, C., Morlot, C., Armand, G. (2018). How rock samples can be representative of in situ condition: A case study of Callovo-Oxfordian claystones. *J. Rock Mech. Geotech. Engin.*, 10(4), 613–623.

Cosserat, E. (1909). *Théorie des Corps Déformables*. Éditions Hermann, Paris.

David, C., Robion, P., Menéndez, B. (2007). Anisotropy of elastic, magnetic and microstructural properties of the Callovo–Oxfordian argillite. *Phys. Chem. Earth, Part A*, 32, 145–153.

Desrues, J. (1998). Localisation patterns in ductile and brittle geomaterials. In *Material Instabilities in Solids*, de Borst, R., van der Giessen, E. (eds). Wiley, New York, 137–158.

Desrues, J. (2004). Experimental characterization of failure, degradation and instability in geomaterials. Progressive deformation and failure observed in laboratory specimens using strain field measurement methods. In *ALERT Doctoral School 2004 – Failure, degradation and instabilities in geomaterials: Rev. Franç. Génie Civ.*, Vardoulakis, I., Mira, P. (eds). 8(5–6), 563–592.

Desrues, J. and Viggiani, G. (2004). Strain localization in sand: an overview of the experimental results obtained in Grenoble using stereophotogrammetry. *Int. J. Numer. Anal. Meth. Geomech.*, 28, 279–321.

Desrues, J., Chambon, R., Mokni, M., Mazerolle, F. (1996). Void ratio evolution inside shear bands in triaxial sand specimens studied by computed tomography. *Géotechnique*, 46(3), 529–546.

Desrues, J., Andò, E., Mevoli, F.A., Debove, L., Viggiani, G. (2018). How does strain localise in standard triaxial tests on sand: Revisiting the mechanism 20 years on. *Mech. Res. Commun.*, 92, 142–146.

Drescher, A., Vardoulakis, I., Han, C. (1990). A biaxial apparatus for testing soils. *Geotech. Test. J.*, 13(3), 226–234.

Flavigny, E., Desrues, J., Palayer, B. (1990). Note technique : le sable d'Hostun 'RF'. *Rev. Franç. Géotech.*, 53, 67–70.

Fleck, N. and Hutchinson, J.W. (1997). Strain gradient plasticity. *Advances Appl. Mech., Academic Press*, 33, 295–361.

Fortin, J., Stanchits, S., Dresen, G., Guéguen, Y. (2009). Acoustic emissions monitoring during inelastic deformation of porous sandstone: Comparison of three modes of deformation. *Pure Appl. Geophys.*, 166(1), 823–841.

Grédiac, M. (2004). The use of full-field measurement methods in composite material characterization: Interest and limitations. *Composites: Part A*, 35, 751–761.

Haimson, B. and Chang, C. (2000). A new true triaxial cell for testing mechanical properties of rock, and its use to determine rock strength and deformability of Westerly granite. *Int. J. Rock Mech. Min. Sci.*, 37(1-2), 285–296.

Hall, S.A. (2012). Digital image correlation in experimental geomechanics. In *ALERT Doctoral School 2012 – Advanced experimental techniques in geomechanics*, Viggiani, G., Hall, S., Romero, E. (eds). Available: http://www.alertgeomaterials.eu/internal/2012/school/2012_ALERT_school.pdf.

Hall, S.A., Bornert, M., Desrues, J., Pannier, Y., Lenoir, N., Viggiani, G., Bésuelle, P. (2010). Discrete and continuum analysis of localised deformation in sand using x-ray μCT and volumetric digital image correlation. *Géotechnique*, 60(5), 315–322.

Hill, R. (1962). Acceleration waves in solids. *J. Mech. Phys. Solids*, 10, 1–16.

Hsieh, J. (2009). *Computed Tomography Principles, Design, Artifacts, and Recent Advances*, 2nd edition. Wiley, New York.

Ketcham, R.A. and Carlson, W.D. (2001). Acquisition, optimization and interpretation of X-ray computed tomographic imagery: Applications to the geosciences. *Comput. Geosci.*, 27, 381–400.

Lanatà, P. (2014). Full-field experimental characterization of mechanical behaviour and failure in a porous rock in plane strain compression: Homogeneous deformation and strain localization. PhD thesis, University of Grenoble, France.

Le Bouil, A., Amon, A., Sangleboeuf, J.-C., Orain, H., Bésuelle, P., Viggiani, G., Chasle, P., Crassous, J. (2014). A biaxial apparatus for the study of heterogeneous and intermittent strains in granular materials. *Granul. Matt.*, 16(1), 1–8.

Lenoir, N. (2006). Comportement mécanique et rupture dans les roches argileuses étudiées par micro-tomographie à rayons X. PhD thesis, University of Grenoble, France.

Lenoir, N., Bornert, M., Desrues, J., Bésuelle, P., Viggiani, G. (2007). Volumetric digital image correlation applied to X-ray microtomography images from triaxial compression tests on argillaceous rock, *Strain*, 43, 193–205.

Mogi, K. (1971). Effect of the triaxial stress system on the failure of dolomite and limestone. *Tectonophysics*, 11, 111–127.

Mühlhaus, H.-B. and Vardoulakis, I. (1987). The thickness of shear band in granular materials. *Géotechnique*, 37(3), 271–283.

Pan, B., Xie, H.-M., Dai, F.-L. (2006). Performance of sub-pixel registration algorithms in digital image correlation. *Meas. Sci. Technol.*, 17, 1615–1621.

Paterson, M.S. and Wong, T.-F. (2005). *Experimental Rock Deformation – The Brittle Field*, 2nd edition. Springer, Berlin.

Peters, W. and Ranson, W. (1982). Digital imaging techniques in experimental stress analysis. *Opt. Eng.*, 21(3), 427–431.

Rice, J. (1976). The localization of plastic deformation. In *Theoretical and Applied Mechanics*, Koiter, W. (ed.). North-Holland Pub. Comp., Delft.

Robinet, J.-C., Sardini, P., Coelho, D., Parneix, J.-C., Prêt, D., Sammartino, S., Boller, E., Altmann, S. (2012). Effects of mineral distribution at mesoscopic scale on solute diffusion in a clay-rich rock: Example of the Callovo-Oxfordian mudstone (Bure, France). *Water Resour. Res.*, 48, W05554.

Sutton, M., Wolters, W., Peters, W., McNiell, S. (1983). Determination of displacements using an improved digital correlation method. *Image Vis. Comput.*, 1, 133–139.

Takemura, T. and Oda, M. (2004). Microstructure observation in deformed geomaterials using microfocus x-ray computed tomography. In *X-ray CT for Geomaterials, Soils, Concrete, Rocks*, Otani, J., Obara, Y. (eds). Balkema, Rotterdam, Kumamoto, 299–304.

Vardoulakis, I. and Aifantis, E.C. (1991). A gradient flow theory of plasticity for granular materials. *Acta Mechanica*, 387, 197–217.

Vardoulakis, I. and Goldscheider, M. (1981). Biaxial apparatus for testing shear bands in soils. In *10th ICSMFE*, 4/61(v), 819–824.

Vardoulakis, I. and Sulem, J. (1995). *Bifurcation Analysis in Geomechanics*. Blackie Academic & Professional, Glasgow.

Vautrin, A., Lee, J., Molimard, J., Surrel, Y. (2002). Full-field optical techniques: applications to strain measurement and mechanical identification. In *Invited keynote at the 10th European conference on composite materials (ECCM10), Bruges (Belgium)*. Available: https://www.nottingham.ac.uk/home/eaziaj/photmechanics/Lee%20Full-field%20optical%20methods%20%20applications%20to%20strain%20measurement%20and%20mechanical%20identification_ECCM10_2002_ keynote.pdf.

Verhulp, E., van Rietbergen, B., Huiskes, R. (2004). A three-dimensional digital image correlation technique for strain measurements in microstructures. *J. Biomech.*, 37, 1313–1320.

Viallon-Galinier, L., Combe, G., Richefeu, V., Atman, A.P.F. (2018). Emergence of shear bands in confined granular systems: Singularity of the q-statistics. *Entropy*, 20, 862–875.

Viggiani, G. and Hall, S.A. (2012). Full-field measurements in experimental geomechanics: Historical perspective, current trends and recent results. In *ALERT Doctoral School 2012 – Advanced experimental techniques in geomechanics*, Viggiani, G., Hall, S., Romero, E. (eds). Available: http://www.alertgeomaterials. eu/internal/2012/school/2012_ALERT_school.pdf.

5

Numerical Modeling of Strain Localization

Panos PAPANASTASIOU[1] **and Antonis ZERVOS**[2]

[1] *Department of Civil and Environmental Engineering, University of Cyprus, Nicosia, Cyprus*

[2] *Faculty of Engineering and Physical Sciences, University of Southampton, UK*

This chapter presents two higher order continua with microstructure, which can be used to regularize the ill-posed mathematical problem of strain-softening material behavior. The continua enable the modeling of progressive localization of deformation in zones of intense shearing, which leads to failure of geomaterials and underground openings. The first continuum is based on a Cosserat plasticity model, which has independent rotation, in addition to the translational degrees of freedom of the classical continuum. The independent rotations introduce in the model curvatures and associated couple stresses. The second higher order continuum belongs to the gradient type and is called gradient elastoplasticity. In this model, higher order deformation strains and internal lengths are present in both the elastic and the plastic regimes. The existence of the higher order strains is treated by numerical discretization based on the displacement formulation with a C1 finite element of high polynomial order. In both models, the introduction of internal length, which can be related to the grain size, improves the computational stability and allows for robust progressive localization modeling in the postbifurcation regime. It is shown through different examples that the models are capable of predicting advanced deformation modes such as surface buckling and localization of deformation in shear bands that may lead to failure of a structure. As an illustration, results of the following applications from geotechnical and petroleum geomechanics are presented: the modeling of failure of biaxial tests and thick walled cylinders under external and internal pressure, and the stability of elliptical shape perforations.

Instabilities Modeling in Geomechanics,

The obtained results clearly show a progressive failure pattern. The calculated modes are in good qualitative agreement with laboratory tests and field observations.

5.1. Introduction

Loading of soil and rock masses can lead progressively to concentration of deformation in narrow bands of intense shearing such as shear bands or shear interface layers. These bands may initiate or intersect the boundary of a free surface leading to failure. It is shown by experiments that near a failure state, the material in these concentration zones, which have a finite thickness (Mühlhaus and Vardoulakis 1987), undergoes significant deformation and degradation, whereas the surrounding material remains inert.

Over the last three decades, many researchers studied localization of deformation in solids using the theory of equilibrium bifurcation. These studies, which were initially analytical for infinite-domain problems with various constitutive models, provided expressions for the critical hardening modulus and the orientation angle of the localization bands at bifurcation (Rudnicki and Rice 1975). Shear-band analyses for complete boundary-value problems and especially for postlocalization analyses that lead to failure can be carried out only computationally.

In order to connect the analytical bifurcation studies with the postfailure analyses, the terminology used in numerical analyses of stability and bifurcation problems is explained here. Figure 5.1 shows the primary branch (I) and the secondary branch (II), which intersects (I) at the bifurcation point. The primary path (I) corresponds to the so-called trivial or homogeneous deformation solution. The secondary path corresponds to the solution that will lead to localized deformation. Examples of analytical studies that are restricted in the primary path and to the finding of the bifurcation point can be found in various studies (Papanastasiou and Durban 1995; Papamichos and van den Hoek 1995; Papanastasiou and Vardoulakis 1989; Vardoulakis and Papanastasiou 1988). The bifurcation point is determined by a singular point where the global stiffness matrix has at least one zero eigenvalue. In order to model the localization of deformation beyond the bifurcation point, the solution must switch from the primary to the secondary path. The method of switching from branch (I) to branch (II) is to perturb the primary solution with the eigenvector of the vanishing eigenvalue at the bifurcation point. Another approach is to introduce a material or a geometrical imperfection to resort to an "imperfect" structure. The solution of the "imperfect" structure will follow the smooth branch (III), which is asymptotic to branches (I) and (II) in the relevant regimes at the bifurcation point (Figure 5.1). Hence, the introduction of a small imperfection transfers the bifurcation problem to a limit problem exhibited by branch III. The limit point is found by the slope of the load-displacement curve, which must be zero. It was found that the Cosserat and gradient models follow smoothly the branch (III) without it being necessary to perturb the primary solution with the eigenvector of the zero eigenvalue (Papanastasiou and Vardoulakis 1992, 1994). In addition, it was shown that the deformation field at the first equilibrium point after bifurcation is consistent with the eigenfield of the lowest eigenvalue.

Turning now to localization analysis, it is suggested that modeling of failure in rocks requires the incorporation of high dilation and significant strength loss in the material constitutive law (Ewy 1993). Material softening can be considered as a macroscopic account of inhomogeneities at the microscopic scale such as microcracks, weak spots and fissures. A simple way to account for these inhomogeneities in a continuum model is by smearing them in a strain-softening constitutive model. The material softening is a necessary physical mechanism to trigger and promote localization. However, a major drawback of strain-softening models is that it leads to a loss of ellipticity of the governing mathematical equations (Vardoulakis and Sulem 1995) resulting in ill-posed boundary value problems, whose finite element analysis is sensitive to mesh refinement and exhibits convergence problems. Furthermore, in classical constitutive models there is no material length and the discretization sets the length scale. This causes the localization region, which is associated with strain softening to depend on the mesh size used for the spatial discretization. Consequently, the thickness and spacing of the shear bands are dictated by the mesh size. From energy consideration, a constitutive theory without internal length often leads to a softening zone of negligible thickness and hence to problematic zero energy dissipation. In other words, numerical computations based on classical continua give spurious or mesh-dependent results in the postpeak softening regime.

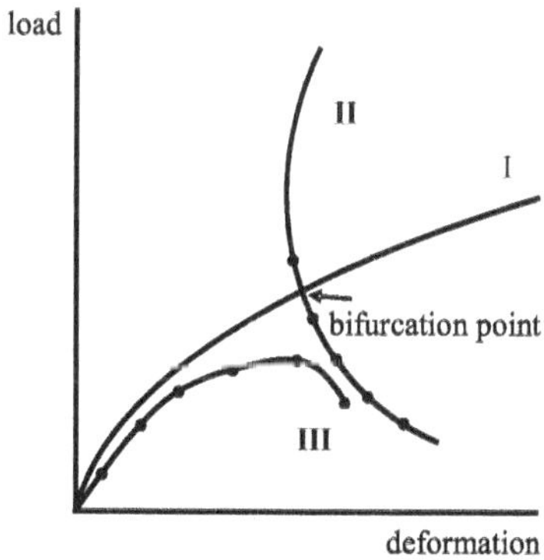

Figure 5.1. *Primary branch (I), secondary branch (II) and branch for the "imperfect" system in a bifurcation problem*

The ill-posed governing equations can be regularized by developing higher order continuum theories. These theories incorporate the microstructure of the deforming solid through additional material parameters and material length. The material length is typically related to the mean grain diameter or to some other, larger characteristic length scale such as the microcrack length. Therefore, the shear band thickness scales with this material length and is no longer indeterminate, allowing for energy dissipation in the plastic region and robust postlocalization computations. Another benefit of the introduction of material lengths is the ability of modeling the scale effect that is often observed in geomechanics problems but cannot be modeled by classical theories.

For regularizing quasi-static, boundary-value problems with strain softening material, different methods were presented in the literature. Higher order theories found in the literature include the Cosserat continuum theory (Cosserat and Cosserat 1909; Mühlhaus and Vardoulakis 1987; Papanastasiou and Vardoulakis 1992; Sulem *et al.* 1995), the Mindlin theory (Mindlin 1964; Zervos 2008), non-local continua (Pijaudier-Cabot and Bazant 1987), gradient plasticity, where higher order derivatives of the plastic strain enter in the yield function or in certain constitutive quantities (Aifantis 1992; Chambon *et al.* 1998; de Borst and Mühlhaus 1992; Sulem *et al.* 1995; Matsushima *et al.* 2002; Vardoulakis and Aifantis 1991; Vardoulakis *et al.* 1992; Vardoulakis and Frantziskonis 1992; Vardoulakis *et al.* 1994; Vardoulakis and Aifantis 1994) and the gradient elastoplasticity (Zervos *et al.* 2001a, 2001b, 2007, 2008), where higher order derivatives of the strain are introduced in both the elastic and the elastoplastic regimes. In the gradient elastoplasticity theory, the introduction of gradient terms in both regimes ensures that the governing equations remain of the same order throughout the whole deformation history. In this way, the need to locate the elastoplastic boundary for imposing internal boundary conditions is avoided. There is no clear agreement as to which one of the higher order continua is preferable, judging from a better agreement of computational results with the experimental observations. Although any approach can regularize to a certain degree the softening behavior, in order to decide among the various competing models one has to look into the physics of the problem. For example, in shear banding in uncemented granular soils or in stability of underground openings embedded in sandstones where there are evidences that initial localization is accompanied by grain rotations (Zervos *et al.* 2000), the Cosserat model appears to be more appropriate. In applications characterized by strong inhomogeneity of the deformation, the gradient models could be a good choice.

This chapter describes the development and implementation of two higher order continua in finite element analysis to treat the ill-posed mathematical problem and the mesh dependence problem of strain softening in order to model localization of deformation and failure in geomechanics. The first model is a Cosserat plasticity continuum, which possesses microstructure (Mühlhaus and Vardoulakis 1987). The second model is of a gradient type and it is called gradient elastoplasticity. The remainder of the chapter is structured as follows. The Cosserat model is presented first in section 5.2. The governing equations are described first in section 5.2.1, the numerical implementation in section 5.2.2, the material parameters in section 5.2.3 two applications of localization modeling near underground openings of cylindrical cavity in section 5.2.4 and elliptical perforations in section 5.2.5. In the second part, the theory and the numerical formulation of gradient elastoplasticity is presented first and then they are used to model the emergence of localization of deformation in shear bands in (a) biaxial tests and (b) expanding cylindrical cavities. The theory and numerical implementation are presented in section 5.3.1. Sections 5.3.2 and 5.3.4 present the numerical results and discussion for biaxial loading and expanding cavities, respectively. Throughout the presented examples, it is demonstrated that the higher order theories and the finite element numerical implementations enable the tracking of triggered instabilities and allow for robust postpeak modeling that leads to the localization of deformation and to the final failure mechanism. The results are summarized and the final conclusions are drawn in section 5.4.

5.2. Cosserat continuum

5.2.1. *Governing equations*

This section briefly describes the basics of the Cosserat model (Cosserat and Cosserat 1909). A full description of Cosserat plasticity theory in geomaterials can be found in Mühlhaus and Vardoulakis (1987) and Vardoulakis and Sulem (1995). A Cosserat continuum possesses at each material point translational degrees of freedom u_i and independent rotational degrees of freedom ω_c (Figure 5.2). The displacements are defined by the motion of the material points and the rotations are defined independently by microrotations represented by the orthogonal tensor. Point rotation and its gradient result in a non-symmetric stress tensor and couple stresses. The non-symmetric part of the stress tensor is in equilibrium with the divergence of the couple stresses (Figure 5.2). A dimensional analysis demonstrates that the microrotation gradients and the couple stresses introduce an internal length in the problem. For a two-dimensional analysis, the state of deformation is described by the four components of the relative deformation as follows:

$$\begin{aligned} &\varepsilon_{11} = u_{1,1} \qquad &&\varepsilon_{12} = u_{1,2} + \omega^c \\ &\varepsilon_{21} = u_{2,1} - \omega^c \qquad &&\varepsilon_{22} = u_{2,2} \end{aligned} \qquad [5.1]$$

and the two components of curvature given by the microrotation gradient:

$$\kappa_1 = \omega^c_{,1} \qquad \kappa_2 = \omega^c_{,2} \tag{5.2}$$

The above six kinematic quantities are conjugate in energy to six stress and couple stress quantities. Ignoring the body forces and body couple forces, force and moment equilibrium result in

$$\sigma_{ij,j} = 0, \quad m_{k,k} + \sigma_{21} - \sigma_{12} = 0 \quad in \quad V, \tag{5.3}$$

$$\sigma_{ij} n_j = t_i, \quad m_i n_i = m \quad on \quad \partial V, \tag{5.4}$$

where t_i and m prescribe tractions and couples on a part ∂V of the boundary of V. The symmetry of the stress tensor follows from equation [5.3] in the special case when the couple stresses or their divergence vanishes. As in classical plasticity theory in a Cosserat plasticity theory, the generalized incremental strains are decomposed into an elastic and a plastic strain:

$$\varepsilon_{ij} = \varepsilon^e_{ij} + \varepsilon^p_{ij} \tag{5.5}$$

The elastic and plastic strains are defined by generalizing the definition of elastic strain-energy potential, as well as the definitions for the yield surface and plastic potential of classical plasticity theory. The relationship between elastic strain and stress increment can be written as:

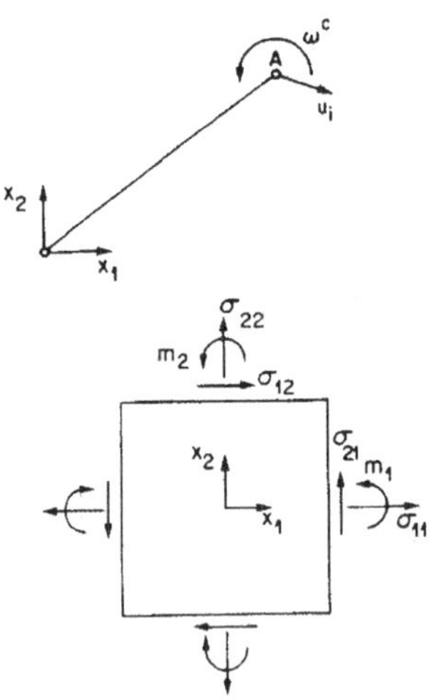

Figure 5.2. *(a) Displacement and rotation fields, and (b) stresses and couple stresses acting in a Cosserat continuum*

$$d\varepsilon_{ij}^{e} = \left\{ 2\left(h_1 \delta_{ik}\delta_{j\ell} + h_2 \delta_{jk}\delta_{i\ell}\right) - \frac{k-1}{2k}\delta_{ij}\delta_{k\ell} \right\} \frac{d\sigma_{kl}}{2G} \, d\kappa_i^{e} = h_3 \frac{dm_i}{GR^2} \qquad [5.6]$$

with $\kappa = K/G = 1/(1-2\nu)$. G, K and ν are the elastic shear modulus, the bulk modulus and Poisson ratio, respectively. R is the internal length, which is defined through constitutive equations and h_i will be defined below.

As in the classical flow theory of plasticity, plastic strains are generated when the yield condition is met. The Drucker–Prager yield criterion can be written as:

$$F = \frac{\tau}{p_0 + p} - \mu = 0, \qquad [5.7]$$

where the material parameters are the mobilized friction coefficient μ and the intercept p_0 of the yield surface with the p-axis, which is related to the material cohesion, $p_0 = c/tan\,\phi$. In equation [5.7], p and τ are invariant measures of average normal and shear contact tractions over the periphery of a macrocell of the "static" Cosserat model medium given in Mühlhaus and Vardoulakis (1987):

$$p = \frac{\sigma_{kk}}{2} \qquad \tau = \sqrt{\left(3s_{ij}s_{ij} - s_{ij}s_{ji}\right)/4 + m_i m_i / R^2}, \qquad [5.8]$$

where s_{ij} is the deviatoric stress, $s_{ij} = \sigma_{ij} + p\delta_{ij}$. In the above definitions, the internal length R can be identified as an equivalent radius of a typical grain or assembly of grains.

In this friction hardening/softening plasticity model, it is assumed that the friction coefficient μ is a function of a plastic hardening parameter, $\mu = \mu\left(\gamma^p\right)$ and the intercept p_0 of the yield surface with the p-axis is a material constant. Hardening is taking place when μ is monotonously increasing and softening when μ is monotonously decreasing with increasing γ^p. The hardening parameter is defined as the integral over the entire loading history of the increment of the generalized plastic shear strain, $\gamma^p = \int d\gamma^p$.

The plastic shear strain increment, $d\gamma^p$, is defined so that it is energy conjugate to the generalized shear stress intensity τ in equation [5.8]:

$$d\gamma^p = \sqrt{\left(3d\varepsilon_{ij}^p \varepsilon_{ij}^p + \varepsilon_{ij}^p \varepsilon_{ji}^p\right)/2 + R^2 d\kappa_i^p d\kappa_i^p}, \qquad [5.9]$$

where $d\varepsilon_{ij}^{p}$ is the deviator of the plastic relative deformation. If the couple stresses m_i and curvatures $d\kappa_i^p$ vanish, the definitions of τ and $d\gamma^p$ in equations [5.8] and [5.9] coincide with those of the classical plasticity.

In analogy of the yield surface, we define a plastic potential to which the normality condition applies as:

$$Q = \frac{\tau}{p_0 + p} - \beta, \qquad [5.10]$$

where $\beta = \beta\left(\gamma^p\right)$ is the mobilized dilatancy coefficient. Plastic strain and plastic curvature increments are derived from the flow rule:

$$d\varepsilon_{ij}^{p} = d\gamma^{p}\frac{\partial Q}{\partial \sigma_{ij}} \qquad [5.11]$$

For the case of $\mu = \beta$, the material obeys an associated flow rule and the incremental plastic strain vector in stress space is normal to the yield surface. In the most general case, geomaterials obey a non-associated flow rule, as the associative flow rule results in excessive plastic dilation. The actual value of proportionality factor $d\gamma^p$ is determined from Prager's consistency condition, $F = 0$ and $dF = 0$.

5.2.2. *Finite element formulation of Cosserat model*

The numerical formulation of the Cosserat model can be easily constructed with simple extension of the classical elastoplasticity algorithms to include the additional terms, as it is explained below (de Borst 1991; de Borst and Mühlhaus 1992).

For finite element analysis, it is convenient to express the equilibrium conditions (3) and (4) in a matrix form through the equivalent principle of virtual work

$$\int_V \{\delta\varepsilon\}^T \{\sigma\} dV = \int_{\partial V} \{\delta u\}^T \{t\} dS, \qquad [5.12]$$

where the vectors for the generalized displacement $\{u\}$, strain $\{\varepsilon\}$, stress $\{\sigma\}$ and tractions $\{t\}$ are defined by:

$$\begin{aligned} &\{u\}^T = \{u_1, u_2, \omega^c\} &&\{\varepsilon\}^T = \{\varepsilon_{11}, \varepsilon_{22}, \varepsilon_{12}, \varepsilon_{21}, \kappa_1 R, \kappa_2 R\} \\ &\{\sigma\}^T = \{\sigma_{11}, \sigma_{22}, \sigma_{12}, \sigma_{21}, m_1 / R, m_2 / R\} &&\{t\}^T = \{t_1, t_2, m\} \end{aligned} \qquad [5.13]$$

The introduction of R in the form of equation [5.13] makes the components of generalized stress and strain vectors have the same dimensions and the same order of magnitude.

In Cosserat plasticity theory, the generalized incremental strains are decomposed into elastic and plastic parts as in classical plasticity:

$$\{d\varepsilon\} = \{d\varepsilon^e\} + \{d\varepsilon^p\} \qquad [5.14]$$

The relationship between elastic strain and stress increment is written in a single matrix equation

$$\{d\sigma\} = \left[D^e\right]\{d\varepsilon^e\}, \qquad [5.15]$$

where the matrix $\left[D^e\right]$ contains the elastic parameters of a two dimensional, linear-elastic, isotropic Cosserat continuum defined by

$$\left[D^e\right] = \begin{bmatrix} K+G & K-G & 0 & 0 & 0 & 0 \\ K-G & K+G & 0 & 0 & 0 & 0 \\ 0 & 0 & G+G^c & G-G^c & 0 & 0 \\ 0 & 0 & G-G^c & G+G^c & 0 & 0 \\ 0 & 0 & 0 & 0 & M & 0 \\ 0 & 0 & 0 & 0 & 0 & M \end{bmatrix}, \qquad [5.16]$$

where G^C is the elastic stiffness of the Cosserat medium with respect to relative rotations and M is the normalized bending stiffness. For the so-called static Cosserat model proposed in Mühlhaus and Vardoulakis (1987):

$$\frac{G^C}{G} = \frac{1}{2}, \qquad \frac{M}{G} = R^2 \qquad [5.17]$$

The plastic strain and plastic curvature increments are derived from the flow rule

$$\{d\varepsilon^p\} = d\gamma^p \left\{\frac{\partial Q}{\partial \sigma}\right\} \qquad [5.18]$$

As mentioned before, the actual value of $d\gamma^p$ is determined from Prager's consistency condition, $F = 0$ and $dF = 0$. The latter equation results in:

$$d\gamma^p = \frac{\left\{\frac{\partial F}{\partial \sigma}\right\}\left(\left[D^e\right]\{d\varepsilon\}\right)}{\left\{\frac{\partial F}{\partial \sigma}\right\}\cdot\left(\left[D^e\right]\left\{\frac{\partial Q}{\partial \sigma}\right\}\right)+\left(p_0+p\right)h_t}, \qquad [5.19]$$

where $h_t = d\mu / d\gamma^p$ is the plastic modulus.

The stress increment $\{d\sigma\}$ can be expressed in terms of the total strain increment

$$\{d\sigma\} = \left[D^{ep}\right]\{d\varepsilon\}, \qquad [5.20]$$

where the elastic-plastic stiffness matrix $\left[D^{ep}\right]$ is given by

$$\left[D^{ep}\right] = \left[D^e\right] - \langle 1 \rangle \frac{\left(\left[D^e\right]\left\{\frac{\partial Q}{\partial \sigma}\right\}\right)\cdot\left(\left[D^e\right]\left\{\frac{\partial F}{\partial \sigma}\right\}\right)^T}{\left\{\frac{\partial F}{\partial \sigma}\right\}\left(\left[D^e\right]\left\{\frac{\partial Q}{\partial \sigma}\right\}\right)+\left(p_0+p\right)h} \qquad [5.21]$$

Loading of the yield surface $F = 0$ takes place when $d\gamma^p > 0$. Accordingly, the switch function $\langle 1 \rangle$ in equation [5.21] is defined as:

$$\langle 1 \rangle = \begin{cases} 1 \quad \text{if} \quad F = 0 \quad \text{and} \quad d\gamma^p > 0 \\ 0 \quad \text{if} \quad F < 0 \quad \text{or} \quad \left\{F = 0 \quad \text{and} \quad d\gamma^p \leq 0\right\} \end{cases} \qquad [5.22]$$

Mesh insensitivity of the results and robustness in the computations for the presented Cosserat model has been demonstrated in Papanastasiou and Vardoulakis (1992). By contrast, computations based on classical elastoplasticity with identical material parameters are mesh sensitive due to the material softening in the constitutive equations.

5.2.3. *Material parameters*

The material parameters for the Mohr–Coulomb elastoplastic constitutive model were derived from triaxial compression tests on Castlegate sandstone and a calibration procedure.

The elastic parameters were found to be $E = 8{,}100\,\text{MPa}$ and $\nu = 0.35$. The parameter related to the material cohesion is $p_0 = 9.81\,\text{MPa}$. The calibration of the friction coefficient, $\mu = sin\phi_m$, as a function of the plastic shearing strain γ^p was based on an interpolation of prepeak triaxial data. Postpeak softening behavior was modeled via the extrapolation of a function, which was fit to the prepeak data. The corresponding curve fit was given by the hyperbolic function

$$\mu = \mu_0 + \frac{\left(1 - c_0\gamma^p\right)\gamma^p}{c_1 + c_2\gamma^p}, \qquad [5.23]$$

where $\mu_0 = 0.463$ is the value of the friction coefficient that defines the state of initial yield; $c_1 = 0.0077$, $c_2 = 7.5$ and $c_0 = 12$ is a control parameter for the rate of softening.

For the dilation coefficient $\beta = sin\psi_m$, an associated flow rule was assumed by considering it equal to the mobilized friction coefficient μ because the Castlegate sandstone exhibits pronounced dilation in the low confinement pressure triaxial tests

$$\beta = \mu \qquad [5.24]$$

The internal length required by the Cosserat model is set equal to $R = 0.2\,\text{mm}$. This value is related to the microstructure of the sandstone, e.g. the grain size or the assembly of grains. We repeat here that the inclusion of the grain size in the constitutive equations provides the Cosserat model with the capability to predict the scale effect that is observed in the thick-walled cylinder tests.

5.2.4. *Failure in thick-walled cylinder test*

In this section, the problem of a thick-walled cylinder under external pressure is modeled. The results will be compared with those of a physical experiment. The main dimensions of the cylinder are the internal radius, $r_a = 1.45\text{cm}$ and the external radius, $r_b = 7.5\text{cm}$. The external pressure was applied incrementally, whereas the internal pressure was maintained at 0. Plane strain conditions are assumed.

Figure 5.3 shows a comparison of the experimental results with model predictions in terms of the applied external pressure versus normalized hole closure. The solid line shows the closure in the direction of breakouts and the dashed line shows the closure in the direction 90° from the breakouts. The initial part of the curve is almost linear, although initial yielding at the hole wall takes place below 10MPa. The first sign of failure in the experiment was observed after the pressure had reached 35 MPa. This prediction compares well with the bifurcation point detected by the break of axisymmetry at pressure 34.75 MPa.

After bifurcation, the deformation is characterized by a warping mode of the finite wave number $m=10$ caused by surface buckling. As mentioned earlier, the Cosserat model switches to the localized branch smoothly, without the need to perturb the primary solution with an imperfection. The computations confirmed that in the primary path the Cosserat rotation remains inactive everywhere, whereas after localization it increases by more than three orders of magnitude, but only on the boundaries of the shear bands due to the intense shearing (Papanastasiou and Vardoulakis 1992). The predicted failure mode is depicted in Figure 5.4 where the contours of the plastic strain have been drawn in the vicinity of the hole after localization of deformation has taken place. Once localization of deformation in shear bands starts, almost all of the remaining region close to the borehole wall unloads elastically. The narrow zones of the forming shear bands continue to plastically deform in the softening regime and progress toward the interior. The shear bands bend further until they finally meet each other, leading to a familiar breakout form observed in experiments and downhole images. Similar computations and extensive analysis can be found in Papanastasiou and Vardoulakis (1992) and Papamichos (2010).

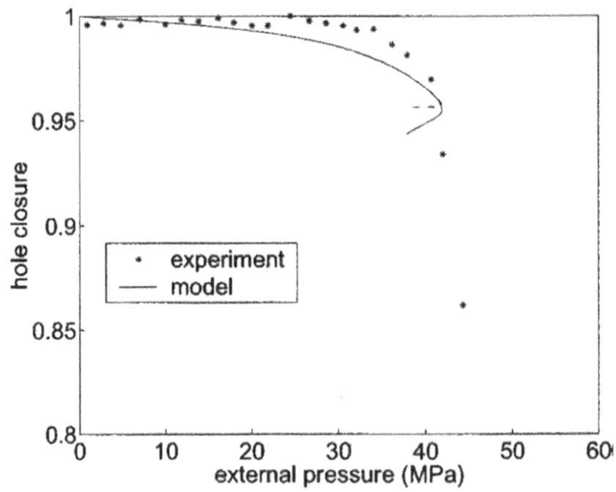

Figure 5.3. *External pressure versus normalized hole closure in a thick walled cylinder experiment*

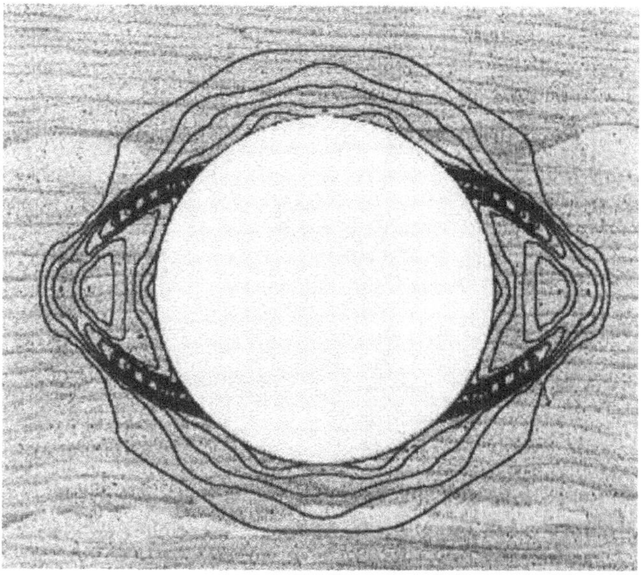

Figure 5.4. *Contours of the shear plastic strain after localization of deformation show the formation of breakouts*

A series of computations was carried out to investigate the issue of scale effect. Experimental results had shown that small holes, compared to large holes, fail at higher external pressure (Haimson and Herrick 1989; Papamichos and van den Hoek 1995; van den Hoek *et al.* 2000; Papamichos 2010). Computations were performed for models with different internal radii but with a fixed thickness ratio 5:1 of the external radius to the internal. The results are shown in Figure 5.5 where the load at which bifurcation takes place (dashed line) and the peak load (solid line) were plotted as a function of the hole radius normalized by the internal length. It is clear that both bifurcation and peak loads increase as the hole radius decreases. For very small holes, bifurcation does not take place and limit point is reached. Smaller holes bifurcate with lower warping modes. The scale effect fades out rapidly with increasing hole size. As the hole gets bigger, the scale effect becomes less pronounced, a tendency also observed in the experiments of Haimson and Herrick (1989), Papamichos and van den Hoek (1995) and Papanastasiou and Vardoulakis (1992). This finding is also supported by experimental results discussed in detail by van den Hoek (2001). Bifurcation in the limit of a large borehole, which approaches a half space geometry, is expected to take place with infinite wavelength mode once the stress state enters the softening regime for an associated material.

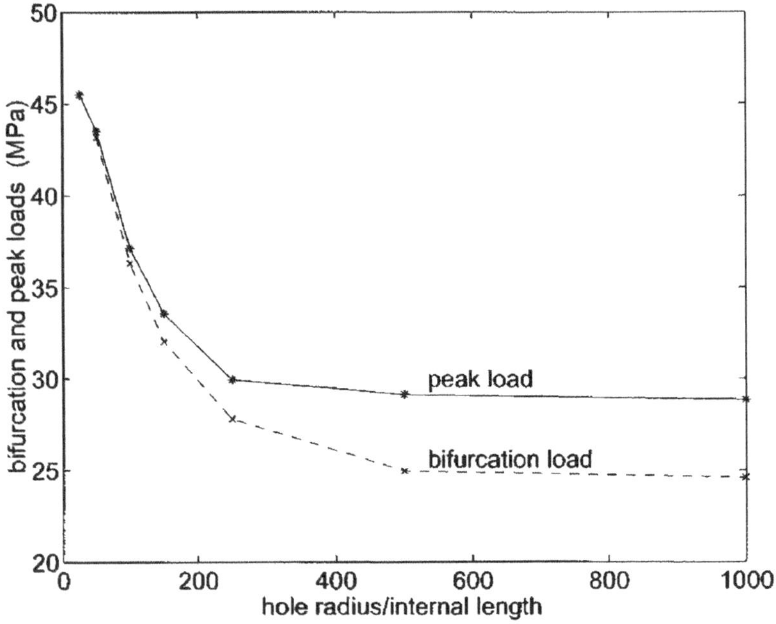

Figure 5.5. *Bifurcation load and peak load versus normalized hole radius predicted by the Cosserat model*

5.2.5. *Stability analysis of elliptical shape perforations*

This section presents another application related to the stability analysis of elliptical shape perforations embedded in a compressive *in situ* stress field. The material parameters used are those of the Castlegate sandstone. The application is related to the stability of perforation that is used for the flow of hydrocarbon in the wellbores. It will be demonstrated that it is essential to carry out advanced localization analysis in order to reach correct results upon which a new perforation shape design can be based. To make this point, it is stressed that classical stress analysis predicts that an elliptical hole suffers less stress concentration than a circular hole when its major axis is aligned with the direction of the major principal stress and its axis ratio is the same as the applied stress ratio. However, this failure analysis based on localization of deformation will show that an elliptical hole is even stronger if its axis ratio is greater than the ratio of the applied stresses.

Figure 5.6 shows the hole closure in terms of a change in cross-sectional area versus applied maximum *in situ* stress ($\sigma_{max}/\sigma_{min} = 1.5$) for a circular hole, $a/b = 1$, and the elliptical holes with axis ratio $a/b = 1.5$ and $a/b = 3$.

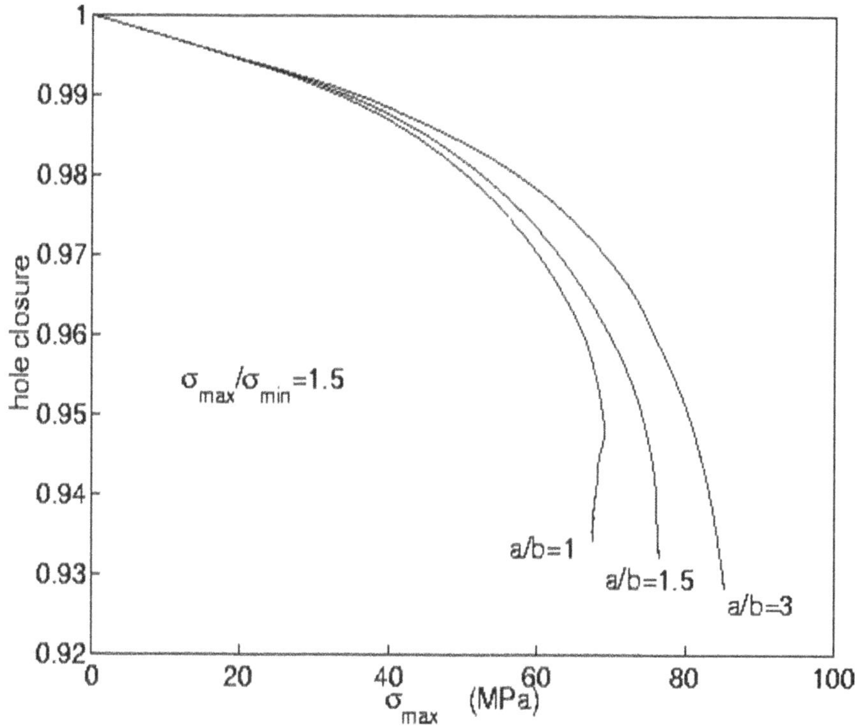

Figure 5.6. *Hole closure versus applied stress for a circular hole and elliptical perforations with different axis ratio a/b*

The predicted failure mode is shown in Figures 5.7 and 5.8 where the displacement fields after localization and the contours of the plastic shear strain were draw for the circular and elliptical holes. Although an elastic stress analysis suggests through the stress concentration near the hole that the more stable ellipse is the one where its axis ratio is the same as the ratio of the applied stresses, this localization analysis has shown that an ellipse with an axis ratio greater than the ratio of the applied *in situ* stress is even stronger. More computations and a detailed analysis can be found in Papanastasiou (2000).

In other applications of the Cosserat plasticity model, Tejchman and Wu (1993) analyzed interface mechanisms in silos filled with granular medium. Adhikary and Dyskin (1997) used the Cosserat model in layered materials. Papamichos (2010) studied the stability of a borehole under an anisotropic stress field and compared the predictions of the Cosserat model with experimental data.

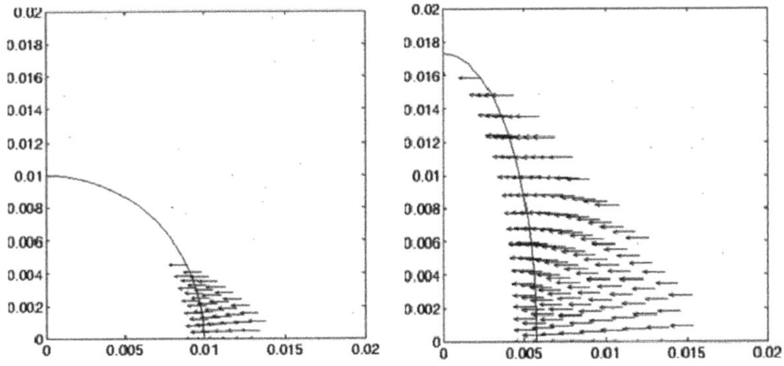

Figure 5.7. *Incremental displacement field after localization of deformation around a circular hole (left) and an elliptical hole (right)*

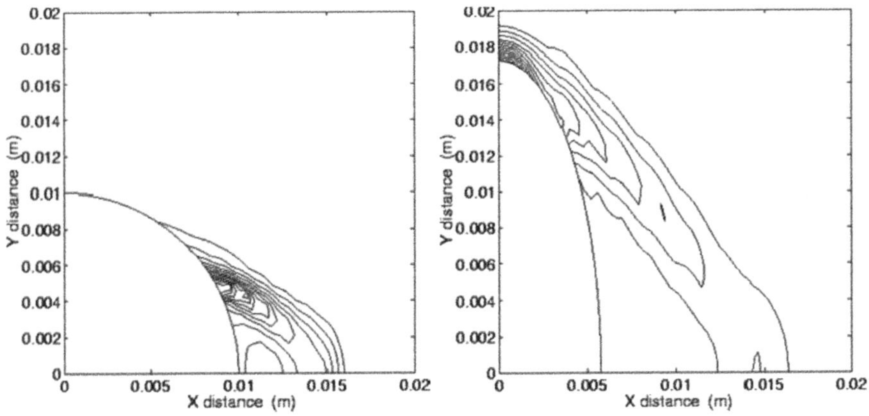

Figure 5.8. *Contours of shear plastic strain around a circular hole (left) and an elliptical hole (right)*

5.3. Gradient elastoplasticity

5.3.1. *Governing equations*

Following the decomposition of the total strain rate $\dot{\varepsilon}_{ij} = \dot{\varepsilon}^{e}_{ij} + \dot{\varepsilon}^{p}_{ij}$ into an elastic part $\dot{\varepsilon}^{e}_{ij}$ and a plastic part $\dot{\varepsilon}^{p}_{ij}$, we define the total (equilibrium) stress rate $\dot{\sigma}_{ij}$ in terms of the elastic strain rate and its Laplacian, as

$$\dot{\sigma}_{ij} = C^{e}_{ijkl}\left(\dot{\varepsilon}^{e}_{kl} - l_e^2 \nabla^2 \dot{\varepsilon}^{e}_{kl}\right), \qquad [5.25]$$

where C^e_{ijkl} is the tensor of elastic moduli and l_e is a material parameter with dimension of length, called the *elastic material length.* The first spatial derivative from equation [5.25] is omitted to avoid directionality in the material response. Plastic deformations develop when the stress state meets the yield condition $F(\tau_{ij},\psi)=0$. The plastic strain rate is defined through the flow rule:

$$\dot{\varepsilon}^p_{ij} = \dot{\psi}\frac{\partial Q}{\partial \tau_{ij}}, \qquad [5.26]$$

where $Q(\tau_{ij},\psi)=0$ is a plastic potential and the scalar quantity $\dot{\psi}$ is the plastic multiplier. In the special case of associative plasticity, $Q \equiv F$.

The yield function F and the plastic potential Q depend on the reduced stress $\dot{\tau}_{ij} = \dot{\sigma}_{ij} - \dot{\alpha}_{ij}$ and on a hardening/softening parameter ψ. The back stress α_{ij} evolves according to the constitutive law

$$\dot{\alpha}_{ij} = -l_p^2 C^e_{ijkl}\nabla^2\dot{\varepsilon}^p_{kl}, \qquad [5.27]$$

where the scalar l_p is another material parameter with dimension of length, called the *plastic material length.* As back stresses are expected to develop due to micro-inhomogeneities in the plastic flow, an assumption that the back stress is proportional to the plastic strain as well as to its Laplacian appears to be more accurate. Back stresses develop only where the deformation becomes sufficiently inhomogeneous, allowing for the region surrounding a material point to contribute to its strength.

The plastic strains are determined using the consistency condition of plasticity,

$$F(\tau_{ij},\psi)=0 \text{ and } \dot{F}(\tau_{ij},\psi)=\frac{\partial F}{\partial \tau_{ij}}\dot{\tau}_{ij}+\frac{\partial F}{\partial \psi}\dot{\psi}=0 \qquad [5.28]$$

which ensures that, during plastic deformation, the stress state remains on the evolving yield surface.

Substituting equations [5.25–5.27] into equation [5.28], using the strain rate decomposition and the definition of the reduced stress, and assuming that $\partial F/\partial \tau_{ij}$ and $\partial Q/\partial \tau_{ij}$ vary slowly in space, the plastic multiplier is expressed as:

$$\left[1-\frac{H_0}{H}\left(l_e^2+l_p^2\right)\nabla^2\right]\dot{\psi}=\frac{1}{H}\frac{\partial F}{\partial \tau_{ij}}C_{ijkl}^e\left(\dot{\varepsilon}_{kl}-l_e^2\nabla^2\dot{\varepsilon}_{kl}\right), \qquad [5.29]$$

where $H=H_0+H_t$, $H_0=\frac{\partial F}{\partial \tau_{ij}}C_{ijkl}^e\frac{\partial Q}{\partial \tau_{kl}}$ and $H_t=-\frac{\partial F}{\partial \psi}$.

An approximate solution of equation [5.29] can be obtained by noting that $H_0/H\simeq 1$ and that both l_e and l_p are small, since they represent some microstructural length scale of the material. The differential operator applied on the left-hand side of equation [5.29] can then be inverted and transferred to the right-hand side. Neglecting the higher order terms, equation [5.29] then yields (Zervos *et al.* 2001a):

$$\dot{\psi}=\frac{1}{H}\frac{\partial F}{\partial \tau_{ij}}C_{ijkl}^e\left(\dot{\varepsilon}_{kl}+l_c^2\nabla^2\dot{\varepsilon}_{kl}\right), \qquad [5.30]$$

where $l_c^2=\frac{H_0}{H}l_p^2-\frac{H_t}{H}l_e^2>0$.

The assumption of the slow spatial variation of $\partial F/\partial \tau_{ij}$ and $\partial Q/\partial \tau_{ij}$, on which the derivation of equation [5.29] is based, is only valid for an initial bifurcation from a relatively homogeneous ground state of plastic straining. In the postbifurcation regime, this assumption progressively breaks down as deformation localizes. Then the terms $\nabla^2\left(\partial F/\partial \tau_{ij}\right)$ and $\nabla^2\left(\partial Q/\partial \tau_{ij}\right)$ are not necessarily negligible. However, the inclusion of such terms would necessitate a rather complicated numerical treatment of the consistency condition and lead to highly nonlinear and expensive calculations.

Here, we use equations [5.29] and [5.30] even in the postlocalization regime, bearing in mind their approximate nature. Numerical results obtained using this approach have been shown to reproduce in a realistic manner localization phenomena observed in experiments (Zervos *et al.* 2001a, 2001b; Zervos *et al.* 2007).

For a total strain increment $\dot{\varepsilon}_{ij}$, equations [5.26] and [5.30] can be used to calculate the plastic part $\dot{\varepsilon}_{ij}^p$. The elastic part is then $\dot{\varepsilon}_{ij}^e=\dot{\varepsilon}_{ij}-\dot{\varepsilon}_{ij}^p$. Substituting to equations [5.25] and [5.27], the constitutive quantities can be expressed in terms of total, rather than elastic and plastic strain increments as follows:

$$\dot{\sigma}_{ij} = C^{ep}_{ijkl}\dot{\varepsilon}_{kl} - C^{m}_{ijkl}\nabla^2\dot{\varepsilon}_{kl} \qquad [5.31]$$

$$\dot{\alpha}_{ij} = -l_p^2 C^{p}_{ijkl}\nabla^2\dot{\varepsilon}_{kl} \qquad [5.32]$$

Here C^{p}_{ijkl} is the known plastic stiffness matrix of classical plasticity

$$C^{p}_{ijkl} = \frac{\langle 1 \rangle}{H} C^{e}_{ijmn} \frac{\partial Q}{\partial \tau_{mn}} \frac{\partial F}{\partial \tau_{st}} C^{e}_{stkl}, \qquad [5.33]$$

where $\langle 1 \rangle = 1$ for loading (i.e. $F=0$ and $\dot{\psi} \geq 0$) and $\langle 1 \rangle = 0$ for unloading (i.e. $F<0$, or $F=0$ and $\dot{\psi}<0$). $C^{ep}_{ijkl} = C^{e}_{ijkl} - C^{p}_{ijkl}$ is the usual elastoplastic stiffness matrix and

$$C^{m}_{ijkl} = l_e^2 C^{ep}_{ijkl} + l_c^2 C^{p}_{ijkl} \qquad [5.34]$$

is a stiffness matrix for the higher order terms.

Due to the approximation introduced to arrive to equation [5.30], the consistency condition is satisfied to $O\left(\left(l_e^2 + l_p^2\right)^2\right)$. In the numerical implementation presented in the next section, errors are prevented from accumulating by checking that the stress state of points that undergo plastic or neutral loading remain on the evolving yield surface and mapping it back if needed (Zervos *et al.* 2001a).

To derive the equation of virtual work, the above equations are reinterpreted within the framework of Mindlin's theory of microstructure (Mindlin 1964). It follows that they describe a restricted Mindlin continuum, i.e. a Mindlin continuum whose microscopic and macroscopic deformations coincide. Hence, equation [5.31] is rewritten as

$$\dot{\sigma}_{ij} = \dot{\sigma}^{(0)}_{ij} + \dot{\sigma}^{(2)}_{ij}, \; \dot{\sigma}^{(0)}_{ij} = C^{ep}_{ijkl}\dot{\varepsilon}_{kl}, \; \dot{\sigma}^{(2)}_{ij} = -C^{m}_{ijkl}\nabla^2\dot{\varepsilon}_{kl} \qquad [5.35]$$

In Mindlin's terminology, $\dot{\sigma}^{(0)}_{ij}$ is the Cauchy stress rate, which is identified as the constitutive stress rate tensor of classical elastoplasticity. The second term is considered as a relative stress rate. We postulate the existence of a double stress rate $\dot{m}_{kij}$, whose gradient equilibrates the relative stress. Then:

$$\dot{\sigma}_{ij}^{(2)} + \dot{m}_{kij,k} = 0 \Rightarrow \dot{m}_{kij} = C_{ijmn}^{m} \dot{\varepsilon}_{mn,k} \quad [5.36]$$

In the absence of body forces and inertial terms, the principle of virtual work is written as (Vardoulakis and Aifantis 1991; Vardoulakis and Sulem 1995; Zervos *et al.* 2001a)

$$\int_{V} \left(\dot{\sigma}_{ij}^{(0)} \delta\dot{\varepsilon}_{ij} + m_{kij} \delta\dot{\varepsilon}_{ij,k} \right) dV = \int_{S_\sigma} \left(t_i \delta\dot{u}_i + \mu_i n_k \delta\dot{u}_{i,k} \right) dS \, , \quad [5.37]$$

where S_σ is the part of the boundary where Neumann-type boundary conditions are applied and n_i is its unit normal. t_i is the applied boundary traction vector, μ_i is the applied boundary double traction vector and $\delta\dot{u}_i$ is the virtual displacement rate vector.

5.3.2. *Finite element formulation*

Since the rate of all constitutive quantities can be expressed in terms of the total strain rate and its derivatives, the usual displacement formulation of the finite element method is suitable for solving boundary value problems of gradient elastoplasticity. However, the existence of strain gradients in the virtual work demands that strains be continuous. Therefore, the interpolation of a displacement field must be C^1-continuous, instead of C^0 which is adequate for classical elastoplasticity.

In the following, we restrict the analysis to 2D plane strain and use the three-noded C^1 triangle with 18 degrees of freedom for each interpolated field (namely the displacement and all its spatial derivatives of first and second order at each node) (Dasgupta and Sengupta 1990). The displacement field varies as a complete quintic with a cubic normal derivative along the element edges. The displacement $u_i \equiv \{u\}$ is interpolated as

$$\{u\} = [N]\{\hat{u}\}, \quad [5.38]$$

where $[N]$ are the shape functions and $\{\hat{u}\}$ the nodal degrees of freedom.

Strain rates, their gradients and their Laplacians are written in vector form as

$$\{\dot{\varepsilon}\} = [L_1][N]\{\dot{\hat{u}}\} = [B_1]\{\dot{\hat{u}}\}, \quad [5.39]$$

$$\{\dot{\varepsilon}_x\} = \frac{\partial}{\partial x}\{\dot{\varepsilon}\} = [L_{2x}][N]\{\hat{u}\} = [B_{2x}]\{\hat{u}\}, \tag{5.40a}$$

$$\{\dot{\varepsilon}_y\} = \frac{\partial}{\partial y}\{\dot{\varepsilon}\} = [L_{2y}][N]\{\hat{u}\} = [B_{2y}]\{\hat{u}\}, \tag{5.40b}$$

$$\nabla^2\{\dot{\varepsilon}\} = [L_3][N]\{\hat{u}\} = [B_3]\{\hat{u}\}, \tag{5.41}$$

where $[L_1]$, $[L_{2x}]$, $[L_{2y}]$ and $[L_3]$ are appropriate differential operators. The constitutive relations of equations [5.32], [5.35] and [5.36] are then written in vector form as

$$\{\dot{\sigma}^{(0)}\} = [C^{ep}]\{\dot{\varepsilon}\}, \tag{5.42}$$

$$\{\dot{m}_x\} = [C^m]\{\dot{\varepsilon}_x\} \qquad \{\dot{m}_y\} = [C^m]\{\dot{\varepsilon}_y\}, \tag{5.43}$$

$$\{\dot{\alpha}\} = -l_p^2[C^p]\nabla^2\{\dot{\varepsilon}\}, \tag{5.44}$$

where $[C^{ep}]$, $[C^m]$ and $[C^p]$ are the tensors C^{ep}_{ijkl}, C^m_{ijkl} and C^p_{ijkl} in matrix form.

Substitution in the virtual work equation [5.37] results in the following linear system of equations for the nodal degrees of freedom:

$$\int_V \left([B_1]^T[C^{ep}][B_1] + [B_{2x}]^T[C^m][B_{2x}] + [B_{2y}]^T[C^m][B_{2y}]\right)dV \cdot \{u\} = \int_{S_\sigma}\left([N]^T\{t\} + \left(n_x\frac{\partial[N]^T}{\partial x} + n_y\frac{\partial[N]^T}{\partial y}\right)\{m\}\right)dS, \tag{5.45}$$

where $\{n\} = (n_x, n_y)$ is the outward unit normal to the boundary, $\{t\}$ is the boundary traction vector and $\{m\}$ is the boundary double traction.

The stiffness matrix on the left-hand side of equation [5.45] is integrated using an $O(h^7)$-accurate, 13-point Gauss quadrature scheme, which is adequate to preserve convergence (Zervos *et al.* 2001a). Equation [5.45] is solved iteratively for the degrees of freedom with the Newton–Raphson method.

5.3.3. *Material model*

We consider that the material behavior is approximated adequately by a nonlinear Mohr–Coulomb yield criterion and plastic potential. To avoid numerical problems related to the apex of the yield function at the point $\tau_1 = \tau_3$, the corner is "rounded" by a circular arc of radius *R*. The arc is taken tangent to the Mohr–Coulomb line, so the derivative of the yield function remains continuous. The modified function is depicted in Figure 5.1 with a solid line and is written as

$$F(\tau_1,\tau_3) = \begin{cases} m\tau_1 - \tau_3 - \sigma_c = 0 & if \quad H(\tau_1,\tau_3) \leq 0 \\ (\tau_1 - \tau^K)^2 + (\tau_3 - \tau^K)^2 - R^2 = 0 & if \quad H(\tau_1,\tau_3) > 0 \end{cases} \qquad [5.46]$$

where $m = (1+sin\phi)/(1-sin\phi)$ quantifies the coefficient of friction and $\sigma_c = 2c\,cos\phi/(1-sin\phi)$ is the *equivalent stress*. ϕ is the angle of internal friction and c is the material cohesion. τ^K defines the center of the arc and $H = (\tau_1 - \tau_1^T) + m(\tau_3 - \tau_3^T) = 0$ is the line normal to the Mohr–Coulomb line, passing from its point of contact with the arc. Geometrical considerations lead to:

$$\tau^K = \frac{1}{1+m}\left[\left(1+m^2\right)\tau_1^T - m\sigma_c\right] \qquad [5.47]$$

$$R^2 = \left(\frac{1}{1+m}\right)^2 \left\{\left(m^2+1\right)\left[\left(m-1\right)\tau_1^T - \sigma_c\right]^2\right\} \qquad [5.48]$$

The arc is defined uniquely by the abscissa τ_1^T of the contact point, which here is taken such that the arc covers 1/10 of the tensile part of the function. Equation [5.22] is also used for the plastic potential, which is defined by the angle of dilatancy ψ in place of the angle of friction ϕ. The condition for associative plasticity is $\psi = \phi$. The hardening/softening behavior is described by assuming dependence of $\sigma_c(\varepsilon_p)$ on the accumulated *equivalent plastic strain* ε_p. The latter is defined as the rate of dissipated work divided by the equivalent stress. The hardening/softening law is:

$$\sigma_c(\varepsilon_p) = \sigma_{c,0} + \frac{\left(1 - C_0\varepsilon_p\right)\varepsilon_p}{C_1 + C_2\varepsilon_p} \qquad [5.49]$$

which is fitted on experimental data. C_1 and C_2 are calibration constants, while $\sigma_{c,0}$ defines the state of initial yield. The constant C_0 controls the rate of softening and it is an open parameter because, due to bifurcations and end restraints, standard laboratory calibration tests cannot determine the "true" material softening. The material lengths l_e and l_p are also open parameters, although they are usually considered to relate to a characteristic micromechanical length scale of the material, such as the average grain size (Mühlhaus and Vardoulakis 1987). As the role of microstructure in the elastic regime is not expected to be significant, extra information on the shear band thickness and the scale effect could, in principle, be inverted using the gradient elastoplasticity model to obtain the unknown softening rate and plastic material length l_p. The elastic material length l_e can then be taken to be a proportion of l_p.

The introduction of two material lengths gives gradient elastoplasticity the advantage that boundary conditions need only be prescribed on the external boundary. In gradient plasticity theories, where only one material length is used, the order of the governing equations changes at the elastoplastic boundary, necessitating a special treatment. In gradient elastoplasticity, this is unnecessary, as gradient terms are introduced in the elastic regime as well. However, it is the plastic material length that dominates the behavior of the material, as deformation gradients are much higher in the plastic than in the elastic regime.

The material parameters used in the computations correspond to weak reservoir sandstone. The elastic parameters were found to be $E = 8{,}100\,\text{MPa}$ and $\nu = 0.35$. The friction angle is considered constant and equal to $\phi = 32.5°$. The calibration constants for the hardening behavior were determined as $C_1 = 1.323 \times 10^{-5}$ and $C_2 = 6.1271 \times 10^{-2}$, while the equivalent stress at initial yield is $\sigma_{c,0} = 25\,\text{MPa}$. The material lengths are set to $l_e = l_p = 0.2\,\text{mm}$. The constant controlling the softening rate is assumed to be $C_0 = 80$.

5.3.4. *Modeling of the biaxial test*

Modeling of the biaxial test has been consistently used in the literature as a "benchmark" for localization computations, as it readily demonstrates the ability of a model to capture localized zones of finite thickness. Although it would be interesting from a practical point of view to compare the predictions of the theory presented here with predictions of other existing higher order theories, meaningful comparisons cannot be made without access to the relevant raw experimental data for calibration purposes. We will therefore consider such comparisons outside the scope of this application.

The geometry of the test and the applied boundary conditions are shown in Figure 5.9. The sample is 5 × 10 cm and is assumed to deform in plane strain. Deviatoric load is applied by prescribing the vertical displacement increment of the top platen. The platen is considered to be rigid and lubricated, allowing the sample to slip without friction. Due to the initial uniformity of the stress and strain fields, an imperfection is needed to trigger localization, so the initial cohesion has been reduced by 10% at a region 0.5 × 0.5 cm at the lower right-hand side corner of the sample. The size of the imperfection does not influence the computed failure mode (Zervos *et al.* 2001a).

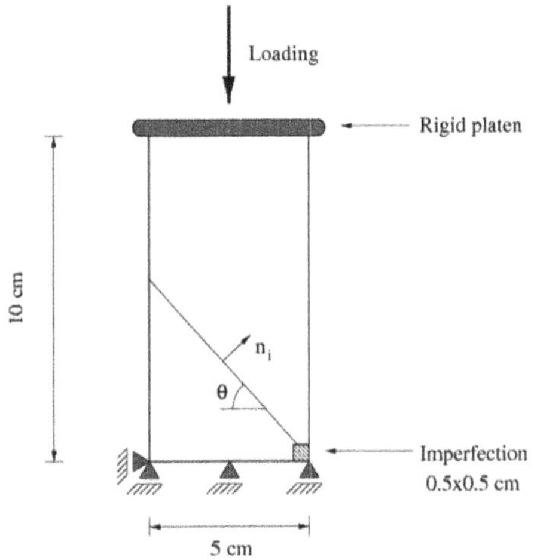

Figure 5.9. *Biaxial geometry, loading and boundary conditions*

A parametric study was conducted to examine the effect of non-associativity on shear band geometry. The dilatancy angle was varied every 5° between $\psi = 0°$ (zero plastic volumetric strain) and $\psi = 32.5°$ (maximum plastic volumetric strain). The results were obtained with a mesh of 21 × 41 = 861 nodes and 10,332 degrees of freedom. It has been shown that this mesh provides converged results (Zervos *et al.* 2001a).

The calculated vertical load versus vertical displacement curves are shown in Figure 5.10. The initial branch, which corresponds to uniform deformation, is identical for all dilatancy angles but the curves differ in the softening branch. In all cases, the observed failure mechanism is a shear banding. In the associative case, the deformation starts localizing at peak load. However, as non-associativity increases, localization starts earlier in the hardening branch. This result agrees with the predictions of bifurcation theory (Rudnicki and Rice 1975; Vardoulakis 1980) and experimental observations (Ord *et al.* 1991). For the non-associative cases, the curve is steeper, suggesting more severe structural softening.

The thickness d of the formed shear bands is shown in Table 5.1. It decreases with a decreasing dilatancy angle. Comparison of the associative case with the case of $\psi = 0°$ reveals a decrease in more than four times; this can also be seen in Figure 5.11, where contours of the equivalent plastic strain are plotted.

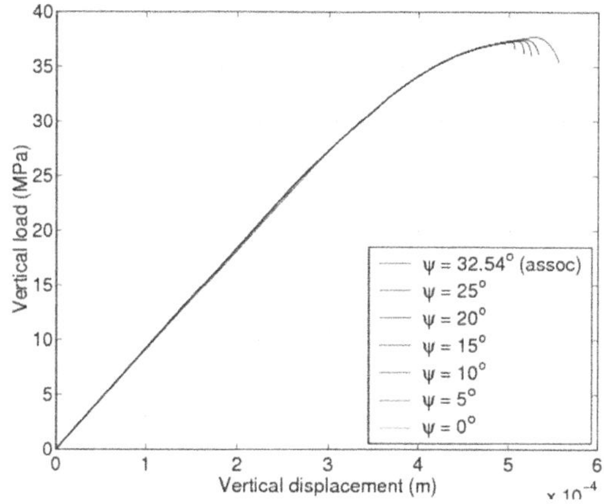

Figure 5.10. *Vertical load versus vertical displacement*

Dilatancy angle Ψ	Band thickness d/l_p
32.50°	56.0
25.00°	45.8
20.00°	42.3
15.00°	35.2
10.00°	31.7
5.00°	24.7
0.00°	14.1

Table 5.1. *Normalized shear band thickness for different dilatancy angles and constant friction angle φ = 32.5°*

The inclination of the shear band (angle θ between the shear band and the x-axis in Figure 5.9) also decreases with a decreasing dilatancy angle. As shown in

Table 5.2, the maximum inclination occurs for the associative case. The effect of non-associativity is in this respect too visible in Figure 5.11: plastic deformation is localized inside the shear band, allowing for the inclination to be graphically determined from the direction of the contour lines. Again, the difference between the associative case (Figure 5.12(a)) and the non-associative ones (Figures 5.11(b–d)) is clear. These results are in close agreement with the predictions of bifurcation analysis, which relate the shear band inclination to the angles of friction and dilatancy (Rudnicki and Rice 1975; Vardoulakis 1980). A comparison is shown in Table 5.2.

Dilatancy angle ψ	Numerical θ	Theoretical θ^{bif}
32.50°	60.0°	61.27°
25.00°	58.0°	59.35°
20.00°	56.5°	58.05°
15.00°	55.0°	56.74°
10.00°	53.5°	55.42°
5.00°	53.5°	54.11°
0.00°	52.5°	52.80°

Table 5.2. *Theoretically predicted and numerically determined shear band inclination, for constant friction angle φ = 32.5° and different values of the dilatancy angle ψ*

In Figure 5.11, we also observe that the inclination of the shear band changes as it approaches the left-hand side boundary of the specimen. Similar shear band reorientations near the boundary have been observed experimentally (Bésuelle *et al.* 2000; Finno *et al.* 1997; Vardoulakis and Sulem 1995). The explanation is that the shear band is influenced by the natural boundary conditions, as it approaches and subsequently intersects the free boundary. The inclination θ^b at the boundary can then be approximated by the Roscoe solution, $\theta^b \simeq \theta^R = \pi/4 + \psi/2$ (Benallal *et al.* 1989). The measured intersection angles are shown in Figure 5.11, at the upper left corner of each plot, and compared in Table 5.3 with the Roscoe solution. The difference is within the error of measurement. In the associative case, the angle remains constant, while for $\psi = 0°$, $\theta^b \simeq 45°$.

Dilatancy angle ψ	Numerical θ^b	Theoretical θ^R
32.50°	60.5°	61.27°
20.00°	54.2°	55.00°
10.00°	49.5°	50.00°
0.00°	44.5°	45.00°

Table 5.3. *Comparison of numerical results (third column) on shear band inclination next to the free boundary with the Roscoe solution (second column), for constant friction angle φ = 32.5° and different dilatancy angles*

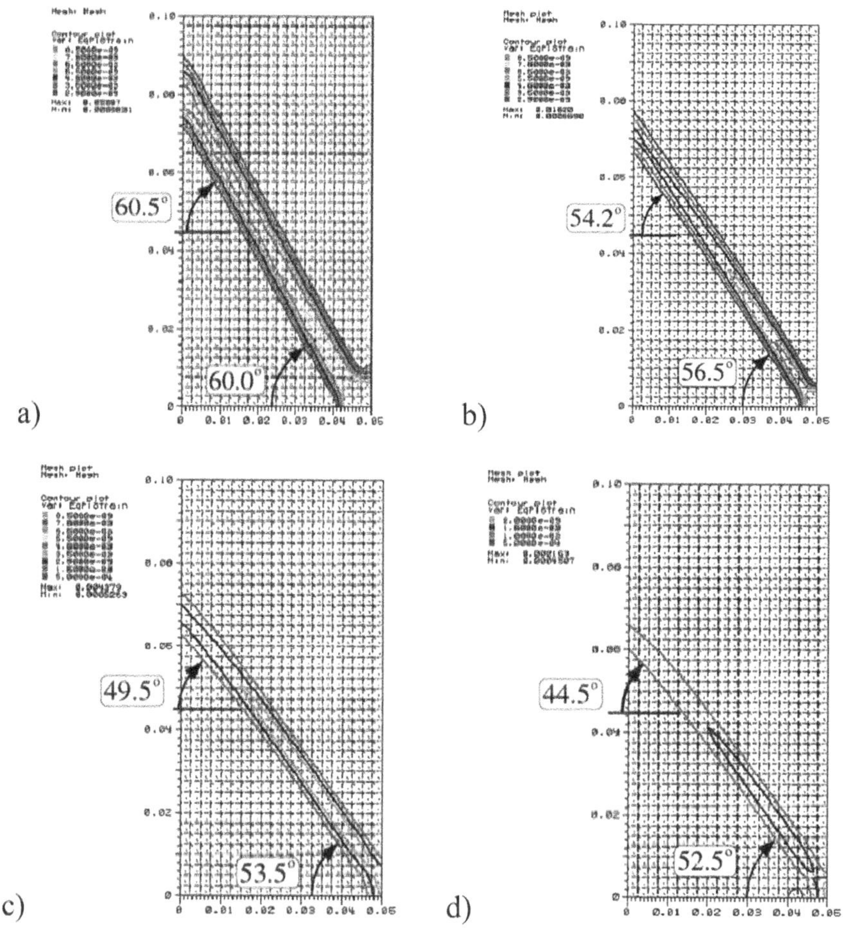

Figure 5.11. *Contours of the equivalent plastic strain: (a) associative case; (b) non-associative, ψ = 20°; (c) non-associative, ψ = 10°; (d) non-associative, ψ = 0°*

5.3.5. *Modeling cavity expansion*

The problem of cylindrical cavity expansion is of significant interest in petroleum engineering in the design of hydraulic fracturing. Hydraulic fracturing is a technique of stimulating oil and gas reservoirs by inducing and propagating fractures through the injection of a high viscosity fluid (Economides and Nolte 2000). The pressure level at which a tensile fracture is initiated, i.e. the breakdown pressure, can be determined assuming that the crack forms when the hoop stress exceeds the tensile strength of the rock. For weaker rocks, however, where plastic deformations develop, analyses predict either higher breakdown pressures than the ones observed, or even that a tensile state of stress cannot be reached during pressurization. Nevertheless, significantly lower breakdown pressures are observed in the experiments (Papanastasiou and Durban 1995). In this section, we investigate the possibility that tensile failure is preceded by the onset of shear localization.

We model internal pressurization of four thick-walled cylinders of weak sandstone, with internal radius R_i of 5, 10, 20 and 40 cm, and an external radius $R_e = 6R_i$. A detailed account can be found in Zervos *et al.* (2008). The material parameters given in section 5.4 are used. Both internal pressure p_i and external pressure p_e are applied: first the cylinders are loaded with $p_i = p_e = 30\,\text{MPa}$ and subsequently p_i is increased under constant p_e. A mesh with a total of 32,640 degrees of freedom provided converged results.

Initially, for $p_i = p_e$, no yielding occurs. In the second stage, as p_i increases, the material near the hole yields first. Deformation is initially axisymmetric, however, as p_i approaches 110 MPa, axisymmetry is lost and the radial displacement increment in the vicinity of the hole switches to the sinusoidal form shown in Figure 5.12. This happens spontaneously, and the relevant discussion in section 5.4 applies here as well. Subsequently, regions near the cavity progressively unload while the deformation localizes into thin bands of softening material that continue to shear. These are shown in Figure 5.12 with black points, while gray points show material still in the hardening regime. As it can be seen, the shear bands initiate from the cavity wall and propagate toward the outer boundary. The development of similar shear bands has been observed experimentally on sand specimens (Alsiny *et al.* 1992). The hoop stress $\sigma_{\theta\theta}$ remains below the tensile strength, excluding the possibility of tensile failure before shear failure.

The corresponding internal pressure versus the hole expansion curve is presented in Figure 5.13, where the trivial deformation path, corresponding to axisymmetric expansion, is also shown for comparison. The two curves coincide up to the bifurcation point, before axisymmetry is lost. Then they slowly separate, as the curve corresponding to the bifurcated solution levels off at a pressure lower than the limit pressure, to which the trivial path asymptotically tends. The mechanism of localized shear failure is thus more critical than the attainment of limit pressure under axisymmetric deformation. From a practical point of view, this suggests that rupture in cavity pressurization may occur at lower internal pressure than the limit pressure, which is used in the interpretation of a pressure-meter test in geotechnical applications or in fracture initiation prediction in weak rock formations.

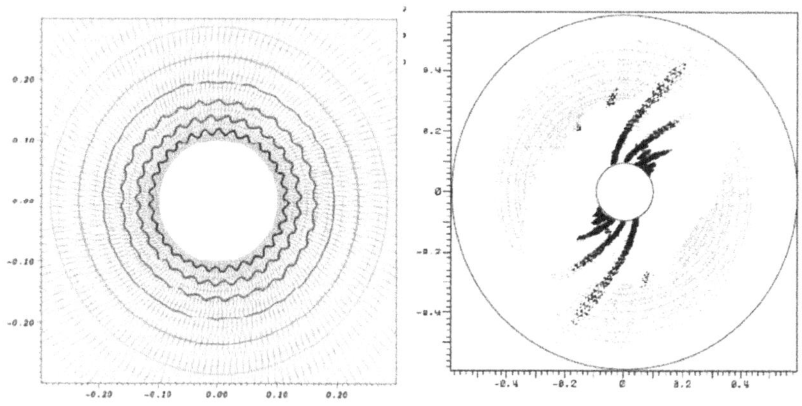

Figure 5.12. *Radial displacement increment at bifurcation (left) and final material state (right) for the Ri = 10 cm model*

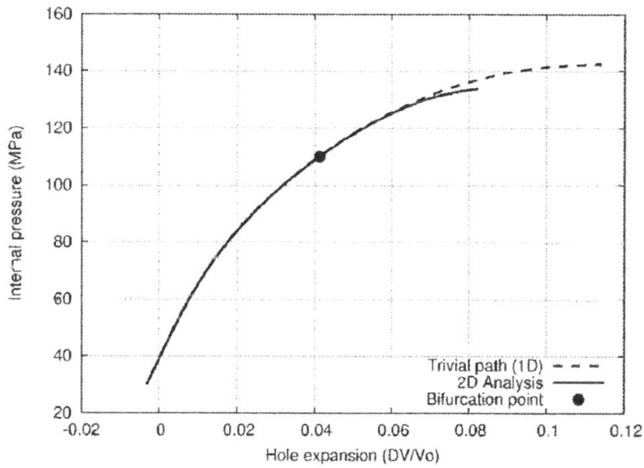

Figure 5.13. *Internal pressure versus hole expansion for the R_i = 10 cm model*

5.4. Conclusion

This chapter presented two advanced theories with higher order continua, one based on a Cosserat continuum and the other on gradient elastoplasticity. Numerical results were presented for different applications and load paths. In all cases, it was shown that the Cosserat continuum and the gradient elastoplasticity can regularize the ill-posedness caused by strain softening. Both models are capable of tracking the spontaneous emergence of instabilities and providing robust predictions of the postlocalization deformation, quantitatively capturing details of the localized mechanism. In other words, both models are capable of modeling the localization of deformation in shear bands, which often leads to failure in geomaterials. The model is also capable of predicting the existence of the scale effect, observed in thick-walled cylinders, which is essential for interpreting the small-scale laboratory tests and for extrapolating the results to the wellbore and tunnels in the field.

The use of localization theories in practical applications is still limited by the issue of material parameters determination. Appropriate procedures or values for the internal lengths and the softening rate must be determined. This as an open issue that requires further research investigation. There are prospects for more applications of these theories with microstructure in the problem of sand prediction as the geometry deals with small holes perforations in reservoir sandstones (van den Hoek 2001). As mentioned earlier, small holes show in experiments a pronounced strength that cannot be predicted by classical plasticity theories. Furthermore, it was shown that localization analysis predicts that an elliptical hole with a long axis parallel to the major *in situ* stress is more stable than a circular hole. This finding can be used in the design of new perforations based on elliptical shape.

5.5. Acknowledgments

A large part of this work was supported by Schlumberger Cambridge Research. We are also grateful to the late Professor Ioannis Vardoulakis for his inspiration and invaluable guidance.

5.6. References

Adhikary, D.P. and Dyskin, A.V. (1997). A Cosserat continuum model for layered materials computers and geotechnics. *Comput. Geotechnics*, 20, 15–45.

Aifantis, E. (1992). On the role of gradients in the localization of deformation and fracture. *Int. J. Eng. Sci.*, 30, 1279–1299.

Alsiny, A., Vardoulakis, I., Drescher, A. (1992). Deformation localization in cavity inflation experiments on dry sand. *Géotechnique*, 42, 395–410.

Benallal, A., Bilardon, R., Geymonat, G. (1989). Conditions de bifurcation à l'intérieur et aux frontières pour une classe de matériaux non standard. *C. R. Acad. Sci. Paris*, 308(II), 893–898.

Bésuelle, P., Desrues, J., Raynaud, S. (2000). Experimental characterisation of the localization phenomenon inside a Vosges sandstone in a triaxial cell. *Int. J. Rock Mech. Min. Sci.*, 37, 1223–1237.

de Borst, R. (1991). Simulation of strain localization: A reappraisal of Cosserat continuum. *Eng. Comput.*, 8, 317–332.

de Borst, R. and Mühlhaus, H.-B. (1992). Gradient-dependent plasticity: Formulation and algorithmic aspect. *Int. J. Num. Meth. Eng.*, 35, 521–539.

Cosserat, E. and Cosserat, F. (1909). *Théorie des Corps Déformables*. Éditions Hermann, Paris.

Chambon, R., Caillerie, D., El Hassan, N. (1998). One-dimensional localization studied with a second grade model. *Eur. J. Mech. A/Solids*, 17(4), 637–656.

Dasgupta, S. and Sengupta, D. (1990). A higher-order triangular plate bending element revisited. *Int. J. Num. Meth. Eng.*, 30, 419–430.

Economides, M.J. and Nolte, K.G. (2000). *Reservoir Stimulation*, 3rd edition. John Wiley & Sons, Chichester, England.

Ewy, R. (1993). Yield and closure of directional and horizontal wells. *Int. J. Rock Mech. Min. Sci. Geomech. Abstr.*, 30(7), 1061–1067.

Finno, R.J., Harris, W.W., Mooney, M.A., Viggiani, G. (1997). Shear bands in plane strain compression of loose sand. *Géotechnique*, 47(1), 149–165.

Haimson, B. and Herrick, C. (1989). Borehole breakouts and in situ stress. In *Proceedings of Drilling Symposium at ATCE-ASME Conference, Houston*, Rowle, J.C. (ed.). American Society of Mechanical Engineers, 22, 49–55.

van den Hoek, P. (2001). Prediction of different types of cavity failure using bifurcation theory. *U.S. Rock Mechanics Symposium*, Washington, DC.

van den Hoek, P., Hertogh, G., Kooijman, A., de Bree, P., Kenter, C., Papamichos, E. (2000). A new concept of sand production prediction: Theory and laboratory experiments. *SPE Drill. Completion*, 15, 261–273.

Matsushima, T., Chambon, R., Caillerie, D. (2002). Large strain finite element analysis of a local second gradient model: Application to localization. *Int. J. Num. Meth. Eng.*, 54, 499–521.

Mindlin, R.D. (1964). Microstructure in linear elasticity. *Arch. Rat. Mech. Anal.*, 10, 51–78.

Mühlhaus, H.-B. and Vardoulakis, I. (1987). The thickness of shear bands in granular materials. *Géotechnique*, 37, 271–283.

Ord, A., Vardoulakis, I., Kajewski, R. (1991). Shear band formation in Gosford sandstone. *Int. J. Rock Mech. Min. Sci. Geomech. Abstr.*, 28(5), 397–409.

Papamichos, E. (2010). Borehole failure analysis in a sandstone under anisotropic stresses. *Int. J. Numer. Analyt. Meth. Geomech.*, 34, 581–603.

Papamichos, E. and van den Hoek, P. (1995). Size dependency of Castlegate and Berea sandstone hollow-cylinder strength on the basis of bifurcation theory. *Proceedings of the 35th U.S. Symposium Rock Mechanics*, 301–306.

Papanastasiou, P. (2000). Localization of deformation and failure around elliptical perforations based on a polar continuum. *Comput. Mech.*, 26, 352–361.

Papanastasiou, P. and Durban, D. (1995). Bifurcation of elastoplastic pressure-sensitive hollow cylinders. *J. Appl. Mech.*, 66(1), 117–123.

Papanastasiou, P. and Durban, D. (1997). Elastoplastic analysis of cylindrical cavity problems in geomaterials. *Int. J. Numer. Analyt. Meth. Geomech.*, 21, 121–132.

Papanastasiou, P. and Vardoulakis, I. (1989). Bifurcation analysis of deep boreholes: I. Scale effect. *Int. J. Numer. Analyt. Meth. Geomech.*, 13, 183–198.

Papanastasiou, P. and Vardoulakis, I. (1992). Numerical treatment of progressive localization in relation to borehole stability. *Int. J. Numer. Analyt. Meth. Geomech.*, 16, 389–424.

Papanastasiou, P. and Vardoulakis, I. (1994). Numerical analysis of borehole stability problem. In *Numerical Analysis and Modelling of Soil Structure Interaction*, Bull, J. (ed.). E&FN Spon, London, 673–711.

Papanicolopulos, S.-A., Zervos, A., Vardoulakis, I. (2009). A three dimensional C1 finite element for gradient elasticity. *Int. J. Num. Meth. Eng.*, 77, 1396–1415.

Pijaudier-Cabot, G. and Bazant, Z.P. (1987). Nonlocal damage theory. *J. Eng. Mech, ASCE*, 113, 1512–1533.

Rudnicki, J.W. and Rice, J.R. (1975). Conditions for the localization of the deformation in pressure sensitive dilatants materials. *J. Mech. Phys. Solids*, 23, 371–394.

Sulem, J., Vardoulakis, I., Papamichos, E. (1995). Microstructure and scale effect in granular rocks. In *Continuum Models for Materials with Microstructure*, Mühlhaus H.-B. (ed.). John Wiley & Sons, New York, pp. 201–237.

Tejchman, J. and Wu, W. (1993). Numerical study on shear band patterning in a Cosserat continuum. *Acta Mech.*, 99, 61–74.

Vardoulakis, I. (1980). Shear band inclination and shear modulus of sand in biaxial tests. *Int. J. Num. Anal. Meth. Geomech.*, 4, 103–119.

Vardoulakis, I. and Aifantis, E.C. (1991). A gradient flow theory of plasticity for granular materials. *Acta Mech.*, 87, 197–217.

Vardoulakis, I. and Aifantis, E.C. (1994). On the role of microstructure in the behavior of soils: Effects of higher order gradients and internal inertia. *Mech. Mater.*, 18, 151–158.

Vardoulakis, I. and Frantziskonis, G. (1992). Micro-structure in kinematic-hardening plasticity. *Eur. J. Mech. A/Solids*, 11(4), 467–486.

Vardoulakis, I. and Papanastasiou, P. (1988). Bifurcation analysis of deep boreholes: I. Surface instabilities. *Int. J. Numer. Analyt. Meth. Geomech.*, 12, 379–399.

Vardoulakis, I., Shah, K.R., Papanastasiou, P. (1992). Modelling of tool-rock shear interfaces using gradient-dependent flow theory of plasticity. *Int. J. Rock Mech. Min. Sci. Geomech. Abstr.*, 29(6), 573–582.

Vardoulakis, I., Papamichos, E., Sulem, J. (1994). A second gradient plasticity model for granular rocks. In *Fracture and Damage in Quasibrittle Structures*, Bazant, Z.P., Bittnar, Z., Mazars, J. (eds). E&FN Spon, London.

Vardoulakis, I. and Sulem, J. (1995). *Bifurcation Analysis in Geomechanics*. Blackie Academic and Professional, Glasgow.

Zervos, A. (2008). Finite elements for elasticity with microstructure and gradient elasticity. *Int. J. Num. Meth. Eng.*, 73(4), 564–595.

Zervos, A., Vardoulakis, I., Jean, M., Lerat, P. (2000). Numerical investigation of granular interfaces kinematics. *Int. J. Num. Analyt. Meth. Geomech.*, 5, 305–324.

Zervos, A., Papanastasiou, P., Vardoulakis, I. (2001a). A finite element displacement formulation for gradient elastoplasticity. *Int. J. Num. Meth. Eng.*, 50, 1369–1388.

Zervos, A., Papanastasiou, P., Vardoulakis, I. (2001b). Modelling of localization and scale effect in thick-walled cylinders with gradient elastoplasticity. *Int. J. Solids Struct.*, 38, 5081–5095.

Zervos, A., Vardoulakis, I., Papanastasiou, P. (2007). Influence of nonassociativity on localization and failure in geomechanics based on gradient elastoplasticity. *Int. J. Geomechanics*, 7(1), 63–74.

Zervos, A., Papanastasiou, P., Vardoulakis, I. (2008). Shear localization in thick walled cylinders under internal pressure based on gradient elastoplasticity. *J. Theor. Appl. Mech.*, 38(1–2), 81–100.

6

Numerical Modeling of Bifurcation: Applications to Borehole Stability, Multilayer Buckling and Rock Bursting

Euripides PAPAMICHOS
Aristotle University of Thessaloniki, Greece, and SINTEF Petroleum Research, Trondheim, Norway

Typical boundary value problems of bifurcation are presented such as (1) borehole stability with applications in the oil and gas industry and in tunneling, (2) multilayer buckling with application in the folding of geological formations and (3) spalling and buckling of surface parallel cracks with application in rock bursting in mining. The formulation of the bifurcation problem in all these cases is described together with the numerical implementation and the main results. If the formulation is in terms of a continuum with microstructure, such as the Cosserat continuum, scale is introduced into the bifurcation problem, which leads to size effect. For example, in borehole failure, smaller diameter holes fail at higher stresses than larger holes.

6.1. Introduction

Three typical bifurcation boundary value problems are formulated and solved numerically. In section 6.2, borehole stability is formulated and analyzed within a finite element method scheme for elastoplastic Cosserat materials, and the results are compared with experimental data. In section 6.3, multilayer buckling in anisotropic

Instabilities Modeling in Geomechanics,
coordinated by Ioannis STEFANOU and Jean SULEM. © ISTE Ltd 2020.

elastic and viscoelastic materials under initial stress is analyzed using the matric transfer method. Finally, in section 6.4 buckling of anisotropic elastic half-spaces with surface parallel cracks under initial stress are analyzed using the displacement discontinuity boundary element method. Conclusions are presented in section 6.5.

6.2. Borehole stability

Borehole failure has been an important geomechanical problem for the assessment of the integrity of tunnels, wellbores and perforations in the field. The most common failure mode for dilatant rocks is the formation of breakouts. A relevant laboratory experiment is the hollow cylinder (HC) test with or without fluid flow. HC experiments under various axial and lateral stress paths in Red Wildmoor sandstone (Papamichos *et al.* 2010) have demonstrated the lateral and axial borehole failure modes shown in Figure 6.1 as suggested by Maury (1993). The first prevails when the tangential stress at the hole is the largest compressive principal stress and failure takes place in the form of shear bands concentrated on two diametrically opposed locations of the borehole. The second prevails when the axial stress at the hole is the largest compressive principal stress and failure takes place in the form of shear bands forming toroids around the hole perimeter.

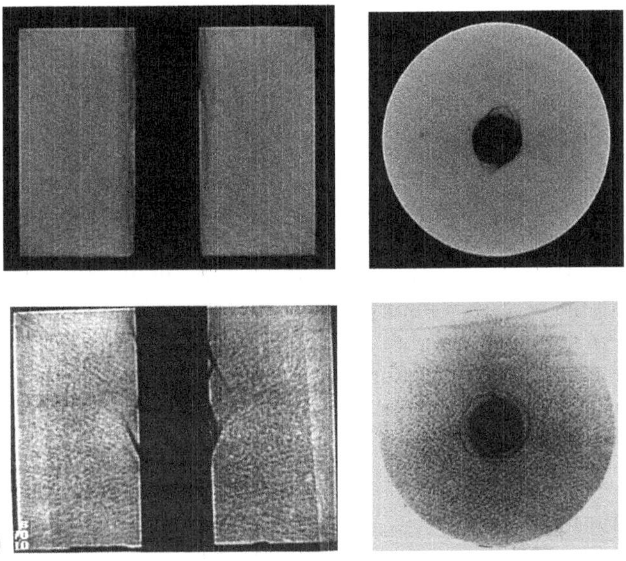

(a) (b)

Figure 6.1. *Vertical (left) and horizontal (right) X-ray CT scan images of hollow cylinder specimens demonstrating (a) lateral borehole failure with breakouts and (b) axial borehole failure with toroids (Papamichos* et al. *2010)*

Analysis based on elasticity and a Mohr–Coulomb failure criterion for the rock has been traditionally used to predict borehole failure. However, comparisons with experimental results have long demonstrated that such an approach greatly underestimates the borehole failure strength. For successful predictions, the borehole failure problem has been treated as a bifurcation phenomenon (Vardoulakis *et al.* 1988) often in connection with continua with microstructure (Cosserat or gradient) (e.g. Vardoulakis and Sulem 1995 and references therein) and elastoplastic constitutive laws. The advantage of this approach is that it describes naturally the observed failure patterns of shear bands and breakouts.

In the following, the Cosserat elastoplastic model is implemented in a finite element scheme for the solution of the HC failure test. Figure 6.2 shows the geometric layout of the problem in the cylindrical (r, θ, z)-coordinate system. A thick wall cylinder of external radius r_e, internal radius r_i and height H is compressed radially with uniform external σ_{rext} and internal σ_{rint} stresses with $\sigma_{rext}/\sigma_{rint} > 1$. Axially, a uniform strain ε_{zz} is applied to simulate the experiment where the axial load is applied through steel (i.e. rigid) loading platens. The strain ε_{zz} is controlled such that it produces a prescribed axial load F_z. The uniform axial strain boundary condition means that the HC test is a two-dimensional problem, i.e. it varies only in two dimensions in the r,θ-plane normal to the hole. The effect of pore pressure on borehole failure is not included here for brevity, but the reader is referred to Papamichos (2010a) for a description of the formulation with steady-state fluid flow. As shown in the experiments, $r_e/r_i = 5$. Section 6.2.1 presents the implementation for the axisymmetric solution of the primary path of cylindrical convergence of the HC hole, and section 6.2.2 formulates the bifurcation condition for borehole failure. Section 6.2.3 presents simulation results and comparison with experimental data. In sections 6.2.1 and 6.2.2, standard mechanics notation of tension positive is employed for the stresses and strains. However, in section 2.3, compressive stresses and strains are considered positive to avoid carrying the negative sign and plotting the results in negative axes.

6.2.1. *Primary loading path*

During the axisymmetric deformations of the primary loading path, shear stresses and Cosserat couple stresses vanish and therefore the solution for the primary loading path can be significantly simplified and become one dimensional with dependence of the quantities only along the radial direction. The finite element formulation is based on the variational equation within a volume V with boundary ∂V, which in axisymmetric problems under uniform and prescribed axial displacement can be written in the cylindrical coordinate system (r,θ,z) as

$$\int_V \left(\sigma_{rr}\delta\varepsilon_{rr} + \sigma_{\theta\theta}\delta\varepsilon_{\theta\theta}\right)dV = \int_{\partial V_\sigma} t_r \delta u_r dS\,, \qquad [6.1]$$

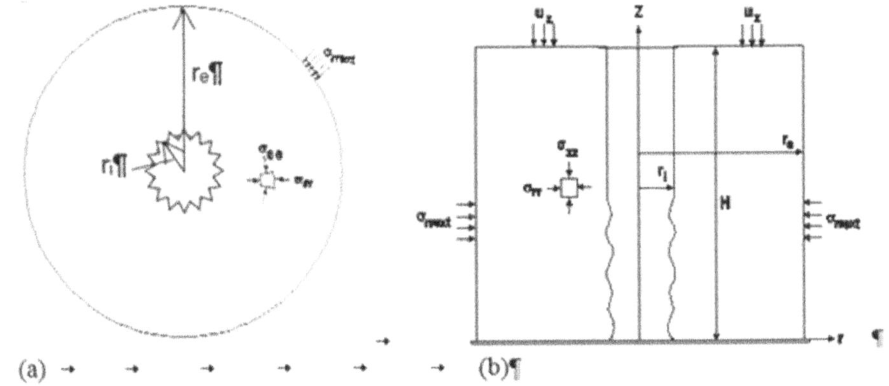

Figure 6.2. *Geometric layout of the thick wall hollow cylinder configuration: (a) plane view with lateral hole warping and (b) vertical section with axial hole warping*

where σ_{rr} is the radial and $\sigma_{\theta\theta}$ the tangential stress, dual in energy to the radial ε_{rr} and tangential $\varepsilon_{\theta\theta}$ strains, respectively, and t_r is a prescribed traction on a part ∂V_σ of the boundary ∂V. Moreover, $\delta\varepsilon_{rr}$, $\delta\varepsilon_{\theta\theta}$ and δu_r are variations of the radial and tangential strains and the radial displacement u_r, respectively. The integrands in equation [6.1] are independent of the θ and z-coordinates and thus equation [6.1] can be integrated with respect to these coordinates and written in matrix form as

$$\int_{r_i}^{r_e} \{\delta\varepsilon\}^T \{\sigma\} r dr = -\delta u_r\big|_{r=r_i} \sigma_{r\,\mathrm{int}} r_i + \delta u_r\big|_{r=r_e} \sigma_{r\mathrm{ext}} r_e\,, \qquad [6.2]$$

where the strain variation $\{\delta\varepsilon\}$ and the stress $\{\sigma\}$ vectors are defined as

$$\{\delta\varepsilon\}^T = \{\delta\varepsilon_{rr}, \delta\varepsilon_{\theta\theta}\}\,, \qquad \{\sigma\}^T = \{\sigma_{rr}, \sigma_{\theta\theta}\} \qquad [6.3]$$

and the stress boundary conditions at the hole and the external surface have been considered in the left-hand side.

In an elastoplastic analysis, incremental and iterative methods are used to solve these equations, where the total external loads are added in increments step by step. At the $(m + 1)$th loading step

$$\int_{r_i}^{r_e} \{\delta\varepsilon\}^T \{d\sigma\} rdr = {}^{m+1}\left[\delta u_r\big|_{r=r_i} \sigma_{r\,\mathrm{int}} r_i + \delta u_r\big|_{r=r_e} \sigma_{rext} r_e \right] - \int_{r_i}^{r_e} \{\delta\varepsilon\}^T\, {}^m\{\sigma\} rdr\,, \quad [6.4]$$

where the first term in the right-hand side of the first equation is the load at the $(m + 1)$th step, ${}^m\{\sigma\}$ is the stress vector at the mth step and $\{d\sigma\}$ the stress increment vector. Using the constitutive relations of Cosserat elastoplasticity (Papamichos *et al.* 2010)

$$\{d\sigma\} = \left[C^{ep}\right]\{d\varepsilon\} \quad [6.5]$$

and taking into account the fact that $d\varepsilon_{zz}$ is constant with r, equation [6.4] can be written as

$$\begin{aligned} \int_{r_i}^{r_e} \{\delta\varepsilon\}^T \left[C_A^{ep}\right]\{d\varepsilon\} rdr = {}^{m+1}\left[\delta u_r\big|_{r=r_i} \sigma_{r\,\mathrm{int}} r_i + \delta u_r\big|_{r=r_e} \sigma_{rext} r_e \right] - \\ - \int_{r_i}^{r_e} \{\delta\varepsilon\}^T\, {}^m\{\sigma\} rdr - d\varepsilon_{zz} \int_{r_i}^{r_e} \{\delta\varepsilon\}^T \left\{C_B^{ep}\right\} rdr \end{aligned}, \quad [6.6]$$

where the constitutive matrix $\left[C_A^{ep}\right]$ and vector $\left\{C_B^{ep}\right\}$ are the following parts of matrix $\left[C^{ep}\right]$ in equation [6.5]

$$\left[C_A^{ep}\right] = \begin{bmatrix} C_{11}^{ep} & C_{12}^{ep} \\ C_{21}^{ep} & C_{22}^{ep} \end{bmatrix}, \qquad \left\{C_B^{ep}\right\} = \begin{Bmatrix} C_{13}^{ep} \\ C_{23}^{ep} \end{Bmatrix} \quad [6.7]$$

The finite element solution is obtained with a one-dimensional discretization along the radius r. A fine discretization is required close to the hole to obtain mesh insensitivity. This is achieved by discretizing the domain in 100 elements, according to a geometric progression with common ratio 1.05. Three-node elements with Lagrange polynomial basis functions of second order are used. Each node has one degree of freedom, the radial displacement u_r.

In nonlinear constitutive models such as elastoplasticity, in contrast with linear elasticity, if a uniform along the radius axial strain is applied, the result will be a non-uniform axial stress σ_{zz} along the HC radius. The experiments in the laboratory are performed by applying a uniform axial strain so that it produces a prescribed axial load F_z. Numerically, this can be accomplished by implementing an axial load servo-control algorithm to control the applied axial strain (Papamichos 2010a).

6.2.2. *Hole failure*

In this section, the bifurcation condition for lateral HC failure is formulated and implemented in the finite element scheme. For axial HC failure, the reader is referred to Papamichos (2010a). The underlying hypothesis of the bifurcation theory is that spalling and shear banding in rocks are the result of material instabilities, termed equilibrium bifurcations. The bifurcation approach associates failure with the occurrence of the instabilities in contrast to classical procedures where failure is usually assumed ad hoc to be an intrinsic material property associated with the elastic-plastic limit. The analysis here focuses on the prediction of the bifurcation point of warping of the hole that subsequently may lead to spalling or shear banding as it has been demonstrated in postbifurcation numerical analyses (e.g. Papanastasiou and Vardoulakis 1992; Papamichos 2010b).

The bifurcation condition is formulated on the basis that in addition to the trivial solution of cylindrical convergence of the hole during the primary loading path, there exists another non-trivial warping solution that fulfills homogeneous boundary conditions. Thus, the bifurcation problem for the HC failure can be formulated with the homogeneous and incremental form of the virtual work equation for a Cosserat continuum

$$\int_V \{\delta d\hat{\varepsilon}\}^T \{d\hat{\sigma}\} dV = 0, \qquad [6.8]$$

where $\{d\hat{\sigma}\}$ and $\{d\hat{\varepsilon}\}$ are the generalized stress and strain increment vectors, respectively, that correspond to the non-trivial bifurcation solution. This solution involves the Cosserat rotations and couple stresses, in contrast to the solution of the primary path, and therefore the bifurcation condition depends on the internal length.

The bifurcation condition for lateral failure that corresponds to warping of the hole on the (r,θ)-plane normal to the HC axis, as shown in Figure 6.2(a), is obtained by solving equation [6.8] for the non-trivial, plane-strain, displacement and microrotation increment field

$$\begin{aligned} d\hat{u}_r(r,\theta) &= V_r(r)\cos(m\theta), & d\hat{\omega}_z^c(r,\theta) &= W_z(r)\sin(m\theta) \\ d\hat{u}_\theta(r,\theta) &= V_\theta(r)\sin(m\theta), & d\hat{\omega}_r^c &= 0 \\ d\hat{u}_z &= 0, & d\hat{\omega}_\theta^c &= 0 \end{aligned} \quad , \qquad [6.9]$$

where m = 1, 2, 3, ... is the wavenumber of the warping mode. The wavelength of the deformation is given as $W = 2\pi r_i/m$. For this displacement and microrotation field, the generalized displacement, strain and stress increment vectors reduce to

$$\begin{aligned}
\{d\hat{u}\}^T &= \{d\hat{u}_r, d\hat{u}_\theta, d\hat{\omega}_z^c\} \\
\{d\hat{\varepsilon}\}^T &= \{d\hat{\varepsilon}_{rr}, d\hat{\varepsilon}_{\theta\theta}, d\hat{\varepsilon}_{r\theta}, d\hat{\varepsilon}_{\theta r}, Rd\hat{\kappa}_{\theta z}, Rd\hat{\kappa}_{rz}\} \\
\{d\hat{\sigma}\}^T &= \{d\hat{\sigma}_{rr}, d\hat{\sigma}_{\theta\theta}, d\hat{\sigma}_{r\theta}, d\hat{\sigma}_{\theta r}, d\hat{\mu}_{\theta z}/R, d\hat{\mu}_{rz}/R\}
\end{aligned} \qquad [6.10]$$

For small deformation analysis, the strain-displacement relation is given by

$$\{d\hat{\varepsilon}\} = [B]\{dU\} \qquad \text{or} \qquad \{\delta d\hat{\varepsilon}\} = [B]\{\delta dU\}, \qquad [6.11]$$

where $\{dU\}$ is the displacement increment vector of nodal points related to the distributed displacement $\{d\hat{u}\}$ by

$$\{du\} = [N]\{dU\} \qquad [6.12]$$

in which $[N]$ is the matrix of shape functions. The strain-displacement matrix $[B]$ is defined as

$$[B] = [L][N] \qquad [6.13]$$

and $[L]$ is a differential operator matrix such that

$$\{d\hat{\varepsilon}\} = [L]\{d\hat{u}\} \qquad [6.14]$$

Following the analysis of axisymmetric problems submitted to non-symmetrical loading (Zienkiewicz and Taylor 1989), a typical term of the $[N]$ matrix of shape functions for the generalized velocity field equation [6.10] is

$$[N_i] = \begin{bmatrix} N_i \cos(m\theta) & 0 & 0 \\ 0 & N_i \sin(m\theta) & 0 \\ 0 & 0 & N_i \sin(m\theta) \end{bmatrix} \qquad [6.15]$$

On substitution of the shape functions [6.15] into equation [6.13], a typical term of $[B]$ is

$$[B_i]=\begin{bmatrix} N_{i,r}\cos(m\theta) & 0 & 0 \\ \frac{1}{r}N_i\cos(m\theta) & \frac{m}{r}N_i\cos(m\theta) & 0 \\ -\frac{m}{r}N_i\sin(m\theta) & -\frac{1}{r}N_i\sin(m\theta) & N_i\sin(m\theta) \\ 0 & N_{i,r}\sin(m\theta) & -N_i\sin(m\theta) \\ 0 & 0 & RN_{i,r}\sin(m\theta) \\ 0 & 0 & \frac{mR}{r}N_i\cos(m\theta) \end{bmatrix} \qquad [6.16]$$

which is conveniently separated to

$$[B_i]=\left[B_i^s\right]\sin(m\theta)+\left[B_i^c\right]\cos(m\theta) \qquad [6.17]$$

By combining equations [6.8] and [6.11] with the incremental constitutive equation [6.5], the governing equation, may be written as

$$\int_V [B]^T [D][B]dV\{dU\}=0\,, \qquad [6.18]$$

where $[D]$ contains only the relevant components of the tangent constitutive matrix $\left[C^{ep}\right]$ in equation [6.5]. Equation [6.18] can be simplified by using equation [6.17] and performing the integration with respect to the θ and z-coordinates reduces to

$$[K]\{dU\}=0\,, \qquad [K]=\int_{r_i}^{r_e}\left\{\left[B^s\right]^T[D]\left[B^s\right]+\left[B^c\right]^T[D]\left[B^c\right]\right\}rdr \qquad [6.19]$$

which results in an eigenvalue problem for the critical bifurcation loads σ_{rint}, σ_{rext} and ε_{zz}. Within a finite element discretization scheme, the solution to this eigenvalue problem is obtained by requiring that the global stiffness matrix $[K]$ becomes singular, i.e.

$$\det[K]=0 \qquad [6.20]$$

which is the bifurcation condition for lateral instability.

The governing equation [6.19] gives significant computational advantages because it reduces the two-dimensional problem to a one-dimensional finite element discretization with respect to the r-coordinate. The finite element solution of the eigenvalue problem equation [6.20] for hole failure is obtained with the same one-dimensional discretization along the radius r of the thick wall HC used for obtaining the solution for the primary path. Each node has now three degrees of freedom. These are the radial V_r and tangential V_θ displacement amplitude and the Cosserat microrotation W_z amplitude. The initial failure bifurcation condition requires that the global stiffness matrix $[K]$ becomes singular. In inhomogeneous problems like the present, failure of local stability conditions does not necessarily imply loss of uniqueness (Papanastasiou and Vardoulakis 1992). In fact, the obtained bifurcation points correspond to loading states where elements close to the hole have entered the softening regime. Although at these elements the local stability criterion is violated, the global stiffness matrix remains positive. The loading stresses at which the bifurcation condition is satisfied depend on the wavenumber m of the bifurcation mode. In a Cosserat continuum, there exists for the HC problem a critical wavenumber m_{cr} that corresponds to the least required loads. It is obtained by solving the bifurcation condition for various m and selecting the m_{cr} corresponding to the least required loads. It maybe, therefore, assumed that the hole would fail under m_{cr}.

6.2.3. *Simulation of hollow cylinder experiments*

Theoretical results from finite element simulations of HC tests under various loading paths are presented and compared to experimental data (Papamichos *et al.* 2010). The bifurcation predictions for failure depend on the Cosserat internal length R. The internal length cannot be calibrated from standard triaxial test data. For this reason, a parametric study with various internal lengths between $R = 2$ and 20 μm was performed to select the internal length R from back calibration of the experimental results on hole failure. Once the internal length is selected, a size effect analysis can be performed for failure predictions of holes of various sizes.

For lateral failure, isotropic, plane-strain and radial loading tests at 3.6, 7.2, 12.5 and 14.5 MPa (average) axial stress were simulated and the results were compared with experimental data. In these tests, the hole remains unsupported while axial and external radial stresses are applied. In the isotropic tests, the applied axial and external radial stresses are the same. In the plane-strain tests, external radial stress is applied under zero axial strain. Finally, in the radial loading test after an initial isotropic loading to 3.6 or 7.2 or 12.5 or 14.5 MPa, the external radial stress is increased while the axial stress is kept constant.

Figure 6.3 shows the bifurcation stresses for various R together with experimental failure stresses. The bifurcation point for the smallest R is found immediately after the material at the hole begins to soften. As R increases, the external stresses for failure increase. The theoretical predictions show that to capture the experimental data, R must be between 10 and 15 μm. This is the case for all loading paths leading to lateral failure, which means that once R is back calibrated on experimental data for one loading path, e.g. isotropic loading, then it can be used to forward predict failure at other loading paths. The average grain diameter d_g of Red Wildmoor sandstone is 107 μm. Thus, the ratio of the back-calibrated R to the grain diameter is $R/d_g = 0.09 - 0.14$.

Figure 6.4 shows the selection of the critical bifurcation mode m_{cr} for isotropic loading where the dependency of the external radial stress for failure with the bifurcation mode m and the internal length R is illustrated. Although the actual evolution of the critical bifurcation mode requires postbifurcation analysis, it can be assumed that high critical modes of small wavelength would lead to spalling all around the hole, whereas low modes would lead to more localized spalling. Experimental and numerical evidence show that although localization may initiate under a high mode, a lower mode may finally evolve. Indeed, experiments show that an initial mode m = 6 localization evolves to an apparent mode m = 3 failure (Haimson and Song 1993). At the same time, postbifurcation computations show that an initial mode m = 12 evolves into a single shear band failure zone (Papanastasiou and Vardoulakis 1992). The results show that the relation between the m_{cr} for lateral hole failure and R is independent of the loading path and that m_{cr} decreases with a power law with R that can be approximate as

$$m_{cr} = 0.074\left(R/r_i\right)^{-0.92} \qquad [6.21]$$

Figure 6.5 compares, as an example, the simulation results for isotropic and plane strain loading with experimental data where the external radial stress is plotted versus the internal and external tangential strains. The internal tangential strain corresponds to the normalized (with its radius) hole closure. In addition, the points of hole failure for to the failure point after which postfailure analysis is required to capture the deformations involved. Theoretical results on the stress profiles along the radius of the HC demonstrate that for isotropic loading the axial stress σ_{zz} is always the intermediate principal stress. Therefore, σ_{zz} does not influence plastic yielding because a Mohr–Coulomb yield surface is employed. However, for plane-strain loading, σ_{zz} is not always the intermediate principal stress but it becomes the least principal stress away from the hole. This leads to a higher stress concentration at the hole for plane-strain and results in earlier hole failure.

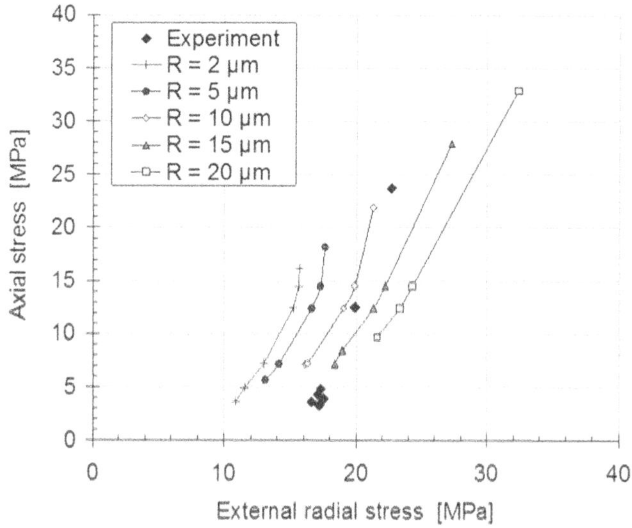

Figure 6.3. *Experimental data and theoretical results for lateral hole failure for various loading paths and internal lengths. For a color version of this figure, see www.iste.co.uk/stefanou/instabilities.zip*

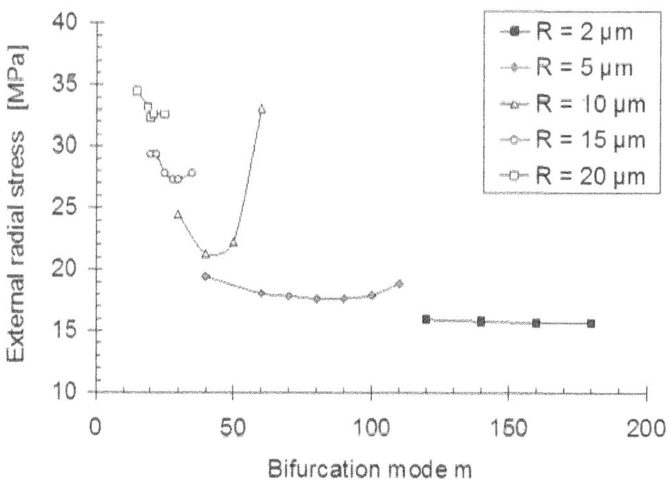

Figure 6.4. *External radial stress at bifurcation failure versus bifurcation mode m and internal length R for isotropic loading. The minimum in each curve leads to the selection of the critical mode m_{cr} corresponding to the least external radial stress. For a color version of this figure, see www.iste.co.uk/stefanou/instabilities.zip*

The value $R = 10$ μm was chosen to construct scale effect diagrams in terms of the hole diameter for the various loading configurations. The results are illustrated in Figure 6.6 where the normalized external radial stress is plotted versus the hole diameter. The normalization is done with the external radial stress for failure of the 20-mm diameter hole, which is considered as the reference hole. The plot includes also the scale effect equation (Papamichos *et al.* 2010)

$$\frac{\sigma_F}{\sigma_{Fref}} = \frac{1}{3} + \frac{2}{3}\left(\frac{D_{ref}}{D}\right)^{2/5}, \qquad D_{ref} = 20 \text{ mm} \qquad [6.22]$$

that was used to approximate experimental results. In equation [6.22], σ_F is the isotropic failure stress of a hole with diameter D, and σ_{Fref} is the isotropic failure stress of a reference hole with diameter $D_{ref} = 20$ mm. The results show that the scale effect is similar for all radial loading paths, while a stronger size effect is predicted for the plane-strain and the isotropic loading paths.

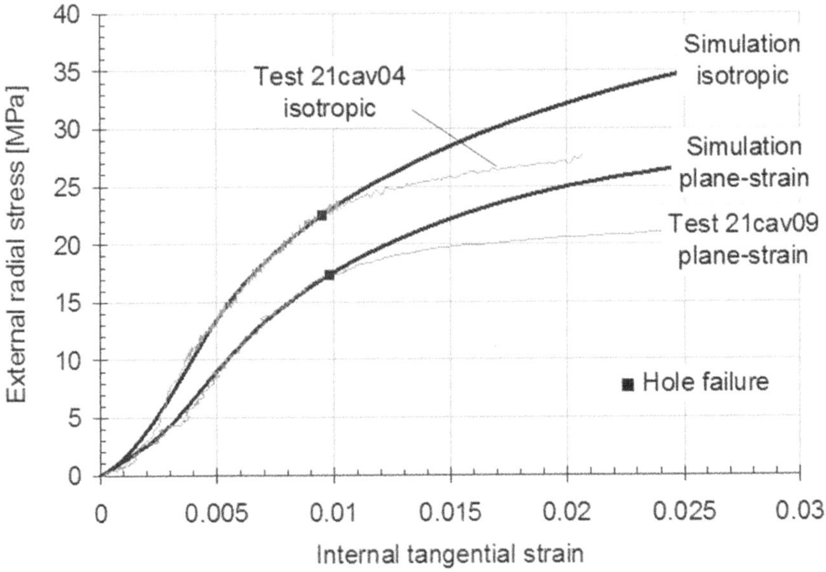

Figure 6.5. *Isotropic and plane-strain loading. Numerical simulations and experimental results for the external radial stress versus internal tangential strain. The hole failure points are indicated. For a color version of this figure, see www.iste.co.uk/stefanou/instabilities.zip*

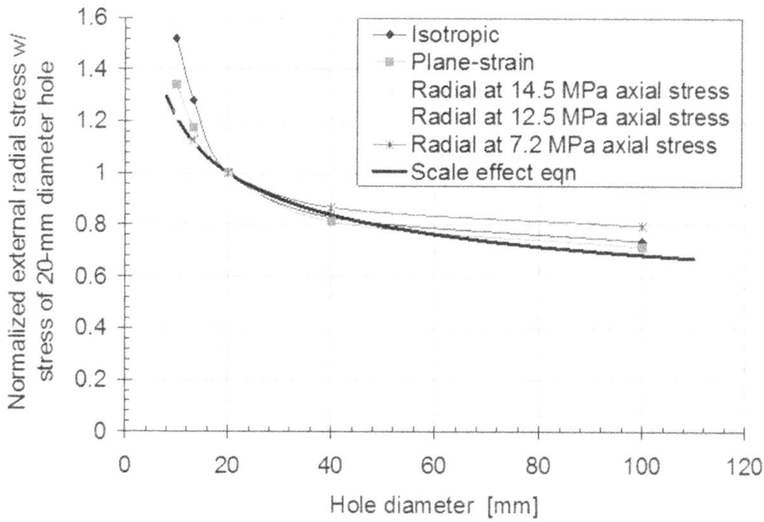

Figure 6.6. *Theoretical predictions of scale effect for lateral hole failure during isotropic, plane-strain and radial loading tests at axial stress 7.2, 12.5 and 14.5 MPa and internal length R = 10 μm. For a color version of this figure, see www.iste.co.uk/stefanou/instabilities.zip*

6.3. Folding of elastic media as a bifurcation problem

The constitutive equations of large strain elasticity theory are utilized to study buckling of elastic layered media. These buckling modes can explain the occurrence of various periodic structures in geology such as folds. The analysis of the behavior of stratified elastic or viscoelastic media under compression is of interest to the structural geologist. Biot has presented an analysis of folding of stratified sedimentary rock in a series of pioneering papers and in his book (Biot 1965 and references therein). The folding mechanism considered in Biot's theory is the spontaneous folding caused by instability under a compressive load acting in a direction parallel to the layers. From the geological viewpoint, a purely elastic theory is not enough to explain folding. Time-dependent phenomena such as viscous behavior must be considered. Biot (1957) developed a general theory of folding of a compressed viscoelastic layer embedded in an infinite medium of another viscoelastic material. He showed that, in general, there exists a lower and a higher critical load between which folding occurs at a finite rate with a dominant wavelength. This is the wavelength whose amplitude increases at the fastest rate. An experimental verification of Biot's theory of folding of stratified viscoelastic media in compression is presented by Biot (1961).

6.3.1. *Buckling of a layer under initial stress*

The problem considered here is the non-homogeneous, plane-strain deformation of a layer of thickness $2h$, due to constant horizontal and vertical compressions σ_1 and σ_2, respectively, as shown in Figure 6.7. The theory in this analysis is based on incremental plane-strain deformations superimposed on the large strain uniform compression. The problem is formulated in terms of the first Piola–Kirchhoff stress π_{ij} with $d\pi_{ij}$ being its increment referred to the deformed initially stressed state. In the case of constant body forces, the equations of equilibrium for the incremental problem take the form $d\pi_{ij,j} = 0$, where a subscript preceded by a comma denotes partial differentiation and repeated indices are summed. The incremental stress boundary conditions are $d\pi_{ij}n_j = d\pi_i$, where $d\pi_i$ are the components of prescribed traction increments referred to the initial stressed state and n_j are the components of the outward unit normal of the boundary. Let du_i be the instantaneous incremental displacement components in the current configuration. The stress increment $d\pi_{ij}$ is related to the Jaumann increments of the corotated Cauchy stress dT_{ij}, the initial stress field σ_{ij} and the incremental strain $d\varepsilon_{ij}$ and spin $d\omega_{ij}$ as follows

$$d\pi_{ij} = dT_{ij} + d\omega_{ik}\sigma_{kj} - \sigma_{ik}d\varepsilon_{kj} + \sigma_{ij}d\varepsilon_{kk} \qquad [6.23]$$

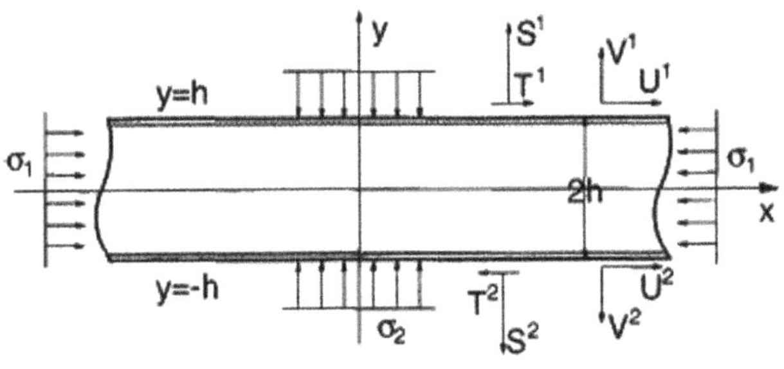

Figure 6.7. *A single layer under initial stress σ_1 and σ_2*

The Jaumann stress increments are related directly to the strain increments through constitutive relations for anisotropic materials

$$\begin{aligned} dT_{11} &= C_{11}d\varepsilon_{11} + C_{12}d\varepsilon_{22} \\ dT_{22} &= C_{21}d\varepsilon_{11} + C_{22}d\varepsilon_{22} \\ dT_{12} &= 2Gd\varepsilon_{12} \end{aligned} \qquad [6.24]$$

Using equation [6.23] and the constitutive equation [6.24], the equations of equilibrium become

$$\begin{aligned} &C_{11}du_{1,11}+(G-\tau)du_{1,22}+(C_{12}+G+\tau)du_{2,12}=0 \\ &(C_{21}+G-\tau)du_{1,12}+(G+\tau)du_{2,11}+C_{22}du_{2,22}=0 \end{aligned}, \qquad [6.25]$$

where $\tau=(\sigma_1-\sigma_2)/2$. For the considered non-homogeneous deformation mode, the displacement field is assumed to be given in terms of two unknown amplitude functions of the dimensionless coordinates $x=x_1/L$, $y=x_2/L$

$$\begin{aligned} &du_1=U(y)\sin(\beta x), \qquad U(y)=Ae^{\gamma y} \\ &du_2=V(y)\cos(\beta x), \qquad V(y)=Be^{\gamma y} \end{aligned}, \qquad [6.26]$$

where L is a reference length associated with the wavelength W and β is a dimensionless wavenumber. The wavelength W of the deformation mode can be written as $W=2\pi L/\beta$. By substituting the displacement field [6.26] in the differential equations [6.25] and letting $Z=\gamma/\beta$, we obtain a system of two linear homogeneous algebraic equations with respect to the integration constants A and B

$$\begin{aligned} &A\left[-C_{11}+(G-\tau)Z^2\right]+B\left[-(C_{12}+G+\tau)Z\right]=0 \\ &A\left[(C_{21}+G-\tau)Z\right]+B\left[-(G+\tau)+C_{22}Z^2\right]=0 \end{aligned} \qquad [6.27]$$

For non-trivial solution in terms of A and B, the determinant of the system of equations [6.27] must vanish. This leads to the following biquadratic equation for Z

$$aZ^4-bZ^2+c=0 \qquad [6.28]$$

where $a=C_{22}(G-\tau)$, $b=C_{11}C_{22}-C_{12}C_{21}-C_{12}(G-\tau)-C_{21}(G+\tau)$ and $c=C_{11}$ $(G+\tau)$. The four roots of equation [6.28] correspond to four solutions U_k and V_k for the displacement field amplitudes. The complete solution for $U(y)$ and $V(y)$ is then given as a linear combination of the function base as

$$U(y)=\sum_{k=1}^{4}a_kU_k, \qquad V(y)=\sum_{k=1}^{4}a_kV_k, \qquad [6.29]$$

where a_k are integration constants. The base functions U_k and V_k are given explicitly in Papamichos (1988).

6.3.2. *Eigen-displacements and tractions at layer boundaries*

The incremental displacement amplitudes at the upper (1) and lower (2) boundaries of the layer are obtained directly from equations [6.29]

$$\begin{aligned} U^1 &= \sum_{k=1}^{4} a_k U_k(h/L), \quad V^1 = \sum_{k=1}^{4} a_k V_k(h/L) \\ U^2 &= \sum_{k=1}^{4} a_k U_k(-h/L), \quad V^2 = \sum_{k=1}^{4} a_k V_k(-h/L) \end{aligned} \qquad [6.30]$$

The boundary tractions $d\pi_i = d\pi_{ij} n_j$ are written through equations [6.23] and [6.24] as

$$d\pi_1 = \frac{\beta}{L} T(y)\sin(\beta x), \quad d\pi_2 = \frac{\beta}{L} S(y)\cos(\beta x), \qquad [6.31]$$

$$\begin{aligned} S(y) &= \sum_{k=1}^{4} a_k S_k, \quad S_k = (C_{21} - \tau - p)U_k + C_{22} V_k' / \beta, \\ T(y) &= \sum_{k=1}^{4} a_k T_k, \quad T_k = (G - \tau)U_k' / \beta + (G - p)V_k, \end{aligned} \qquad [6.32]$$

where $p = (\sigma_1 + \sigma_2)/2$ and $(\cdot)' = d/dy$. The stress amplitudes S and T at the upper (1) and lower (2) boundaries of the layer, in accordance with equation [6.30], are

$$\begin{aligned} S^1 &= \sum_{k=1}^{4} a_k S_k(h/L), \quad T^1 = \sum_{k=1}^{4} a_k T_k(h/L) \\ S^2 &= \sum_{k=1}^{4} a_k S_k(-h/L), \quad T^2 = \sum_{k=1}^{4} a_k T_k(-h/L) \end{aligned} \qquad [6.33]$$

In the matrix form, the general solution for the upper $(i = 1)$ or lower $(i = 2)$ boundary displacement and traction amplitudes is written as

$$\begin{Bmatrix} U^i \\ V^i \\ S^i \\ T^i \end{Bmatrix} = \begin{bmatrix} U_1^i & U_2^i & U_3^i & U_4^i \\ V_1^i & V_2^i & V_3^i & V_4^i \\ S_1^i & S_2^i & S_3^i & S_4^i \\ T_1^i & T_2^i & T_3^i & T_4^i \end{bmatrix} \begin{Bmatrix} a_1 \\ a_2 \\ a_3 \\ a_4 \end{Bmatrix} \qquad [6.34]$$

6.3.3. *Buckling of a layer system – the transfer matrix technique*

Buckling of a layer system can be analyzed and solved numerically using the solution for a single layer and the transfer matrix technique (Bufler 1965). Figure 6.8 shows a system of n layers of different materials and different initial stresses parallel to the layer axis, with a global coordinate system located at the top layer.

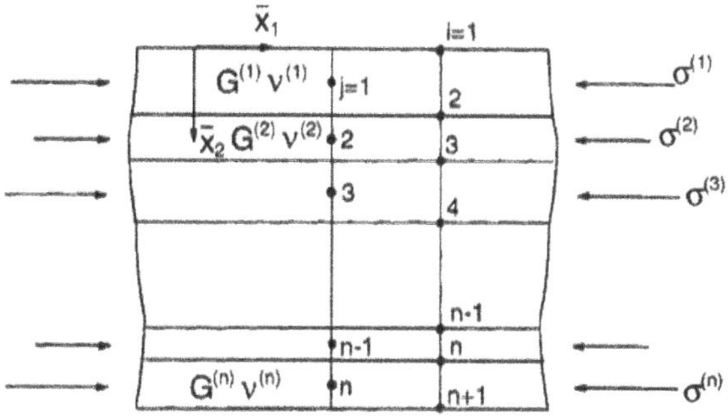

Figure 6.8. *Multilayered medium under compressive initial stresses*

This medium is obtained by superposing adhering layers, each of which can be viewed individually as a single layer. The layers are numbered from 1 to n starting at the top. By assuming perfect adherence at the interfaces, the incremental stresses and displacements are continuous along the interfaces. Under these conditions, the equations for the buckling of the system of layers are derived immediately from the results of section 6.3.2 provided that all local coordinates are expressed in the global coordinate system. From equation [6.34], the amplitude of the incremental stresses and displacements for the ith interface of the jth layer can be assembled in the matrix form as

$$\begin{Bmatrix} U^{ij} \\ V^{ij} \\ S^{ij} \\ T^{ij} \end{Bmatrix} = \begin{bmatrix} U_1^{ij} & U_2^{ij} & U_3^{ij} & U_4^{ij} \\ V_1^{ij} & V_2^{ij} & V_3^{ij} & V_4^{ij} \\ S_1^{ij} & S_2^{ij} & S_3^{ij} & S_4^{ij} \\ T_1^{ij} & T_2^{ij} & T_3^{ij} & T_4^{ij} \end{bmatrix} \begin{Bmatrix} a_1^j \\ a_2^j \\ a_3^j \\ a_4^j \end{Bmatrix} \quad \text{or} \quad \{X^{ij}\} = [F^{ij}]\{A^j\} \qquad [6.35]$$

By requiring continuity of the incremental displacements and tractions at all interfaces, the integration constants of every layer are linked to the integration constants of the top layer as follows

$$\{A^n\} = [F]\{A^1\}, \quad [F] = [F^n]\cdots[F^2], \quad [F^k] = [F^{kk}]^{-1}[F^{k(k-1)}] \qquad [6.36]$$

To formulate the eigenvalue problem, we must consider boundary conditions only at the upper and lower boundary surfaces of the layered medium. As an example, the case of zero tractions at the upper boundary surface ($i = 1, j = 1$) and zero displacements at the lower boundary surface ($i = n+1$, $j = n$) is considered. These boundary conditions can be written in matrix form as

$$\left[Y^1\right]\left\{A^1\right\}=0, \quad \left[Y^n\right]\left\{A^n\right\}=0, \qquad [6.37]$$

where $\left[Y^1\right]$ contains the last two rows of $\left[F^{11}\right]$ and $\left[Y^n\right]$ the first two rows of $\left[F^{(n+1)n}\right]$. By taking into account equation [6.36], the matrix equation [6.37] can be assembled in a homogeneous algebraic system of equations for the integration constants $\left\{A^1\right\}$

$$\begin{bmatrix} \left[Y^n\right] & [F] \\ \left[Y^1\right] & \end{bmatrix}\left\{A^1\right\}=0 \quad \text{or} \quad [Y]\left\{A^1\right\}=0 \qquad [6.38]$$

The resulting homogeneous system of equations has non-trivial solutions in terms of the integration constants involved only if the determinant of the system is zero, i.e. $\det([Y])=0$. This provides an equation whose roots give the corresponding eigenvalues. In this equation, we fix the wavenumber β and we consider it as an equation for the load. When monotonic loading is assumed, then the lowest loading level associated with the least eigenvalue provides the critical buckling load.

6.3.4. *Buckling of layered half-space*

The buckling condition for a homogeneous half space is independent of the wavelength of the considered mode. This is because no length appears in this problem and consequently the various modes corresponding to different wavelengths cannot be differentiated. The introduction of length leads to the selection of a buckling mode. An example is the buckling of a layer on top of a half space due to a horizontal homogeneous strain field, as shown in Figure 6.9(a). Both media are assumed isotropic and compressible with elastic parameters G, ν and G_L, ν_L for the half space and the layer, respectively. Figure 6.9(b) presents the results for the critical buckling stress. The wave number is given as $\beta = 2\pi h/W$, where h is the thickness of the layer and W is the wavelength of the deformation field. For the two limiting cases of vanishingly short wavelength ($\beta \to \infty$) and infinitely large wavelength ($\beta \to 0$), the buckling stress corresponds to that of a homogeneous half space. For short wavelengths, the layer behaves like a half-space, whereas large wavelengths cannot "see" the layer. If the layer is stiffer than the half-space, then there is a dominant buckling mode, the wavelength of which depends on the stiffness contrast G_L/G.

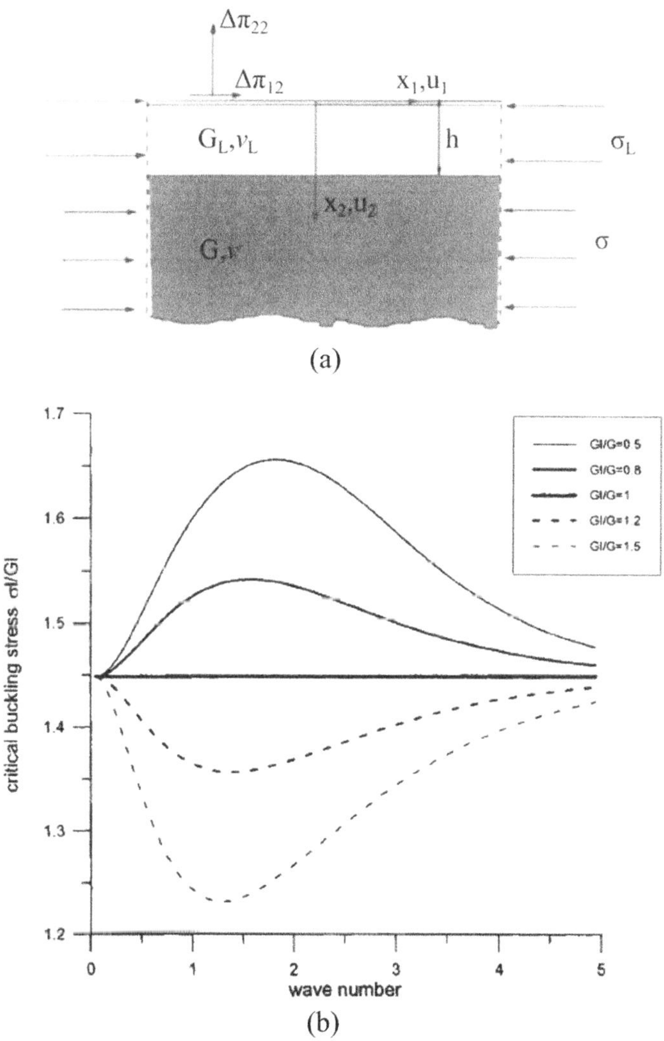

Figure 6.9. *(a) Buckling of a layer on top of a half-space, and (b) critical buckling stress as a function of wavenumber for various values* G_L/G

For viscoelastic behavior for the layer and the half-space, there exists a dominant wavelength for fastest growth (Papamichos 1988). Figure 6.10 plots the critical buckling stress as a function of wavelength β for various rates of growth and relaxation constant $r/r_L = 0.01$ Maxwell-type viscoelastic materials are elastic for fast deformations ($p \rightarrow \infty$) but in slow deformations ($p \rightarrow 0$) exhibit Newtonian viscosity. In all cases, the stiffness of the layer is sufficiently higher than that of the half-space and both are incompressible, i.e. $v = v_L = 0.5$. Figure 6.10 shows that for a given load there exists a wavenumber of fastest rate of growth, i.e. highest value of p. The possibility of buckling with multiple wavelengths is also possible. If, for example, more than one different load is applied, the dominant wavelength of each load will be present and will be superposed. This fact is of great interest in the mechanics of folding and can reveal the stress states that the formations have experienced in the past.

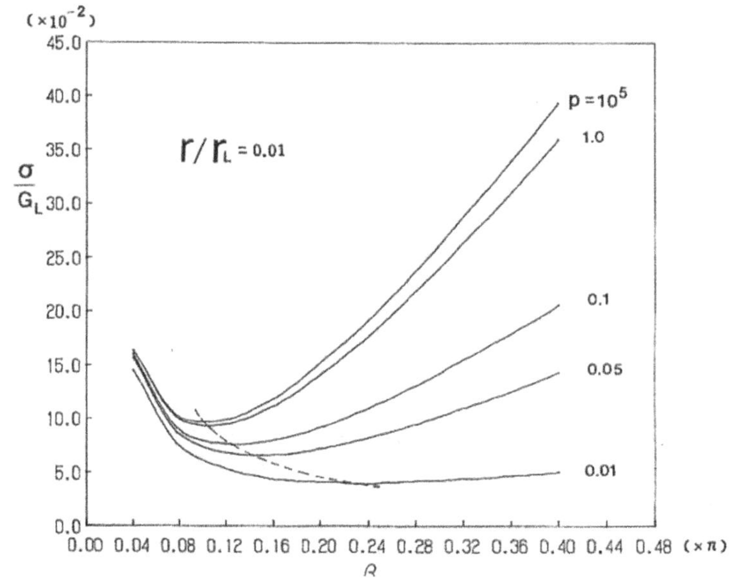

Figure 6.10. *Critical buckling stress of a viscoelastic layer on a viscoelastic half space with relaxation constant ratio $r/r_L = 0.01$ and rate of growth p*

6.4. Axial splitting and spalling

Axial splitting in uniaxial loading, spalling of a free surface and rock bursting are common phenomena in brittle materials, like rock and concrete, under compressive stresses parallel to a free surface. A micromechanically motivated mechanism is presented here that views axial splitting and spalling because of the interaction between surface instabilities and surface parallel Griffith cracks. According to this mechanism, surface instabilities in a uniformly stressed half space produce secondary tensile stresses, which, for material points close to a free surface, remain unbalanced in the direction normal to the surface. These tensile stresses cause latent, surface parallel cracks to open and thus magnify the effect of diffuse bifurcation. Tensile stress concentrations develop at the crack tips resulting in unstable crack growth and finally axial splitting and spalling of the material.

Unstable crack growth and axial splitting correspond to some critical crack surface density (distributed material damage), which manifests itself in the mean spacing of forming spalls. The fractured material becomes much weaker than the intact one due to the lower buckling stress of individual columns. Keeping this model in mind, we may consider that rock bursting is the result of buckling of flaked rock surfaces, the flaking itself being triggered by surface instabilities. To investigate this mechanism, the buckling problem under plane-strain conditions due to a uniform compression of a semiinfinite medium containing surface parallel Griffith cracks is considered. Analytical solutions for this type of problems have been presented by Keer *et al.* (1982), who solved the buckling problem in solids containing a periodic array of coplanar cracks, and Nazarenko (1986), who considered a single crack. Vardoulakis and Papamichos (1991) and Papamichos and Vardoulakis (1992) have developed a numerical boundary element solution based on the displacement discontinuity method for solving the problems of an arbitrary number and/or geometry of surface parallel cracks in elastic, anisotropic media. In the following, the weakening of a medium due to the presence of periodic crack arrays is investigated.

6.4.1. *Buckling of a half-space with surface parallel cracks*

The buckling problem of a half-space containing a single array or periodic arrays of coplanar cracks is solved using a numerical solution for the eigen-displacements of the two crack faces and the free surface. The problem is formulated using the boundary element method for crack problems in semiinfinite anisotropic media under initial horizontal stress (Vardoulakis and Papamichos 1991). The method was constructed by modifying the displacement discontinuity method Crouch and Starfield (1983) to account for the existence of the initial stress field. It is based on the analytical solution to the problem of a constant displacement discontinuity over a finite line segment in a semiinfinite, anisotropic medium. For a line segment, the displacement discontinuity D_i is defined as the difference in displacement between the two sides of the segment. The formulation provides the functions that allow the determination of the displacement and first Piola–Kirchhoff stress increments at any point in a half-space due to a displacement discontinuity increment D_i over the line segment.

Figure 6.11 shows the deformed crack shape and the free surface due to buckling of a single crack for three values of the ratio h/a, where h is the distance of the crack from the free surface and a the crack half length. At each point along the crack, the relative displacements between the two faces are the eigen-displacement discontinuities of the crack. The two sides of the crack deform by different amounts and, as expected, the disturbed region of the free surface localizes with the crack approaching the free surface. For a periodic crack array (Figure 6.12), the infinitely extended crack array is approximated with 10 25-element cracks. Such an approximation is considered satisfactory since it was found that an increase in the number of cracks does not affect significantly the results. Figure 6.13 presents the results where the normalized critical buckling stress $\xi = -\sigma/G$ is plotted as a function of the dimensionless depth h/L and the dimensionless crack length α/L, where L is the spacing between the cracks. Poisson's ratio has the value of 0.3 in all calculations. This is the solution for the first bifurcation mode, which is antisymmetric, and corresponds to cracks tending to open and close alternately, as shown in Figure 6.13(a). The numerical results of the advanced solution for the symmetric mode of all open cracks are shown in Figure 6.13(b). Higher stress is required for the all open mode, and in fact, the critical stress in this case increases with decreasing spacing.

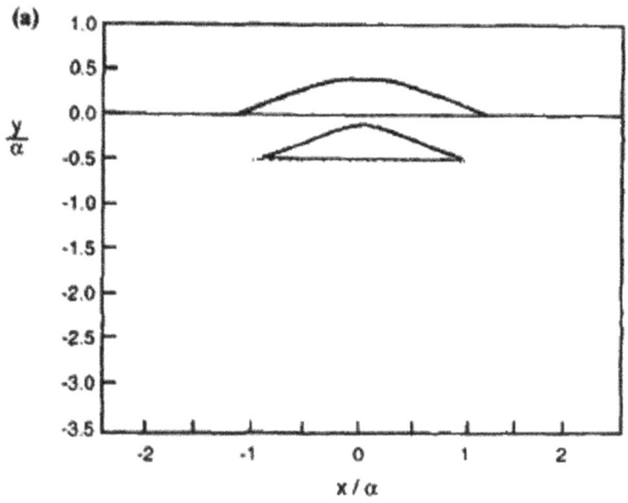

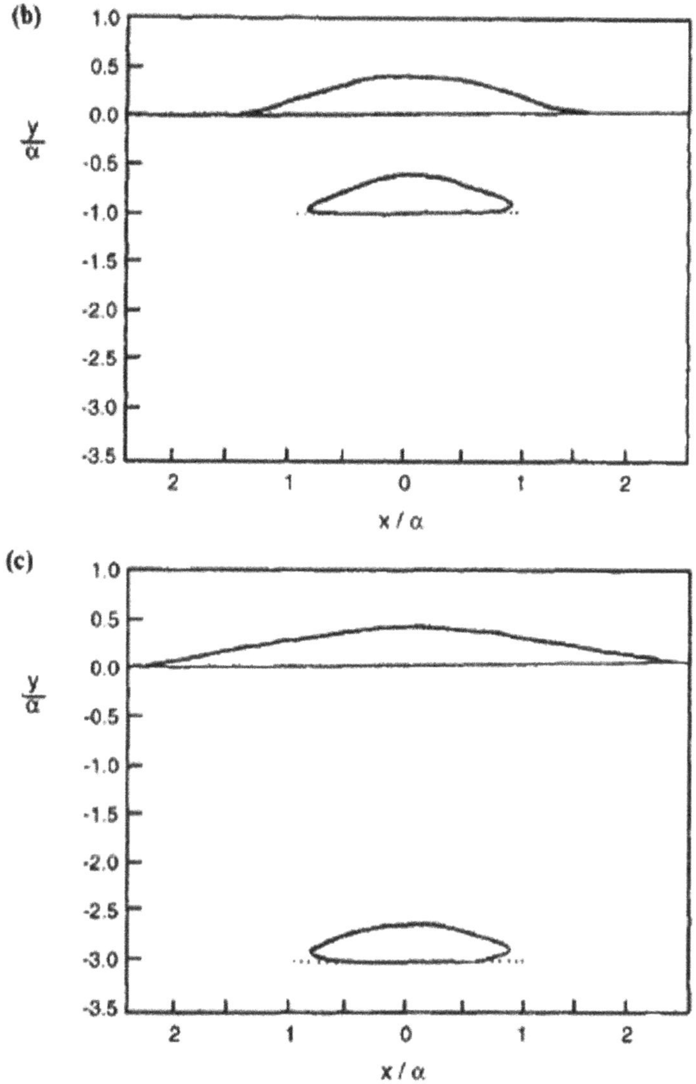

Figure 6.11. *Deformed shape of crack and free surface due to buckling of the half space for h/α = (a) 0.5, (b) 1 and (c) 3 (boundary element solution, ν = 0.3)*

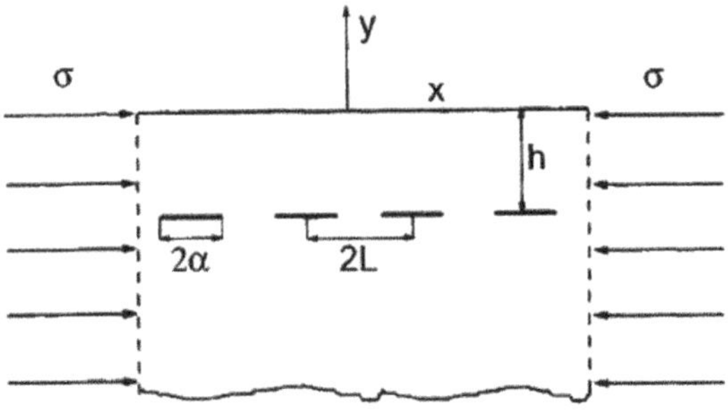

Figure 6.12. *Half-space with periodic array of collinear cracks compressed by uniform horizontal stress*

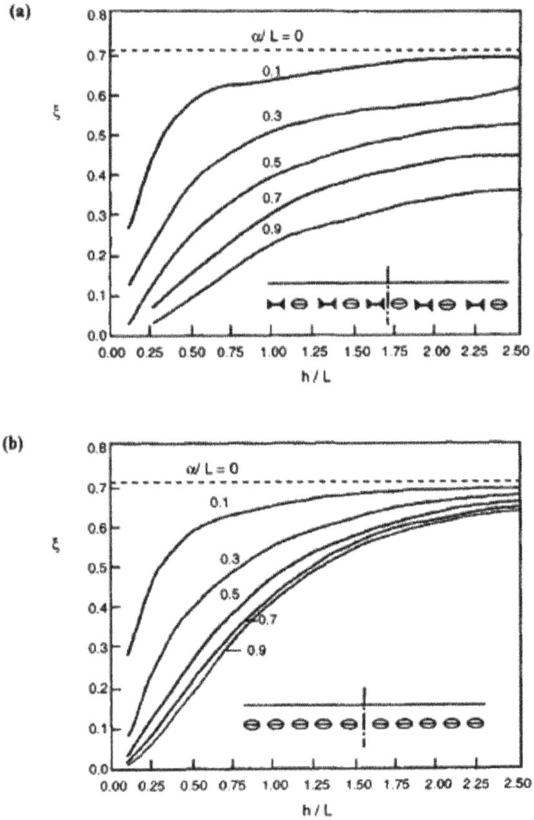

Figure 6.13. *Critical buckling stress* $\xi = -\sigma/G$ *of a half-space with a periodic array of collinear cracks for (a) alternate open mode and (b) all open mode (ν = 0.3)*

Once the periodic crack array problem is solved, additional crack arrays can be placed in the half-space to investigate the weakening of the medium due to the presence of additional crack arrays. The crack arrays were approximated with 10 20-element cracks. For $\alpha/L = 0.5$, h/a, between 0.5 and 5 and the lowest buckling mode, the second crack array reduces the critical buckling stress by approximately 6%, while the third array reduces it by an additional 1.5%. This means that the buckling load is mainly determined by the cracks closer to the free surface.

Furthermore, the eigen-displacements at the critical state show that the buckling of the half-space affects primarily the crack array closer to the free surface, suggesting a progressive spalling behavior that starts close to the free surface and subsequently progresses deeper into the material. This spalling advancement was observed experimentally, for example, in Berea sandstone specimens tested in the surface instability detection apparatus (Papamichos *et al.* 1994).

6.5. Conclusion

A three-dimensional Cosserat nonlinear elastic, Mohr–Coulomb elastic-plastic model was formulated and calibrated based on generalizations of classical continuum models. The model was calibrated on Red Wildmoor sandstone data and applied to the bifurcation problem of borehole failure. The Cosserat internal length was calibrated on back analysis of borehole failure data from HC tests. Once calibrated, it can be used for forward predictions under different loading paths and fluid flows. The lateral borehole failure mode was analyzed. This mode prevails when the tangential stress at the borehole is the largest compressive principal stress.

In elastic, semiinfinite media containing surface parallel Griffith cracks, the resistance to crack propagation diminishes as the critical stress for surface instability is approached. In particular, the effective stress intensity factor increases rapidly near the surface instability condition and therefore we may assume that any crack at this stress level will propagate. Furthermore, the critical buckling stress of the media under consideration decreases rapidly with increases relative to the depth and length of the cracks. This suggests that in the case of a constant far-field stress (e.g. load control conditions in a laboratory setting), cracks close to the free surface will tend to become unstable once they begin to grow, leading to spalling of the surface. For stable crack growth, a reduction in the value of the far-field stress is necessary (e.g. displacement control conditions). Thus, a material with surface parallel cracks could be modeled as a softening material with respect to the crack length and spacing.

6.6. Acknowledgments

This work was supported by the Research Council of Norway and Aker BP, Anadarko, DEA Norway and Hess [Petromaks 2 program, Project number 268159, 2017], and the sponsors of SINTEF's Sand production project, ConocoPhillips, Total, Eni, Statoil and Shell.

6.7. References

Biot, M.A. (1957). Folding instability of a layered viscoelastic medium under compression. *Proc. Roy. Soc. Lond., Ser. A*, 242, 444–454.

Biot, M.A. (1961). Theory of folding of stratified viscoelastic media and its implications in tectonics and orogenesis. *Geol. Soc. Am. Bull.*, 72, 1595–1620.

Biot, M.A. (1965). *Mechanics of Incremental Deformations*. John Wiley, New York.

Bufler, H. (1965). Die Drucksstabilität rechteckiger Verbundplatten. *Ingenieur Archiv*, 34, 109–128.

Crouch, S.L. and Starfield, A.M. (1983). *Boundary Element Methods in Solid Mechanics*. Allen & Unwin, London.

Haimson, B.C. and Song, I. (1993). Laboratory study of borehole breakouts in Cordova Cream: A case of shear failure mechanism. *International Journal for Rock Mechanics and Mining Sciences & Geomechanics Abstracts*, 30(7), 1047–1056.

Keer, L.M., Nemat-Nasser, S., Oranratnachai, A. (1982). Surface instability and splitting in compressed brittle elastic solids containing crack arrays. *Journal of Applied Mechanics*, 49, 761–767.

Maury, V. (1993). An overview of tunnel, underground excavation and borehole collapse mechanisms. In *Comprehensive Rock Engineering, Vol 4: Principles, Practice & Projects*, J.A. Hudson (ed.). Pergamon Press, Oxford, 369–412.

Nazarenko, V.M. (1986). Theory of fracture of materials in compression along near surface cracks under plane-strain conditions. *Soviet Appl. Mech.*, 22(12), 1192–1199.

Papamichos, E. (1988). Numerical analysis of buckling phenomena in layered rock media. Thesis, University of Minnesota, Minneapolis.

Papamichos, E. (2010a). Analysis of borehole failure modes and pore pressure effects. *Computers and Geotechnics*, 37, 141–152.

Papamichos, E. (2010b). Borehole failure analysis in a sandstone under anisotropic stresses. *Intl Journal for Numerical and Analytical Methods in Geomechanics*, 34, 581–603.

Papamichos E. and Vardoulakis I. (1989). The coupling effect of surface instabilities and surface-parallel Griffith cracks in rock. In *Intl Symp, Rock Mechanics and Rock Physics at Great Depth*, V. Maury and D. Fourmaintraux (eds). Pau, France, 1, 481–487.

Papamichos, E., Labuz, J.F., Vardoulakis, I. (1994). A surface instability detection apparatus. *Rock Mechanics Rock Engineering*, 27(1), 37–56.

Papamichos, E., Tronvoll, J., Skjærstein, A., Unander, T.E. (2010). Hole stability of Red Wildmoor sandstone under anisotropic stresses and sand production criterion. *Journal of Petroleum Sciences and Engineering*, 72, 78–92.

Papanastasiou, P. and Vardoulakis, I. (1992). Numerical treatment of progressive localization in relation to borehole stability. *International Journal for Numerical and Analytical Methods in Geomechanics*, 16, 389–424.

Vardoulakis I. and Papamichos E. (1991). Surface instabilities in elastic anisotropic media with surface-parallel Griffith cracks. *Intl Journal for Rock Mechanics and Mining Sciences & Geomechanics Abstracts*, 28, 2/3, 163–173. Available at: http://dx.doi.org/10.1016/0148-9062(91)92164-T.

Vardoulakis, I. and Sulem, J. (1995). *Bifurcation Analysis in Geomechanics*. Blackie Academic & Professional, Glasgow.

Vardoulakis, I., Sulem, J., Guenot, A. (1988). Borehole instabilities as bifurcation phenomena. *International Journal of Rock Mechanics and Mining Sciences & Geomechanics Abstracts*, 25(3), 159–170.

Zienkiewicz, O.C., Taylor, R.L. (1989). *The Finite Element Method, Vol. 2, Solid and Fluid Mechanics Dynamics and Non-linearity*, 4th edition. McGraw-Hill, London.

7

Numerical Modeling of Multiphysics Couplings and Strain Localization

Frédéric COLLIN[1], Panagiotis KOTRONIS[2] and Benoît PARDOEN[3]

[1] *ArGEnCo, University of Liège, Belgium*

[2] *GeM, Ecole Centrale de Nantes, France*

[3] *iMMC, Université catholique de Louvain, Belgium*

Rupture in geomaterials is often preceded by a localization of the deformations within thin bands. The strain localization is therefore an important process, which has been studied both experimentally and theoretically. This chapter summarizes the main observations on localized phenomena and proposes numerical tools to characterize localization processes. To deal with interactions occurring between the different phases of porous media, a regularization technique based on the second gradient model has been extended to multiphysic couplings.

7.1. Introduction

Since the material behavior and rupture are of importance regarding the design of geotechnical works for which the materials can be subjected to strong solicitations, failure has been widely investigated in geomechanics. Experimental observations on geomaterials clearly indicate the appearance of localized ruptures (Desrues 1984). Theoretically, the concept of a rupture surface is one of the oldest cases of material localized failure and was already used in the design of works and structures a few

Instabilities Modeling in Geomechanics,
coordinated by Ioannis STEFANOU and Jean SULEM.

centuries ago (Coulomb 1773). In some cases, a diffuse mode of failure can also be observed and it corresponds to homogeneous failure in laboratory tests (Khoa *et al.* 2006). Nowadays, it is commonly assumed that localized deformation and damage can appear in materials prior to the rupture in many situations. In rock material, a stress redistribution can engender damage that can first be diffused and then localized. Once the damage threshold is reached, microcracks initiate, then grow, accumulate and propagate within the material. If the microcracks coalesce, the distributed damage can further lead to strain localization and to the initiation of interconnected fractures by the onset of macrocracks, which provokes a sudden material rupture (Diederichs 2003).

The fracturing process instigates discontinuities in the material that can be represented theoretically and numerically by various approaches. Two main categories exist: the continuous and discrete descriptions of fracture. The continuous description includes material damage and strain localization, while the discrete description explicitly represents cracks. In fracture mechanics, the different fractures can be in tensile or opening mode (mode I), in sliding shear mode (mode II), in tearing shear mode (mode III), or in mixed mode (modes I–II) (Jenq and Shah 1988). The modeling of shear strain localization is a continuous approach that does not explicitly reproduce fractures and their discontinuities. Nevertheless, it generally induces the appearance of shear bands and a non-uniform strain distribution that may engender a displacement discontinuity between the materials located on the two sides of a shear band.

Furthermore, soils and rocks are porous materials, where the porous volume is filled with one or several fluids (water, gas, oil, etc.). The general behavior of the medium depends not only on the skeleton response (solid phase) to a given loading path, but also on the interactions occurring between the different phases of the medium. Capillary effects, temperature variations and chemical reactions induce specific behaviors, which have to be modeled by multiphysical constitutive laws. The numerical tools for modeling strain localization problems thus have to be extended to this multiphysic context to deal with applications related, for instance, to nuclear waste disposal and concrete behavior under severe loading. Then, new questions arise concerning the interactions between the localization process and the physical process (like liquid diffusion, for example). The answer to these new questions can only be given by experimental evidences.

Section 7.2 of this chapter summarizes the main observations on the localization phenomenon, coming from experimental results. Section 7.3 describes the regularization methods used to properly model the strain localization process. Section 7.4 presents the description of the coupled second gradient model in saturated conditions. A biaxial compression test is modeled in order to show the ability of the second gradient model to correctly represent the post-peak behavior. In section 7.5, the second gradient model is extended to unsaturated conditions. The example of a gallery excavation is proposed in section 7.6 to evidence the influence of hydromechanical couplings in saturated and unsaturated conditions on the strain localization process. The chapter ends with some conclusions in section 7.7.

7.2. Experimental evidences of strain localization

Strain localization is frequently observed prior to material rupture. Starting from a homogeneous deformation state, strain localization consists of a brutal accumulation of strain in a limited zone that can lead to cracks and failure (rupture lines). In geomaterials like soils and rocks, it is often considered as a shear strain accumulation in band mode (Desrues 2005). Nevertheless, the type of localization may be of different nature for other materials.

Geomaterials have low tensile strength, thus tensile rupture is arduous to characterize. However, plenty of small-scale compression laboratory tests are dedicated to strain localization (Alshibli *et al.* 2003; Finno *et al.* 1996, 1997; Han and Drescher 1993; Vardoulakis *et al.* 1978) and allow us to characterize the compression material behavior up to the rupture. They are generally realized on an axisymmetric triaxial or plane-strain biaxial compression apparatus and involve special techniques, such as stereophotogrammetry (Desrues 1984; Desrues and Viggiani 2004), X-ray microtomography and three dimensional digital image correlation (Lenoir *et al.* 2007) to study the evolution of the strain localization process. The advantage of biaxial compression experiments is that the localization process is clearly evidenced, whereas it can remain hidden inside the sample in triaxial compression tests.

Under a compressive regime, rupture is governed essentially by shear failure and these experimental studies generally highlight shear strain localization in band mode (Desrues 2005). It is commonly accepted that the shear band establishment corresponds to a peak stress in the stress–strain global response curve of the specimen (Desrues 2005; Mokni and Desrues 1999).

The experimental localization studies mostly analyze the behavior of sand and only a few are actually available on rocks (Bésuelle *et al.* 2000). Analyzing the formation of fractures and strain localization bands in rocks is quite challenging due to their high resistance and brittle behavior (quasi-brittle material), thence the development of appropriate apparatus designed to test this type of material is necessary (Desrues and Viggiani 2004).

7.3. Regularization methods

The next step is to define an appropriate and robust method that allows us to properly model strain localization and shear banding with the finite element method, leading finally to rupture in the localized mode. Local descriptions of failure with classical finite element methods are not efficient in the reproduction of strain localization because they suffer from a mesh dependency (to mesh size and orientation) as indicated by Collin *et al.* (2009b), Pietruszczak and Mróz (1981), Wu and Wang (2010), Zervos *et al.* (2001b). This pathological problem is due to the properties of the underlying mathematical problem.

The dependence to the finite element discretization can be solved by employing a proper regularization technique. Such a method has to introduce an internal length scale in the problem to correctly model the post-localization behavior. Two principal categories of enhanced models exist: one consists of the enrichment of the constitutive law with, for instance, non-local (Bazant *et al.* 1984; Guy *et al.* 2012; Pijaudier-Cabot and Bazant 1987; Peerlings *et al.* 1996b) or gradient plasticity (Aifantis 1984; de Borst and Mühlhaus 1992; Peerlings *et al.* 1996a), and the other consists of approaches in the enrichment of the continuum kinematics with microstructure effects. For this second category, the microkinematics are characterized at microscale in addition to the classical macrokinematics (Cosserat and Cosserat 1909; Germain 1973; Mindlin 1964; Toupin 1962). However, enhanced models restore mesh objectivity but not the uniqueness of the solution.

7.3.1. *Enrichment of the constitutive law*

In this approach, an internal length scale is introduced at the level of the constitutive model. Advanced analyses of localization phenomena have indicated that constitutive equations with internal length are one solution to model a strain localized pattern properly.

The internal length scale is introduced by developing a non-local definition of internal variables involved in the material behavior. The non-local variable $\mathring{\mathrm{v}}$ at a material point x_i can be defined as an averaging value of the local variable v in a considered region Ω near that point (Peerlings *et al.* 2001; Pijaudier-Cabot and Bazant 1987), as illustrated in Figure 7.1.

The non-local integral method gives:

$$\mathring{\mathrm{v}}(x_i) = \frac{1}{V} \int_{\Omega} \Psi \, \mathrm{v}(y_i) \, d\Omega, \qquad [7.1]$$

$$V = \int_{\Omega} \Psi \, d\Omega, \qquad [7.2]$$

where x_i is the coordinate vector of the material point where the non-local variable is considered, Ω is a representative volume centered in x_i, y_i is the coordinate vector of the infinitesimal volume $d\Omega$ and Ψ is a weight function scaling $\mathring{v}$ to $\mathring{v} = v$ for a homogeneous distribution of the variable. It is generally defined with a Gaussian distribution:

$$\Psi = \frac{1}{(2\pi)^{3/2}\, l_c^3}\, exp\left(-\frac{||x_i - y_i||^2}{2\, l_c^2}\right) \qquad [7.3]$$

which depends on the distance $||x_i - y_i||$ and on a characteristic length parameter l_c. This length parameter, or this internal length scale, defines the material volume that significantly contributes to the non-local variable and is consequently related to the microstructure.

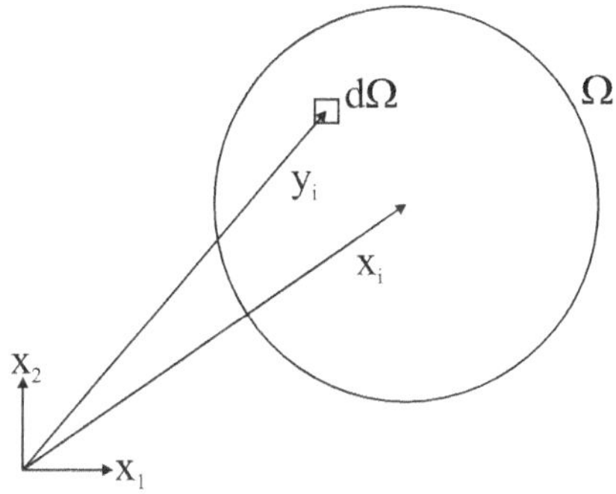

Figure 7.1. *Non-local approach on a representative material volume*

The regularized variable can also be defined explicitly from the local variable $v(x_i)$ and its gradient. In his pioneering works, Aifantis (1984) introduced such a gradient in the constitutive equations. The explicit gradient formulation is given as

$$\mathring{v} = v + \bar{l}\, \frac{\partial^2 v}{\partial x_i \partial x_i} \qquad [7.4]$$

where the dependence of v and $\mathring{v}$ on the coordinate vector x_i is dropped for simplicity and $\bar{l}$ has the dimension of length squared, so $\sqrt{\bar{l}}$ can be related to the internal length scale l_c introduced to regularize the model. Because the gradient term is a local quantity, the spatial interaction of the material points located in the vicinity of $\mathring{v}$ is infinitesimal and the explicit gradient model is therefore local. This is a main difference with the non-local integral formulation of equation [7.1] where the interaction distance is finite and related to the weight function. Moreover, the explicit gradient formulation can be derived from the non-local integral formulation by introducing the gradient of the internal variable, expanding the local variable $\mathrm{v}(y_i)$ into a Taylor series (Bazant *et al.* 1984; Lasry and Belytschko 1988; Peerlings *et al.* 1996a), using the weight function definition of equation [7.3] and neglecting the terms above the second order (approximation).

The definition of equation [7.4] is less suitable in the context of numerical analyses, such as the finite element method, because of the explicit dependence of $\mathring{\mathrm{v}}$ with its local (second) gradient. This dependence leads to a continuity requirement for the internal variable, which has to be a continuously differentiable function (class C^1 function whose derivative is continuous). To avoid this drawback, an alternative implicit gradient formulation, introducing an approximation of equation [7.1] similar to equation [7.4], can be expressed as follows (Peerlings *et al.* 1996a, 2001):

$$\mathring{\mathrm{v}} - \bar{l}\,\frac{\partial^2 \mathring{\mathrm{v}}}{\partial x_i \partial x_i} = \mathrm{v} \qquad [7.5]$$

and enables a continuous definition of v (class C^0 function). For the implicit gradient model, the non-local internal variable is an additional unknown, which is a solution of the Helmoltz differential equation [7.5]. The solution of this equation can only be found provided that the additional boundary condition on $\mathring{\mathrm{v}}$ is specified. The following condition is usually assumed (Lasry and Belytschko 1988):

$$\frac{\partial \mathring{\mathrm{v}}}{\partial x_i}\, n_i = 0, \qquad [7.6]$$

where n_i is the normal unit vector to the external boundary. This condition enables $\mathring{\mathrm{v}} = \mathrm{v}$ for homogeneous distribution. In contrast to the explicit formulation, the non-local variable $\mathring{\mathrm{v}}$ is implicitly given as the solution of equations [7.5] and [7.6], and the spatial interaction has a finite distance that implies a non-local character. The solution is of the same form of the non-local equation [7.1] with $\Psi = Gr$ and $V = 1$, Gr being the Green's function (Zauderer 1989):

$$\mathring{\mathrm{v}}(x_i) = \int_\Omega Gr\ \mathrm{v}(y_i)\, d\Omega \qquad [7.7]$$

$$Gr = \frac{1}{4\pi\,\bar{l}\,||x_i - y_i||}\ exp\left(-\frac{||x_i - y_i||}{\sqrt{\bar{l}}}\right) \qquad [7.8]$$

The implicit gradient model is therefore a special case of the non-local model.

Non-local quantities as well as the gradient of internal variables can finally be introduced in constitutive models. Among other authors, Bazant *et al.* (1984) and Pijaudier-Cabot and Bazant (1987) proposed a family of constitutive models derived from the non-local damage theory, in which a non-local internal variable is used instead of the local one. For instance, a non-local damage energy release rate obtained by equation [7.1] is introduced in the damage loading function. Other variables such as non-local equivalent strain are usually used in a damage model (Peerlings *et al.* 2001).

7.3.2. *Enrichment of the kinematics*

The previous approaches (enrichment of the constitutive law) introduce the effect of microstructure with non-local or gradient terms but the microstructure itself is not explicitly defined. To this end, the classical kinematics of a continuous medium can be enriched with additional description of the microstructure kinematics, leading to a microstructure continuum medium also called enriched medium.

For a classical continuous medium, a material particle of volume Ω is defined at macroscopic scale by its (macro) displacement field u_i. The classical kinematic fields are the macrodeformation field:

$$F_{ij} = \frac{\partial u_i}{\partial x_j} \qquad [7.9]$$

corresponding to the gradient of the displacement field, the macrostrain field:

$$\varepsilon_{ij} = \frac{1}{2}\left(F_{ij} + F_{ji}\right) \qquad [7.10]$$

corresponding to the symmetric part of F_{ij}, and the macrorotation field:

$$r_{ij} = \frac{1}{2}\left(F_{ij} - F_{ji}\right) \qquad [7.11]$$

corresponding to the antisymmetric part of F_{ij}. Their rate forms are also commonly used; the velocity gradient field:

$$L_{ij} = \frac{\partial \dot{u}_i}{\partial x_j} \qquad [7.12]$$

the strain rate field:

$$\dot{\varepsilon}_{ij} = \frac{1}{2}\left(L_{ij} + L_{ji}\right) \qquad [7.13]$$

and the spin rate field:

$$\omega_{ij} = \frac{1}{2}\left(L_{ij} - L_{ji}\right) \tag{7.14}$$

The first and most famous enhanced model was developed by the Cosserat brothers (Cosserat and Cosserat 1909) who introduced local rotation degrees of freedom r_i^c in addition to the displacements of classical continua u_i (Figure 7.2). The Cosserat (or micropolar) elastic continuum theory is mostly suitable for the kinematic description of granular materials. Accordingly, additional kinematic fields are introduced (Vardoulakis and Sulem 1995). The deformation due to the particle rotation, also called microrotation (antisymmetric tensor), becomes:

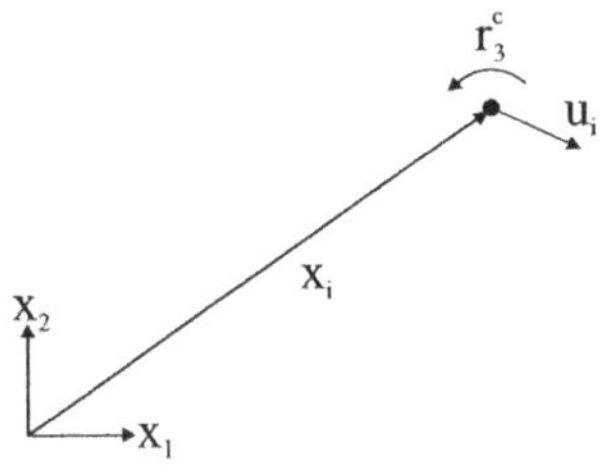

Figure 7.2. *Kinematic degrees of freedom of the Cosserat elastic continuum theory*

$$r_{ij}^c = e_{ijk}\, r_k^c, \tag{7.15}$$

where e_{ijk} is the alternating tensor, and the gradient of the particle rotation, also called curvature:

$$K_{ij}^c = \frac{\partial r_i^c}{\partial x_j} \tag{7.16}$$

A relative strain is deduced as the difference between macrodeformation and microrotation:

$$\bar{\varepsilon}_{ij} = F_{ij} - r_{ij}^c \tag{7.17}$$

whose symmetric part coincides with the macrostrain ε_{ij} and its antisymmetric part with the difference between the macro- and microrotation $r_{ij} - r_{ij}^c$. The latter characterizes the relative rotation of a material point with regard to the rotation of its neighborhood. A couple stress (torques) tensor associated with the rotations is thus added, introducing bending and torsion at the material point. This results in a moment equilibrium equation involving the couple stresses that come in addition to the classical (local) momentum balance equation involving the stress field σ_{ij}. Moreover, supplementary elastic constants are considered in the constitutive equations, which consist of internal length scale parameters related to the microstructure (Vardoulakis and Sulem 1995).

In the 1960s, Toupin (1962) and Mindlin (1964) defined materials with microstructure. A macrovolume Ω is composed of smaller microscale particles that can be represented by a microvolume Ω^m, embedded in the material volume Ω (Figure 7.3). A microdisplacement field u_i^m is defined independently of the macrodisplacement u_i and its gradient leads to a microdeformation field:

$$v_{ij} = \frac{\partial u_i^m}{\partial x_j} \qquad [7.18]$$

which is homogeneous in the microvolume Ω^m but non-homogeneous in the macrovolume Ω.

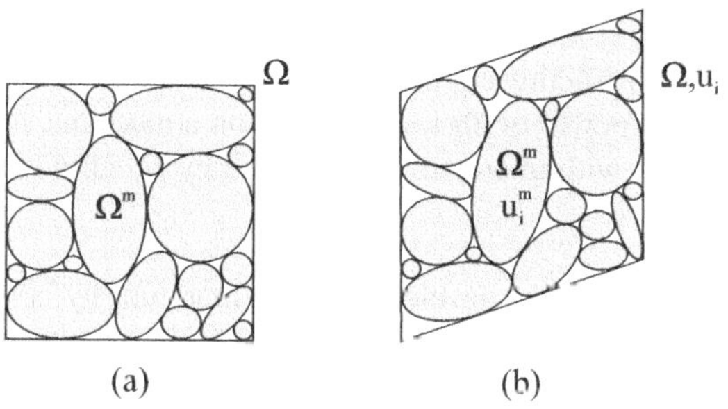

Figure 7.3. *Kinematics of a microstructure continuum: (a) initial configuration and (b) configuration after external solicitation with relative displacement of the microstructure*

The symmetric and antisymmetric parts of v_{ij} correspond to the microstrain and microrotation:

$$\varepsilon_{ij}^m = \frac{1}{2}\left(v_{ij} + v_{ji}\right) \qquad [7.19]$$

$$r_{ij}^m = \frac{1}{2}\left(v_{ij} - v_{ji}\right) \qquad [7.20]$$

with the microrotation corresponding to the rotation components of the Cosserat model $r_{ij}^m = r_{ij}^c$ (equation [7.15]). The Cosserat model is in fact a particular case of a microstructure medium. Moreover, the micro second gradient is defined as:

$$h_{ijk} = \frac{\partial v_{ij}}{\partial x_k} = \frac{\partial^2 u_i^m}{\partial x_j \partial x_k} \qquad [7.21]$$

The relative deformation of the microstructure is defined as the difference between the macro- and the microdeformation fields:

$$\bar{\varepsilon}_{ij} = F_{ij} - v_{ij} \qquad [7.22]$$

whose symmetric part coincides with the difference between the macro- and the microstrain $\varepsilon_{ij} - \varepsilon_{ij}^m$ and its antisymmetric part with the difference between the macro- and microrotation $r_{ij} - r_{ij}^m$. Similar to the Cosserat's continuum description, additional stresses are introduced: the microstress, an additive stress field associated with the microstructure and the double stress.

Later, Germain (1973) introduced the virtual power principle to provide a global framework for the microstructure continuum formulation. This principle states that, by equilibrium, the virtual power of all forces acting on a mechanical system is null. In the following, materials with microstructure defined by Mindlin (1964) and Germain (1973) will be considered.

A large panel of models are developed in the literature by adding mathematical constraints to microstructure media. Among them, the second gradient model developed in Grenoble (Chambon *et al.* 1998, 2001a) will be most particularly developed hereafter. Yet, the following conclusions can be generalized with other regularization techniques.

7.4. Coupled local second gradient model for microstructure saturated media

The coupled local second gradient model is developed for enriched continua including microstructure effects (Chambon *et al.* 2001a). This model was extended from monophasic to biphasic porous media (solid and fluid) by Collin *et al.* (2006) to highlight the possible interactions of the fluid (liquid water) with the strain localization process and with the internal length introduced by the model. The developments proposed by Collin *et al.* (2006) are recalled in this section. They account for a medium with incompressible solid grains under saturated and isothermal conditions. The solid and fluid phases are considered as immiscible and phase changes, like evaporation and dissolution, and are therefore not taken into account.

As for a classical continuum, the material is considered as a porous medium and the balance equations are based on mixture theories. The unknowns of the coupled problem are the macrodisplacement u_i, the microdeformation field v_{ij} (or the microdisplacement field u_i^m by equation [7.18]) and the pore water pressure p_w. An additional unknown field of Lagrange multipliers λ_{ij} will be added for the finite element method implementation.

7.4.1. *Balance equations for microstructure poromechanics*

7.4.1.1. *Local second gradient model for a monophasic medium*

The class of virtual kinematics introduced in the virtual work principle for the classical kinematic theory can be extended in the framework of a microstructure continuum theory by adding a description of the microstructure kinematics to the classical displacement field. According to equation [7.18], the kinematics at microscale are described by a microkinematic gradient field v_{ij}. With respect to classical continuum mechanics, additional terms are added in the internal virtual work of a given body (Germain 1973). The following expression holds for any virtual quantities:

$$W_{int}^{*} = \int_{\Omega} \left(\sigma_{ij}\, F_{ij}^{*} - \tau_{ij}\, \bar{\varepsilon}_{ij}^{*} + \Sigma_{ijk}\, h_{ijk}^{*}\right) d\Omega, \qquad [7.23]$$

where $\bar{\varepsilon}_{ij}^{*}$ is the virtual relative deformation of the microstructure:

$$\bar{\varepsilon}_{ij}^{*} = F_{ij}^{*} - v_{ij}^{*} \qquad [7.24]$$

where τ_{ij} is an additional stress associated with the microstructure also called the microstress, $h_{ijk}^{*} = \frac{\partial v_{ij}^{*}}{\partial x_k}$ is the virtual micro second gradient and Σ_{ijk} is the double stress dual of h_{ijk}^{*}, which needs an additional constitutive law introducing the internal length scale. The external virtual work can be defined as follows:

$$W_{ext}^{*} = \int_{\Omega} \rho\, g_i\, u_i^{*}\, d\Omega + \int_{\Gamma_\sigma} \left(\bar{t}_i\, u_i^{*} + \overline{P}_{ij}\, v_{ij}^{*}\right) d\Gamma, \qquad [7.25]$$

where $\overline{P}_{ij}$ is an additional external double surface traction acting on a part Γ_T of the boundary Γ (Figure 7.4) and $\Gamma_\sigma = \{\Gamma_t \cup \Gamma_T\}$ regroups the classical and additional external solicitations. The virtual work principle assumes the equality between internal and external virtual works and leads to the weak form of the momentum balance equation:

$$\int_{\Omega} \left(\sigma_{ij}\, F^{*}_{ij} - \tau_{ij}\, \left(F^{*}_{ij} - v^{*}_{ij}\right) + \Sigma_{ijk}\, h^{*}_{ijk}\right) d\Omega$$

$$= \int_{\Omega} \rho\, g_i\, u^{*}_i\, d\Omega + \int_{\Gamma_\sigma} \left(\bar{t}_i\, u^{*}_i + \overline{P}_{ij}\, v^{*}_{ij}\right) d\Gamma \qquad [7.26]$$

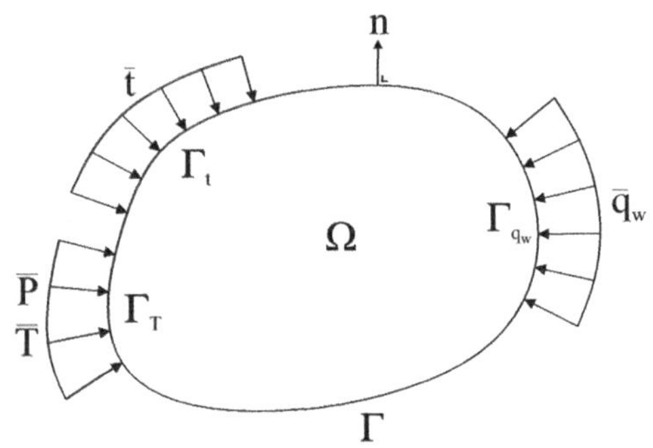

Figure 7.4. *Material system with current configuration Ω and boundary conditions for the second gradient model*

The local equilibrium equations are formulated for the macro- and the microquantities; the local momentum balance equations are:

$$\frac{\partial\left(\sigma_{ij} - \tau_{ij}\right)}{\partial x_j} + \rho\, g_i = 0 \qquad [7.27]$$

$$\frac{\partial \Sigma_{ijk}}{\partial x_k} - \tau_{ij} = 0 \qquad [7.28]$$

and the boundary conditions are:

$$\bar{t}_i = \left(\sigma_{ij} - \tau_{ij}\right) n_j \qquad [7.29]$$

$$\overline{P}_{ij} = \Sigma_{ijk}\, n_k \qquad [7.30]$$

The boundary conditions for the mixture are also enriched with microstructure effects, which leads to non-classical boundary conditions.

In the specific case of the local second gradient model used in the following, a kinematic constraint is added in order to obtain a local second gradient continuum medium. No relative deformation of the microstructure is assumed $\bar{\varepsilon}_{ij} = 0$, meaning that the microkinematic gradient is equal to the macrodeformation:

$$v_{ij} = F_{ij} \quad [7.31]$$

As a result:

$$v^*_{ij} = F^*_{ij} \quad [7.32]$$

for the virtual fields. Therefore, the principle of virtual work can be rewritten as follows:

$$\int_{\Omega} \left(\sigma_{ij} \frac{\partial u^*_i}{\partial x_j} + \Sigma_{ijk} \frac{\partial^2 u^*_i}{\partial x_j \partial x_k} \right) d\Omega = \int_{\Omega} \rho\, g_i\, u^*_i\, d\Omega + \int_{\Gamma_\sigma} \left(\bar{t}_i\, u^*_i + \overline{T}_i\, Du^*_i \right) d\Gamma \quad [7.33]$$

where $\overline{T}_i$ is the additional external double force per unit area on Γ_T (Figure 7.4) and the notation Da denotes the normal derivative of any quantity a:

$$Da = \frac{\partial a}{\partial x_i}\, n_i \quad [7.34]$$

with:

$$Du^*_i = \frac{\partial u^*_i}{\partial x_j}\, n_j = F^*_{ij}\, n_j = v^*_{ij}\, n_j \quad [7.35]$$

The local momentum balance equation reads:

$$\frac{\partial \sigma_{ij}}{\partial x_j} - \frac{\partial^2 \Sigma_{ijk}}{\partial x_j \partial x_k} + \rho\, g_i = 0 \quad [7.36]$$

and the boundary conditions are:

$$\bar{t}_i = \sigma_{ij}\, n_j - n_k\, n_j\, D\Sigma_{ijk} - \frac{D\Sigma_{ijk}}{Dx_k}\, n_j - \frac{D\Sigma_{ijk}}{Dx_j}\, n_k + \frac{Dn_l}{Dx_l}\, \Sigma_{ijk}\, n_j\, n_k - \frac{Dn_j}{Dx_k}\, \Sigma_{ijk}, \quad [7.37]$$

$$\overline{T}_i = \overline{P}_{ij}\, n_j = \Sigma_{ijk}\, n_j\, n_k, \qquad [7.38]$$

where $\frac{Da}{Dx_i}$ is the tangential derivative of any quantity a:

$$\frac{Da}{Dx_i} = \frac{\partial a}{\partial x_i} - \frac{\partial a}{\partial x_j}\, n_j\, n_i \qquad [7.39]$$

The local second gradient possesses the advantage that the constitutive equations remain local, with the stress fields σ_{ij} and Σ_{ijk} being local quantities. A second gradient extension can henceforth be formulated for any classical continuum mechanics constitutive law.

7.4.1.2. *Second gradient constitutive equation*

Similarly to classical media for which a constitutive equation links the stress to the kinematic history, an additional constitutive law has to be defined between the double stress and the microkinematics. In this chapter, the latter is assumed to be decoupled of the classical first gradient part (a 1D numerical study considering the two coupled laws is presented in Kotronis (2008)). However, only a little information is available on the relation existing between the double stress and the microdeformation. A linear elastic mechanical law is chosen hereafter for simplicity reasons with the purpose of introducing as few additional parameters as possible. It consists of an isotropic linear relationship involving five independent parameters derived by Mindlin (1965):

$$\tilde{\Sigma}_{ijk} = D_{ijklmn}\, \frac{\partial \dot{v}_{lm}}{\partial x_n} \qquad [7.40]$$

giving the Jaumann double stress rate:

$$\tilde{\Sigma}_{ijk} = \dot{\Sigma}_{ijk} + \Sigma_{ljk}\, \omega_{li} + \Sigma_{imk}\, \omega_{mj} + \Sigma_{ijp}\, \omega_{pk} \qquad [7.41]$$

as a function of the micro second gradient rate $\dot{h}_{ijk}$. Because the physical meaning of the material parameters composing D_{ijklmn} is not well established, a simplified version introducing only one parameter has been proposed (Matsushima *et al.* 2002). For two-dimensional problems, it reads:

$$\begin{bmatrix} \tilde{\Sigma}_{111} \\ \tilde{\Sigma}_{112} \\ \tilde{\Sigma}_{121} \\ \tilde{\Sigma}_{122} \\ \tilde{\Sigma}_{211} \\ \tilde{\Sigma}_{212} \\ \tilde{\Sigma}_{221} \\ \tilde{\Sigma}_{222} \end{bmatrix} = D \begin{bmatrix} 1 & 0 & 0 & 0 & 0 & \frac{1}{2} & \frac{1}{2} & 0 \\ 0 & \frac{1}{2} & \frac{1}{2} & 0 & -\frac{1}{2} & 0 & 0 & \frac{1}{2} \\ 0 & \frac{1}{2} & \frac{1}{2} & 0 & -\frac{1}{2} & 0 & 0 & \frac{1}{2} \\ 0 & 0 & 0 & 1 & 0 & -\frac{1}{2} & -\frac{1}{2} & 0 \\ 0 & -\frac{1}{2} & -\frac{1}{2} & 0 & 1 & 0 & 0 & 0 \\ \frac{1}{2} & 0 & 0 & -\frac{1}{2} & 0 & \frac{1}{2} & \frac{1}{2} & 0 \\ \frac{1}{2} & 0 & 0 & -\frac{1}{2} & 0 & \frac{1}{2} & \frac{1}{2} & 0 \\ 0 & \frac{1}{2} & \frac{1}{2} & 0 & 0 & 0 & 0 & 1 \end{bmatrix} \begin{bmatrix} \frac{\partial \dot{v}_{11}}{\partial x_1} \\ \frac{\partial \dot{v}_{11}}{\partial x_2} \\ \frac{\partial \dot{v}_{12}}{\partial x_1} \\ \frac{\partial \dot{v}_{12}}{\partial x_2} \\ \frac{\partial \dot{v}_{21}}{\partial x_1} \\ \frac{\partial \dot{v}_{21}}{\partial x_2} \\ \frac{\partial \dot{v}_{22}}{\partial x_1} \\ \frac{\partial \dot{v}_{22}}{\partial x_2} \end{bmatrix} \qquad [7.42]$$

The constitutive elastic parameter D represents the physical microstructure. The internal length scale relevant for the shear band width is related to this parameter (Chambon *et al.* 1998; Collin *et al.* 2009a; Kotronis *et al.* 2005, 2007, 2008; Kotronis 2008; Soufflet *et al.* 2018).

7.4.1.3. *Coupled local second gradient model*

The second gradient theory was extended from monophasic to biphasic medium by Collin *et al.* (2006). As for a monophasic medium, microstructure effects have to be introduced in the balance equations of classical poromechanics.

The linear momentum balance equation is identical to equation [7.33]:

$$\int_\Omega \left(\sigma_{ij} \frac{\partial u_i^*}{\partial x_j} + \Sigma_{ijk} \frac{\partial^2 u_i^*}{\partial x_j \partial x_k} \right) d\Omega$$

$$= \int_\Omega \rho\, g_i\, u_i^*\, d\Omega + \int_{\Gamma_\sigma} \left(\overline{t}_i\, u_i^* + \overline{T}_i\, Du_i^* \right) d\Gamma \qquad [7.43]$$

The water mass balance equation is written, in a weak form, in a similar way as the momentum balance equation. A kinematically admissible virtual pore water pressure field p_w^* is considered and is involved, as well as its first derivative, in the internal and external virtual quantities. The water mass balance equation reads:

$$\int_\Omega \left(\dot{M}_w\, p_w^* - f_{w,i} \frac{\partial p_w^*}{\partial x_i} \right) d\Omega = \int_\Omega Q_w\, p_w^*\, d\Omega - \int_{\Gamma_{q_w}} \overline{q}_w\, p_w^*\, d\Gamma, \qquad [7.44]$$

where M_w is the water mass inside Ω, $f_{w,i}$ is the water mass flow, Q_w is a sink term of water mass, and $\overline{q}_w$ is the input water mass (positive for inflow) per unit area on a part Γ_{q_w} of Γ (Figure 7.4).

According to the previous assumptions, the momentum balance equation [7.33] remains valid provided ρ and σ_{ij} are defined, knowing that the medium is a mixture of a solid phase and one fluid.

The mixture homogenized mass density is given by:

$$\rho = \rho_s\,(1 - \Phi) + \rho_w\, \Phi \qquad [7.45]$$

and the effective stress is defined according to Terzaghi's postulate:

$$\sigma_{ij} = \sigma'_{ij} + p_w\, \delta_{ij} \qquad [7.46]$$

Furthermore, it is assumed that the pore fluid does not have an influence at microscale; therefore, pore water pressure variations do not generate microkinematic gradients. This hypothesis was formulated by Ehlers (Ehlers and Volk 1998) on a Cosserat model for a biphasic medium. Second gradient effects are only assumed for the solid phase and the water mass balance equation (equation [7.44]) of classical poromechanics is conserved. The governing equations of the coupled problem are therefore equations [7.33] and [7.44].

As already mentioned for the classical poromechanics, the effect of water on the total stress is defined according to the effective stress postulate (equation [7.46]), while on the contrary the double stress Σ_{ijk} is independent of the pore water pressure. The double stress is only related to the solid phase.

The water mass M_w inside Ω and the water mass flow $f_{w,i}$ are defined in the following equations:

$$M_w = \rho_w \, \Phi \, \Omega \qquad [7.47]$$

$$f_{w,i} = -\rho_w \, \frac{k_w}{\mu_w} \left(\frac{\partial p_w}{\partial x_i} + \rho_w \, g_i \right) \qquad [7.48]$$

The definitions of the phase density variations and of the porosity evolution are:

$$\frac{\dot{\rho}_w}{\rho_w} = \frac{\dot{p}_w}{\chi_w} \qquad [7.49]$$

$$\dot{\rho}_s = 0 \qquad [7.50]$$

$$\dot{\Phi} = (1 - \Phi) \, \frac{\dot{\Omega}}{\Omega} \qquad [7.51]$$

The latter lead to the time derivative of the water mass per unit mixture volume:

$$\dot{M}_w = \rho_w \left(\frac{\dot{p}_w}{\chi_w} \, \Phi + \frac{\dot{\Omega}}{\Omega} \right) \qquad [7.52]$$

7.4.2. *Coupled finite element formulation*

7.4.2.1. *Numerical implementation*

The virtual work formulation allows us to implement the second gradient models in a finite element code. To implement the momentum balance equation of equation [7.33], the displacement field has to be a continuously differentiable function because second order derivatives of the displacement field are involved (Zervos *et al.* 2001b). To avoid the use of the C1 function, the kinematic restrictions $v_{ij} = F_{ij}$ and $v^*_{ij} = F^*_{ij}$ are introduced in the momentum balance equation through a field of Lagrange multipliers λ_{ij} related to a weak form of the constraint (Chambon *et al.* 1998) (see Soufflet *et al.* 2018 for some numerical problems related to this choice). The field equations of the numerical coupled problem are:

$$\int_{\Omega^t} \left(\sigma^t_{ij} \frac{\partial u^*_i}{\partial x^t_j} + \Sigma^t_{ijk} \frac{\partial v^*_{ij}}{\partial x^t_k} \right) d\Omega^t - \int_{\Omega^t} \lambda^t_{ij} \left(\frac{\partial u^*_i}{\partial x^t_j} - v^*_{ij} \right) d\Omega^t$$

$$= \int_{\Omega^t} \rho^t \, g_i \, u^*_i \, d\Omega^t + \int_{\Gamma^t_\sigma} \left(\bar{t}^t_i \, u^*_i + \overline{T}^t_i \, v^*_{ik} \, n^t_k \right) d\Gamma^t \quad [7.53]$$

$$\int_{\Omega^t} \lambda^*_{ij} \left(\frac{\partial u^t_i}{\partial x^t_j} - v^t_{ij} \right) d\Omega^t = 0 \quad [7.54]$$

$$\int_{\Omega^t} \left(\dot{M}^t_w \, p^*_w - f^t_{w,i} \frac{\partial p^*_w}{\partial x^t_i} \right) d\Omega^t = \int_{\Omega^t} Q^t_w \, p^*_w \, d\Omega^t - \int_{\Gamma^t_{q_w}} \bar{q}^t_w \, p^*_w \, d\Gamma^t \quad [7.55]$$

where the notation a^t corresponds to the current value of any quantity a for a given time t. For boundary conditions problems, the virtual quantities included in the above equations depend on the boundary conditions history. Thus, the governing equations and the constitutive equations have to hold at any time t.

7.4.2.2. *Linearization of the field equations*

Solving the loading process of a boundary conditions problem consists of determining the unknown fields u_i, v_{ij}, λ_{ij} and p_w for which the equilibrium equations [7.53–7.55] are valid. Since this system of nonlinear equations is *a priori* not verified for any instant t, the problem is numerically solved by iterative procedure. It involves a time discretization over finite time steps Δt:

$$\tau = t + \Delta t \quad [7.56]$$

and an implicit scheme of finite differences for the rate of any quantity a:

$$\dot{a}^{\tau} = \frac{a^{\tau} - a^{t}}{\Delta t} \tag{7.57}$$

A full Newton–Raphson method is used to find a solution for the new fields u_i, v_{ij}, λ_{ij} and p_w at the end of each time step, which is in equilibrium with the boundary conditions.

Following the approach of Borja and Alarcón (1995), the method aims to define a linear auxiliary problem deriving from the continuum one. A first configuration Ω^t in equilibrium with the boundary conditions at a given time t is assumed to be known and another Ω^τ in equilibrium at the end of the time step $\tau = t + \Delta t$ has to be found. The aim of the iterative numerical procedure is to determine this new configuration at the end of the time step. First, a configuration which is close to the solution but not at equilibrium is guessed and denoted as $\Omega^{\tau 1}$. Both configurations at time t and $\tau 1$ are assumed to be known and non-equilibrium forces for the three considered equations, i.e. the residuals $\Delta_1^{\tau 1}$, $\Delta_2^{\tau 1}$ and $\Delta_3^{\tau 1}$ are defined. The objective is to find another configuration $\Omega^{\tau 2}$ close to $\Omega^{\tau 1}$ for which the non-equilibrium forces vanish. To obtain the linear auxiliary problem, the field equations for $\Omega^{\tau 2}$ are subtracted from the field equations in configuration $\Omega^{\tau 1}$, after being rewritten in configuration $\Omega^{\tau 1}$ by using the Jacobian matrix of the transformation between the two configurations:

$$\overline{F}_{ij} = \frac{\partial x_i^{\tau 2}}{\partial x_j^{\tau 1}} \tag{7.58}$$

and its Jacobian determinant:

$$det(F) = \left| \frac{\partial x_i^{\tau 2}}{\partial x_j^{\tau 1}} \right| \tag{7.59}$$

Assuming that g_i, $\bar{t}_i$, $\bar{q}_w$ and Q_w are independent of the different unknown fields (displacement and pore water pressure) and that $\overline{T}_i$ vanishes give:

$$\begin{aligned}
&\int_{\Omega^{\tau 1}} \frac{\partial u_i^*}{\partial x_l^{\tau 1}} \left(\sigma_{ij}^{\tau 2} \frac{\partial x_l^{\tau 1}}{\partial x_j^{\tau 2}} det(F) - \sigma_{il}^{\tau 1} \right) + \frac{\partial v_{ij}^*}{\partial x_l^{\tau 1}} \left(\Sigma_{ijk}^{\tau 2} \frac{\partial x_l^{\tau 1}}{\partial x_k^{\tau 2}} det(F) - \Sigma_{ijl}^{\tau 1} \right) d\Omega^{\tau 1} \\
&- \int_{\Omega^{\tau 1}} \frac{\partial u_i^*}{\partial x_l^{\tau 1}} \left(\lambda_{ij}^{\tau 2} \frac{\partial x_l^{\tau 1}}{\partial x_j^{\tau 2}} det(F) - \lambda_{il}^{\tau 1} \right) - v_{ij}^* \left(\lambda_{ij}^{\tau 2} \, det(F) - \lambda_{ij}^{\tau 1} \right) d\Omega^{\tau 1} \\
&- \int_{\Omega^{\tau 1}} u_i^* \left(\rho^{\tau 2} \, det(F) - \rho^{\tau 1} \right) g_i \, d\Omega^{\tau 1} = -\Delta_1^{\tau 1}
\end{aligned} \tag{7.60}$$

$$\int_{\Omega^{\tau 1}} \lambda^*_{ij} \left(\left(\frac{\partial u_i^{\tau 2}}{\partial x_k^{\tau 1}} \frac{\partial x_k^{\tau 1}}{\partial x_j^{\tau 2}} \, det(F) - \frac{\partial u_i^{\tau 1}}{\partial x_j^{\tau 1}} \right) - \left(v_{ij}^{\tau 2} \, det(F) - v_{ij}^{\tau 1} \right) \right) d\Omega^{\tau 1}$$
$$= -\Delta_2^{\tau 1} \qquad [7.61]$$

$$\int_{\Omega^{\tau 1}} p^*_w \left(\dot{M}_w^{\tau 2} \, det(F) - \dot{M}_w^{\tau 1} \right) - \frac{\partial p^*_w}{\partial x_l^{\tau 1}} \left(f_{w,i}^{\tau 2} \frac{\partial x_l^{\tau 1}}{\partial x_i^{\tau 2}} \, det(F) - f_{w,l}^{\tau 1} \right) d\Omega^{\tau 1}$$
$$= -\Delta_3^{\tau 1} \qquad [7.62]$$

By making the two configurations tend toward each other, the variations between them can be defined for any quantity a as:

$$da^{\tau 1} = a^{\tau 2} - a^{\tau 1} \qquad [7.63]$$

The balance equations can be rewritten by taking into account these variations. The complete development of the linearization of the field equation system and of the resulting linear auxiliary problem is exposed by Collin *et al.* (2006).

7.4.2.3. *Spatial discretization*

In finite element methods, each continuum body is discretized by finite elements and the above field equations are spatially discretized. For the second gradient model, the discretization is realized by two-dimensional plane-strain isoparametric finite elements. These elements are composed of eight nodes for the displacement field u_i and the pore water pressure p_w, four nodes for the microkinematic gradient field v_{ij} and one node for the Lagrange multiplier field λ_{ij} (Figure 7.5). Quadratic serendipity shape functions (Zienkiewicz and Taylor 2000) are used for the u_i and p_w interpolations, linear shape functions are used for v_{ij}, whereas λ_{ij} is assumed constant.

The balance equations of the coupled finite element problem (linear auxiliary problem) have to be rewritten in matricial form to define the local stiffness matrix of an element:

$$\int_{\Omega^{\tau 1}} \left[U^{*,\tau 1}_{(x_1,x_2)} \right]^T \left[E^{\tau 1} \right] \left[dU^{\tau 1}_{(x_1,x_2)} \right] d\Omega^{\tau 1} = -\Delta_1^{\tau 1} - \Delta_2^{\tau 1} - \Delta_3^{\tau 1}, \qquad [7.64]$$

where $\left[dU_{(x_1,x_2)}^{\tau 1}\right]$ is the vector of the unknown increments of nodal variables in the current element configuration:

$$\left[dU_{(x_1,x_2)}^{\tau 1}\right]_{25\times 1} = \left[\frac{\partial du_1^{\tau 1}}{\partial x_1^{\tau 1}}\ \frac{\partial du_1^{\tau 1}}{\partial x_2^{\tau 1}}\ \frac{\partial du_2^{\tau 1}}{\partial x_1^{\tau 1}}\ \frac{\partial du_2^{\tau 1}}{\partial x_2^{\tau 1}}\ du_1^{\tau 1}\ du_2^{\tau 1}\ \frac{\partial dp_w^{\tau 1}}{\partial x_1^{\tau 1}}\ \frac{\partial dp_w^{\tau 1}}{\partial x_2^{\tau 1}}\ dp_w^{\tau 1}\ \frac{\partial dv_{11}^{\tau 1}}{\partial x_1^{\tau 1}} \right.$$
$$\frac{\partial dv_{11}^{\tau 1}}{\partial x_2^{\tau 2}}\ \frac{\partial dv_{12}^{\tau 1}}{\partial x_1^{\tau 1}} \dots \frac{\partial dv_{22}^{\tau 1}}{\partial x_2^{\tau 1}}\ dv_{11}^{\tau 1}\ dv_{12}^{\tau 1}\ dv_{21}^{\tau 1}\ dv_{22}^{\tau 1}\ d\lambda_{11}^{\tau 1}\ d\lambda_{12}^{\tau 1} \quad [7.65]$$
$$\left. d\lambda_{21}^{\tau 1}\ d\lambda_{22}^{\tau 1}\right]^T$$

$\left[U_{(x_1,x_2)}^{*,\tau 1}\right]$ is a vector having the same structure with the corresponding virtual quantities:

$$\left[U_{(x_1,x_2)}^{*,\tau 1}\right]_{1\times 25} = \left[\frac{\partial u_1^*}{\partial x_1^{\tau 1}}\ \frac{\partial u_1^*}{\partial x_2^{\tau 1}}\ \frac{\partial u_2^*}{\partial x_1^{\tau 1}}\ \frac{\partial u_2^*}{\partial x_2^{\tau 1}}\ u_1^*\ u_2^*\ \frac{\partial p_w^*}{\partial x_1^{\tau 1}}\ \frac{\partial p_w^*}{\partial x_2^{\tau 1}}\ p_w^*\ \frac{\partial v_{11}^*}{\partial x_1^{\tau 1}}\ \frac{\partial v_{11}^*}{\partial x_2^{\tau 2}} \right. \quad [7.66]$$
$$\left. \frac{\partial v_{12}^*}{\partial x_1^{\tau 1}} \dots \frac{\partial v_{22}^*}{\partial x_2^{\tau 1}}\ v_{11}^*\ v_{12}^*\ v_{21}^*\ v_{22}^*\ \lambda_{11}^*\ \lambda_{12}^*\ \lambda_{21}^*\ \lambda_{22}^*\right]$$

and $\left[E^{\tau 1}\right]$ is the current element stiffness (tangent) matrix defined as follows:

$$\left[E^{\tau 1}\right]_{25\times 25} = \begin{bmatrix} E^{\tau 1}_{1_{4\times 4}} & 0_{4\times 2} & K^{\tau 1}_{WM_{4\times 3}} & 0_{4\times 8} & 0_{4\times 4} & -I_{4\times 4} \\ G^{\tau 1}_{1_{2\times 4}} & 0_{2\times 2} & G^{\tau 1}_{2_{2\times 3}} & 0_{2\times 8} & 0_{2\times 4} & 0_{2\times 4} \\ K^{\tau 1}_{MW_{3\times 4}} & 0_{3\times 2} & K^{\tau 1}_{WW_{3\times 3}} & 0_{3\times 8} & 0_{3\times 4} & 0_{3\times 4} \\ E^{\tau 1}_{2_{8\times 4}} & 0_{8\times 2} & 0_{8\times 3} & D^{\tau 1}_{8\times 8} & 0_{8\times 4} & 0_{8\times 4} \\ E^{\tau 1}_{3_{4\times 4}} & 0_{4\times 2} & 0_{4\times 3} & 0_{4\times 8} & 0_{4\times 4} & I_{4\times 4} \\ E^{\tau 1}_{4_{4\times 4}} & 0_{4\times 2} & 0_{4\times 3} & 0_{4\times 8} & -I_{4\times 4} & 0_{4\times 4} \end{bmatrix} \quad [7.67]$$

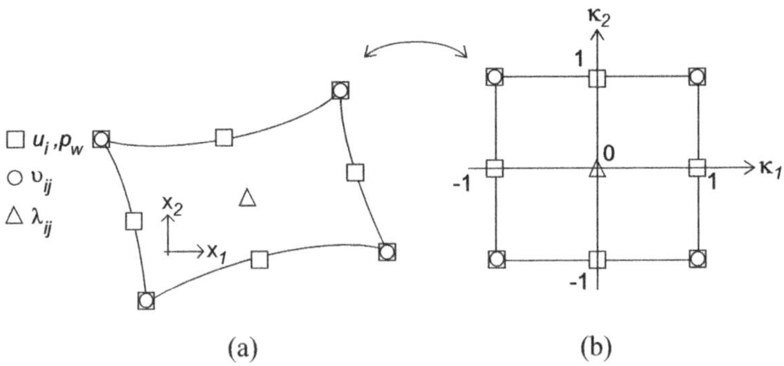

Figure 7.5. *Finite element used for the spatial discretization of the coupled local second gradient model: (a) current quadrilateral element and (b) parent element (Collin* et al. *2006)*

The matrices $\left[E_1^{\tau 1}\right]$, $\left[E_2^{\tau 1}\right]$, $\left[E_3^{\tau 1}\right]$, $\left[E_4^{\tau 1}\right]$ and $\left[D^{\tau 1}\right]$ are the same as the ones used in the local second gradient model for monophasic medium by Chambon and Moullet (2004) ($\left[D^{\tau 1}\right] = D_{ijklmn}$ in equations [7.40] and [7.42]). $\left[K_{WW}^{\tau 1}\right]$ is the classical stiffness matrix of a flow problem, and $\left[K_{MW}^{\tau 1}\right]$ and $\left[K_{WM}^{\tau 1}\right]$ are matrices of the coupling between the flow and the mechanical problems detailed by Collin *et al.* (2006). Moreover, $\left[G_1^{\tau 1}\right]$ and $\left[G_2^{\tau 1}\right]$ are related to the contribution of gravity volume force.

The finite element spatial discretization of the linear auxiliary problem is introduced in equation [7.64] by using transformation matrices $\left[T^{\tau 1}\right]$ and $[B]$ that connect the current element vector $\left[dU_{(x_1,x_2)}^{\tau 1}\right]$ to the parent element vector $\left[dU_{(\kappa_1,\kappa_2)}^{\tau 1}\right]$ and to the nodal variables $\left[dU_{Node}^{\tau 1}\right]$:

$$\left[dU_{(x_1,x_2)}^{\tau 1}\right] = \left[T^{\tau 1}\right]\left[dU_{(\kappa_1,\kappa_2)}^{\tau 1}\right] = \left[T^{\tau 1}\right][B]\left[dU_{Node}^{\tau 1}\right] \quad [7.68]$$

The matrices $[B]$ and $\left[T^{\tau 1}\right]$ contain the interpolation functions and their derivatives. Moreover, the vector $\left[U_{(x_1,x_2)}^{*,\tau 1}\right]$ is related to $\left[U_{Node}^{*,\tau 1}\right]$ in the same manner.

The integration in equation [7.64] can be expressed for each parent element as follows:

$$\int_{\Omega^{\tau 1}} \left[U_{(x_1,x_2)}^{*,\tau 1}\right]^T \left[E^{\tau 1}\right]\left[dU_{(x_1,x_2)}^{\tau 1}\right] d\Omega^{\tau 1} = \left[U_{Node}^{*,\tau 1}\right]^T \left[k^{\tau 1}\right]\left[dU_{Node}^{\tau 1}\right], \quad [7.69]$$

where $\left[k^{\tau 1}\right]$ is the local element stiffness matrix:

$$\left[k^{\tau 1}\right] = \int_{-1}^{1}\int_{-1}^{1} [B]^T \left[T^{\tau 1}\right]^T \left[E^{\tau 1}\right]\left[T^{\tau 1}\right][B]\, det(J^{\tau 1})\, d\kappa_1\, d\kappa_2 \quad [7.70]$$

with $det(J^{\tau 1})$ the determinant of the Jacobian matrix of the transformation between the parent (κ_1, κ_2) and the current (x_1, x_2) elements:

$$det(J^{\tau 1}) = \left|\frac{\partial x_i^{\tau 1}}{\partial \kappa_j}\right| \quad [7.71]$$

The residual terms are also computed locally for each element and define the elementary out of balance force vector $\left[f_{OB}^{\tau 1}\right]$:

$$-\Delta_1^{\tau 1} - \Delta_2^{\tau 1} - \Delta_3^{\tau 1} = \left[U_{Node}^{*,\tau 1}\right]^T \left[f_{OB}^{\tau 1}\right] \quad [7.72]$$

7.4.2.4. *Global solution*

Once the elementary stiffness matrices and out of balance force vectors are computed, they are assembled to obtain the global stiffness matrix $\left[K^{\tau 1}\right]$ and the global out of balance force vector $\left[F_{OB}^{\tau 1}\right]$ of the whole continuum. The linear auxiliary system is solved by computing:

$$\left[K^{\tau 1}\right]\left[\delta U_{Node}^{\tau 1}\right] = -\left[F_{OB}^{\tau 1}\right], \qquad [7.73]$$

where $\left[\delta U_{Node}^{\tau 1}\right]$ is the global correction vector of the nodal degrees of freedom. The current configuration is actualized by adding the corrections to their respective current values. The new current configuration is closer to the well-balanced configuration and its equilibrium is checked, leading to a new iteration or to the end of the loading step of the iterative procedure.

7.4.3. *Two-dimensional specimen under compression*

A finite element modeling of two-dimensional plane-strain compression tests is first considered. These tests have been widely reproduced on small-scale specimens to emphasize the strain localization effects.

Among various authors, the results obtained by Collin *et al.* (2009b) for a uniaxial compression are principally developed hereafter. A sketch of the boundary value problem in plane-strain state is illustrated in Figure 7.6. The vertical displacement u_a of the sample upper surface (smooth and rigid boundary) is progressively increased during the test with a constant loading strain rate to model the vertical compression. The vertical displacement of the bottom surface is blocked (rigid boundary) and the displacement of the central node is blocked in both directions to avoid rigid body displacement.

7.4.3.1. *Classical medium*

A mechanical modeling is presented hereafter for a classical medium (without a regularization method). The (first gradient) constitutive law is an elastoplastic strain-softening model in an associated softening plasticity framework ($\varphi = \psi$, $F^p = G^p$): a Drucker–Prager yield criterion is considered with no hardening of the friction angle, and a cohesion softening function (Collin *et al.* 2009b).

A homogeneous response of the specimen is first studied. The global response is detailed in Figure 7.7(a) where one can observe a linear elastic behavior, then a nonlinear plastic behavior before the peak stress including cohesion softening, and finally a plastic behavior with decrease in the global load response. Concerning the orientation of the shear band, the Rice criterion gives the orientation of the first possible shear band occurrence (Rudnicki and Rice 1975; Rice 1976). This criterion can be used for a single-mechanism elastoplastic model such as the studied problem. The Rice criterion $det(\wedge_{jk})$ evolution is presented during the increasing loading history in Figure 7.7(b) as a function of $tan\Theta$, where $\wedge_{jk}$ is the acoustic tensor and Θ is the orientation of the shear band normal to the loading vertical axis, i.e. the shear band orientation with the horizontal direction. The criterion is positive $det(\wedge_{jk} > 0)$ as long as the behavior is elastic and even for an elastoplastic loading until the first bifurcation is predicted $(det(\wedge_{jk}) = 0)$. For a certain load, the bifurcation criterion is met at every material point and two symmetric (conjugate) bifurcation directions are predicted with an orientation of $\Theta = \pm 60°$. This bifurcation point corresponds to the peak stress on the global response curve and to the start of the load response reduction in associated plasticity. A range of possible orientations is predicted for an increasing load corresponding to a multitude of possible solutions $(det(\wedge_{jk}) <= 0)$.

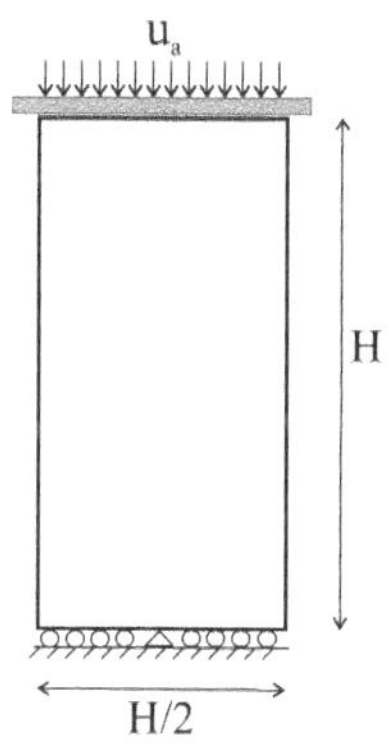

Figure 7.6. *Sketch of the plane-strain compression test*

For a perfect sample, although softening plasticity is considered, the strain localization is not automatically triggered and the numerical solution may remain homogeneous even after the bifurcation criterion is met. In reality, the localization process is generated because geomaterials exhibit heterogeneities. Different numerical procedures are available to force the occurrence of strain localization. The most used one is the introduction of an imperfection, such as disturbing force, material imperfection, or geometrical defect (Charlier *et al.* 1997; Matsushima *et al.* 2002; Zhang *et al.* 2001). The modification of numerical parameters, such as time step size and sequences, can also be performed (Marinelli *et al.* 2014; Sieffert *et al.* 2009). A third method that will be discussed later is the random initialization of variables (Chambon *et al.* 2001b).

Among these procedures, Collin *et al.* (2009b) introduced a material imperfection in the bottom left finite element of the sample under compression. Initially, the strain field in the sample is homogeneous, and once the bifurcation criterion is met, the imperfection instigates the development of a shear band across the specimen. The strain localization as well as its dependency to the mesh size is illustrated in Figure 7.8 for a classical medium. The localized solution is therefore non-homogeneous, with the shear band under plastic loading and the outer material under elastic unloading. The global sample response is detailed in Figure 7.7(a), where a rapid decrease in the global reaction is observed once the shear band is established. The shear band appearance corresponds therefore to the curve peak load (or peak stress) as concluded from laboratory evidences in section 7.2. The latter also indicate that a material inclusion can act as a strain localization attractor, which is confirmed by the numerical results.

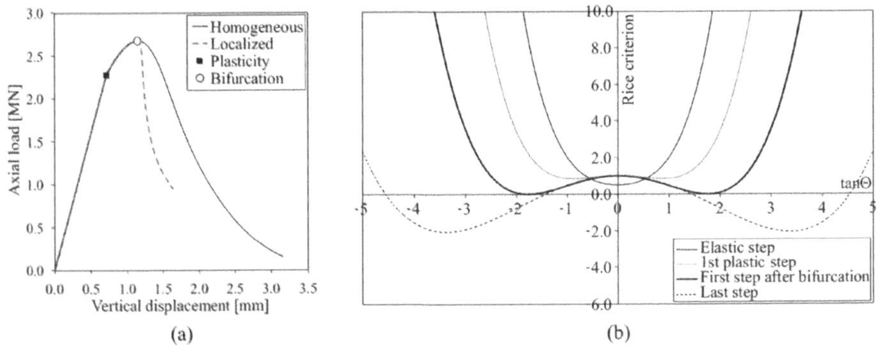

Figure 7.7. *Uniaxial compression: (a) global specimen response curves and (b) Rice criterion at several loading steps for the homogeneous solution (Collin* et al. *2009b)*

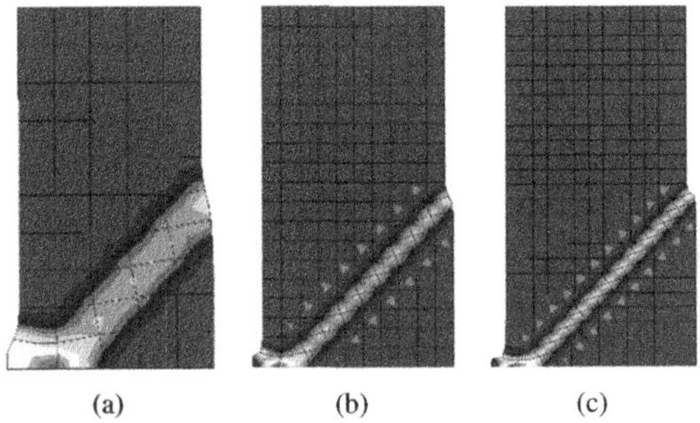

Figure 7.8. *Localized patterns represented by the deviatoric deformation for classical medium: finite element meshes of (a) 50, (b) 190 and (c) 325 elements. For a color version of this figure, see www.iste.co.uk/stefanou/instabilities.zip*

7.4.3.2. *Microstructure enhanced medium*

To fix the pathological mesh sensitivity, an enhanced microstructure medium is used. More specifically, the local second gradient medium is adopted, with the second gradient constitutive law given by equation [7.42]. The strain localization pattern induced by the imperfection is illustrated in Figure 7.9 with the Gauss integration points under softening plastic loading shown as red squares. This allows us to measure the shear band width and to notice that it stays constant no matter the element size, implying that the shear strain localization is mesh independent. Thus, the strain localization is correctly regularized because of the internal length scale introduced by the second grade model. This is also the case for unstructured mesh (Bésuelle *et al.* 2006) and for a biphasic porous medium under saturated conditions, using the coupled local second gradient model (Collin *et al.* 2006).

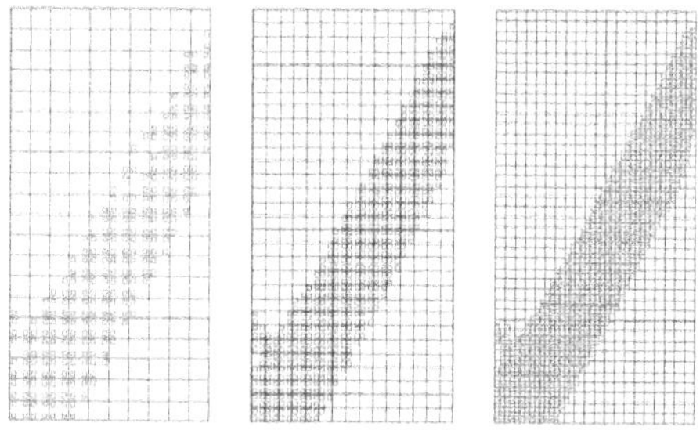

(a) (b) (c)

Figure 7.9. *Localized patterns represented by the plastic zone for a second gradient medium: finite element meshes of (a) 200, (b) 450 and (c) 800 elements. For a color version of this figure, see www.iste.co.uk/stefanou/instabilities.zip*

However, the regularization of the strain localization process is satisfactory, provided that the second gradient elastic modulus D is calibrated properly to represent the shear bands properly. As already mentioned, the internal length scale inherent to the second gradient mechanical law is related to this constitutive parameter. The value of D should therefore be evaluated based on experimental measurements of shear band thickness for the considered material. From a modeling point of view, a better numerical prediction of the post-localization plastic behavior within the bands is obtained if at least three elements compose the shear band width (Bésuelle *et al.* 2006). This remark is valid for any regularization technique including the second gradient model but also gradient plasticity and non-local models.

7.4.3.3. *Non-uniqueness of the solution*

The non-uniqueness of the post-peak solution of an initial boundary value problem can be studied using special numerical techniques. In fact, instead of using a material imperfection, localized solutions can be found for a homogeneous material using a random initialization of the strain rate field (nodal velocities) or of material characteristics at the beginning of the iterative procedure. This technique has been mainly proposed by Chambon *et al.* (2001b), who developed an algorithm to search several possible localized solutions by random initialization. This algorithm has been adapted to the second-gradient models by Chambon and Moullet (2004).

Numerical modeling of compression tests performed with the second gradient model illustrates the non-uniqueness of localized solutions of the same initial boundary value problem (Bésuelle *et al.* 2006). The random initialization is adopted for the increment of nodal quantities $\left[dU_{Node}^{\tau 1}\right]$ (equation [7.68]) related to the values obtained at the end of the preceding time step. The obtained non-homogeneous solutions are detailed in Figure 7.10(a), where the different solutions exhibit one to three bands with a possible reflection on the top and bottom faces of the sample because of the imposed vertical displacement. The results indicate that the band thickness is reproducible even if the localization pattern is different in terms of bands position and number.

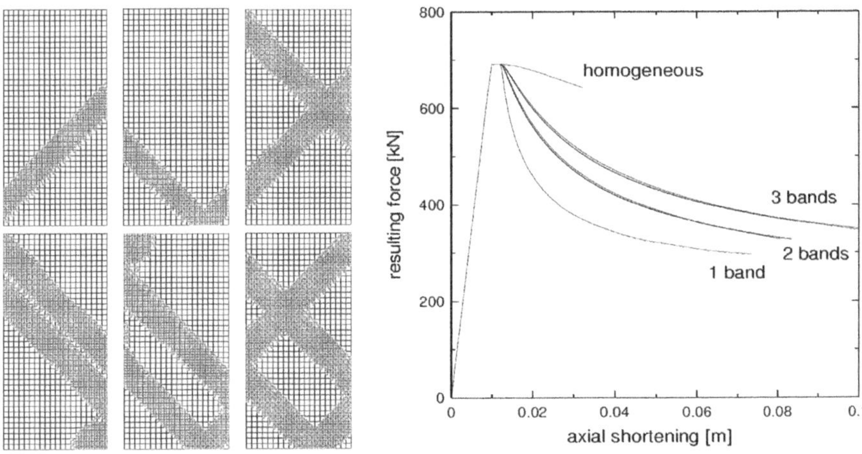

(a) (b)

Figure 7.10. *Example of localized solutions for a compression test obtained after a random initialization: (a) plastic zone and (b) global response curve (Bésuelle* et al. *2006)*

As before, the strain localization occurring at the bifurcation point is due to the strain softening behavior and possible elastic unloading. Figure 7.10(b) illustrates the global response curves that are different to those in Figure 7.7 because a different first gradient law is used. These curves are grouped in packages characterized by the number of deformation bands. It is evident that the higher the band number, the closer the curves are to the homogeneous plastic case. A similar conclusion was drawn from 1D problems (e.g., bar in traction) by Chambon *et al.* (1998), Jouan *et al.* (2014), and Kotronis *et al.* (2008).

7.4.3.4. *Bifurcation criterion for the second gradient model*

A bifurcation analysis applied to the second gradient model is proposed by Bésuelle *et al.* (2006). The authors indicate that the bifurcation criterion of the second gradient model is, as for a classical medium, a necessary but not sufficient condition for the localization onset and that it is met after the bifurcation criterion of the classical medium is verified. Thus, the bifurcation analysis reduces to an analysis on the classical part of the constitutive model.

7.5. Coupled local second gradient model for an unsaturated medium

The procedure to extend the local second gradient model in saturated conditions to other multiphysical contexts is more or less the same: additional balance equations have to be considered to model the other processes. The main issue is not a numerical one but rather a physical one. What are the possible interactions between the second gradient model and the thermal diffusion, the suction or the chemical reaction? These questions should be first addressed by experimental campaigns. Concerning the internal length introduced by the second gradient model, it comes as the ratio of two constitutive moduli: the one related to the second gradient constitutive law and the one related to the classical law (Chambon *et al.* 1998). Considering that the classical constitutive moduli are influenced by the different processes (chemical, thermal, suction, etc.), the conclusions should be that the internal length scale should be modified by these latter processes (under the condition that second gradient law is not influenced by the processes). This should again be confirmed by some experimental evidences!

The coupled local second gradient model developed by Collin *et al.* (2006) for a biphasic porous medium can be extended to unsaturated materials with compressible solid grains. Hereafter, the unsaturated conditions are taken into account and the compressibility of the solid grains is introduced through Biot's coefficient. Additionally, the permeability anisotropy is incorporated in the model to better represent the water flows, even if the mechanical behavior remains isotropic.

7.5.1. *Partial saturation conditions*

For unsaturated conditions, the water mass inside a porous material volume Ω corresponds to:

$$M_w = \rho_w \, \Phi \, S_{r,w} \, \Omega \qquad [7.74]$$

and its time derivative corresponds to:

$$\dot{M}_w = \dot{\rho}_w \; \Phi \; S_{r,w} \; \Omega + \rho_w \; \dot{\Phi} \; S_{r,w} \; \Omega + \rho_w \; \Phi \; \dot{S}_{r,w} \; \Omega + \rho_w \; \Phi \; S_{r,w} \; \dot{\Omega} \qquad [7.75]$$

This amount of water, which depends on the degree of water saturation $S_{r,w}$, leads to the following mixture homogenized mass density:

$$\rho = \rho_s \; (1 - \Phi) + S_{r,w} \; \rho_w \; \Phi \qquad [7.76]$$

and the water advective flow for anisotropic hydraulic permeability is given by Darcy's law:

$$f_{w,i} = -\rho_w \; \frac{k_{w,ij} \; k_{r,w}}{\mu_w} \; \left(\frac{\partial p_w}{\partial x_j} + \rho_w \; g_j \right), \qquad [7.77]$$

where $k_{r,w}$ is the water relative permeability.

The fluid mass and fluid flows are mostly governed by the water retention property of the material and by its hydraulic permeability. Both of them are related to the partial water saturation and a relative permeability coefficient is introduced in the generalized Darcy law. Among various possible analytical expressions, the water retention and relative permeability curves are given by van Genuchten's and Mualem's models (Mualem 1976; van Genuchten 1980):

$$S_{r,w} = S_{res} + (S_{max} - S_{res}) \left(1 + \left(\frac{p_c}{P_r} \right)^{\frac{1}{1-\mathcal{M}}} \right)^{-\mathcal{M}}, \qquad [7.78]$$

$$k_{r,w} = \sqrt{S_{r,w}} \left(1 - \left(1 - S_{r,w}{}^{\frac{1}{\mathcal{M}}} \right)^{\mathcal{M}} \right)^2, \qquad [7.79]$$

where P_r is the air entry pressure, S_{max} and S_{res} are the maximum and residual water degrees of saturation, $\mathcal{M}$ is a model coefficient and p_c is the capillary pressure.

7.5.2. *Anisotropy of the intrinsic permeability*

The advective flow of water (equation [7.77]) depends on the anisotropic characteristics of the material through the anisotropic intrinsic permeability. For anisotropic materials and by symmetry of the tensor, the intrinsic permeability tensor $k_{w,ij}$ requires six components to describe the flow characteristics. However, materials commonly exhibit limited forms of anisotropy and stratified geomaterials only require two parameters for the description of the water flow. For horizontal layering in the plane (x_1,x_3), the intrinsic hydraulic permeability tensor is defined with the horizontal and vertical permeabilities, $k_{w,h}$ and $k_{w,v}$, as follows:

$$k_{w,ij} = \begin{bmatrix} k_{w,h} & 0 & 0 \\ 0 & k_{w,v} & 0 \\ 0 & 0 & k_{w,h} \end{bmatrix} \quad [7.80]$$

7.5.3. *Compressibility of the solid grains*

The material compressibility is defined within the scope of poroelasticity (Detournay and Cheng 1993) and is based on the different compressibilities of a porous material. These are as follows: the compressibility of the bulk material C (solid skeleton), the compressibility of the pores C_p and the compressibility of the solid phase C_s (rock matrix) with $C_s < C$. The different types of compressibility induce different behaviors of the rock matrix and of the porous material. They can deform differently and the porous material may enter plastic state while the solid grains remain elastic. In the general Biot framework (Biot 1941), Biot's coefficient is expressed by:

$$b = \frac{\Phi \, C_p}{C} = 1 - \frac{K}{K_s} \quad [7.81]$$

as a function of the drained bulk modulus of the material K and the bulk modulus of the solid phase K_s. This coefficient represents the relative deformability of the solid grains with regard to the solid skeleton (Biot 1941; Biot and Willis 1957; Skempton 1960). Biot proposed for the effective stress definition to use b as a scaling factor that reduces the effect of p_w on σ_{ij} due to a reduction of pore compressibility. Biot's stress definition can be formulated under unsaturated conditions presuming that the assumptions on compressibility hold under these conditions (Nuth and Laloui 2008):

$$\sigma_{ij} = \sigma'_{ij} - b \, S_{r,w} \, p_w \, \delta_{ij} \quad [7.82]$$

The latter expression includes the effect of partial saturation on the effective stress field (tensile stress is positive).

For the solid phase behavior, the isotropic solid density variation is linked to the variations of pore water pressure and mean effective stress by (Coussy 2004; Detournay and Cheng 1993):

$$\frac{\dot{\rho}_s}{\rho_s} = \frac{(b - \Phi) \, S_{r,w} \, \dot{p}_w - \dot{\sigma}'}{(1 - \Phi) \, K_s} \quad [7.83]$$

The time derivative of the porosity is obtained by solid mass conservation $\dot{M}_s = 0$ and reads:

$$\dot{\Phi} = (1-\Phi)\left(\frac{\dot{\rho}_s}{\rho_s} + \frac{\dot{\Omega}}{\Omega}\right) = (1-\Phi)\left(\frac{(b-\Phi)\, S_{r,w}\, \dot{p}_w - \dot{\sigma}'}{(1-\Phi)\; K_s} + \frac{\dot{\Omega}}{\Omega}\right) \quad [7.84]$$

Furthermore, the time derivative of the water mass in equation [7.75] becomes, by including the fluid compressibility, the porosity variation, and by considering a unit mixture volume:

$$\dot{M}_w = \rho_w\left(\frac{\dot{p}_w}{\chi_w}\,\Phi\, S_{r,w} + \frac{\dot{p}_w}{K_s}\,(b-\Phi)\; S_{r,w}^2 + \left(\frac{\dot{\Omega}}{\Omega} - \frac{\dot{\sigma}'}{K_s}\right)\; S_{r,w} + \Phi\, \dot{S}_{r,w}\right) \quad [7.85]$$

The above expressions can be rewritten under poroelastic assumption:

$$\dot{\sigma}' = K\, \dot{\varepsilon}_v = K\, \frac{\dot{\Omega}}{\Omega} \quad [7.86]$$

and using Biot's coefficient expression of equation [7.81]. The equations become:

$$\frac{\dot{\rho}_s}{\rho_s} = \frac{(b-\Phi)\, S_{r,w}\, \dot{p}_w - K\, \dfrac{\dot{\Omega}}{\Omega}}{(1-\Phi)\; K_s} \quad [7.87]$$

$$\dot{\Phi} = (b-\Phi)\left(\frac{S_{r,w}}{K_s}\,\dot{p}_w + \frac{\dot{\Omega}}{\Omega}\right) \quad [7.88]$$

$$\dot{M}_w = \rho_w\left(\frac{\dot{p}_w}{\chi_w}\,\Phi\, S_{r,w} + \frac{\dot{p}_w}{K_s}\,(b-\Phi)\; S_{r,w}^2 + b\,\frac{\dot{\Omega}}{\Omega}\, S_{r,w} + \Phi\, \dot{S}_{r,w}\right) \quad [7.89]$$

Biot's theory and the equations of poroelasticity are valid only for an elastic behavior. Extending these equations to poroplasticity (Coussy 1995) with permanent changes in fluid mass content and in porosity requires us to include the plastic material behavior, which is complex to implement and is not included in this chapter.

According to the previous assumptions, the momentum balance equation [7.33] and the water mass balance equation [7.44] remain valid provided that the different variables included in these two equations are adapted to unsaturated conditions (ρ, σ_{ij}, $\dot{M}_w$, etc.).

7.6. Modeling of a gallery excavation

The processes of underground drilling and induced shear strain localization are investigated in detail. Papanastasiou and Vardoulakis (1992) were the first to present a numerical analysis of progressive localization around an excavated cavity in rock with a Cosserat microstructure. In the following, a gallery excavation is considered in a clayey rock. In the rock mass, the fractured structure around the galleries develops preferentially in the horizontal or vertical direction depending on the anisotropy of both the stress state and material properties (Armand *et al.* 2014; Pardoen *et al.* 2015b). As a first large-scale approach including strain localization, an isotropic mechanical model is used with the objective of analyzing if the appearance of fractures during the drilling of galleries is governed by the anisotropy of the *in situ* stress state (Pardoen *et al.* 2015a). So far, the numerical modeling of gallery drilling with the second gradient model has highlighted strain localization but was essentially limited to mechanical analyses with isotropic initial stress state (Fernandes 2009; Sieffert *et al.* 2009).

Moreover, during the operational phases of underground openings, an air ventilation is performed inside the galleries to control the air relative humidity and temperature. This ventilation induces fluid transfers and a desaturation of the rock that must be taken into account in the coupled second gradient model.

Many studies have been performed with two-dimensional isotropic mechanical models. Our purpose is to investigate if this type of model can reproduce the *in situ* observations and measurements by incorporating the fracture modeling with strain localization. The zone that develops around the gallery is called the excavation fractured zone, and it is related to the irreversible hydromechanical property changes.

It should be pointed out that regularization techniques have already been used for this type of problem. They generate results that are mesh independent but these theories do not restore the uniqueness of the solution for the gallery excavation problem (Fernandes 2009; Sieffert *et al.* 2009). These remarks are valid for all regularization methods.

7.6.1. *Numerical model*

A hydromechanical modeling of a gallery excavation is performed in two-dimensional plane strain state. The modeled gallery corresponds to the GED gallery of the Andra's URL oriented parallel to the minor horizontal principal total stress σ_h and having a radius of 2.3 m. The initial pore water pressure and anisotropic stress state are:

$$\sigma_{x,0} = \sigma_H = 1.3\,\sigma_h = 15.6\ \text{MPa}$$

$$\sigma_{y,0} = \sigma_v = 12 \text{ MPa}$$

$$\sigma_{z,0} = \sigma_h = 12 \text{ MPa}$$

$$p_{w,0} = 4.5 \text{ MPa}$$

A schematic representation of the models, the meshes and the boundary conditions is detailed in Figure 7.11. Two meshes are used: a full gallery and a quarter of a gallery. The mesh extension of the full gallery is 120 m, both horizontally and vertically, and the spatial discretization is performed with a total of 29,040 nodes and 7,440 elements. Assuming symmetry along the x and y-axes, only one quarter of the gallery can be discretized. In this case, the mesh extension is 60 m, both horizontally and vertically, and the discretization is performed with a total of 9,801 nodes and 2,480 elements. For both meshes, the initial stresses and pore water pressure are imposed at the mesh external boundary (drained boundary) and the meshes have a more refined discretization close to the gallery. To establish the symmetry, the normal displacements and the normal water flows are blocked to a value of zero along the symmetry axes, which are therefore impervious. Nonetheless, as mentioned by Zervos *et al.* (2001a), special care must be brought to the kinematic boundary conditions required to establish the symmetry. Due to the existence of gradient terms in the equilibrium equations, higher order constraints have to be characterized in addition to the classical boundary condition on the normal displacements. This second kinematic condition requires that the radial displacement u_r must be symmetric on both sides of the symmetry axes. This implies that the normal derivative of u_r, with respect to the tangential (orthoradial) direction θ, has to cancel:

$$\frac{\partial u_r}{\partial \theta} = 0 \qquad [7.90]$$

which is equivalent to:

$$x - axis: \quad \frac{\partial u_x}{\partial y} = 0 \qquad [7.91]$$

$$y - axis: \quad \frac{\partial u_y}{\partial x} = 0 \qquad [7.92]$$

Furthermore, natural boundary conditions for the double forces, $\overline{T}_i = 0$, are assumed on the different boundaries and gravity is not taken into account.

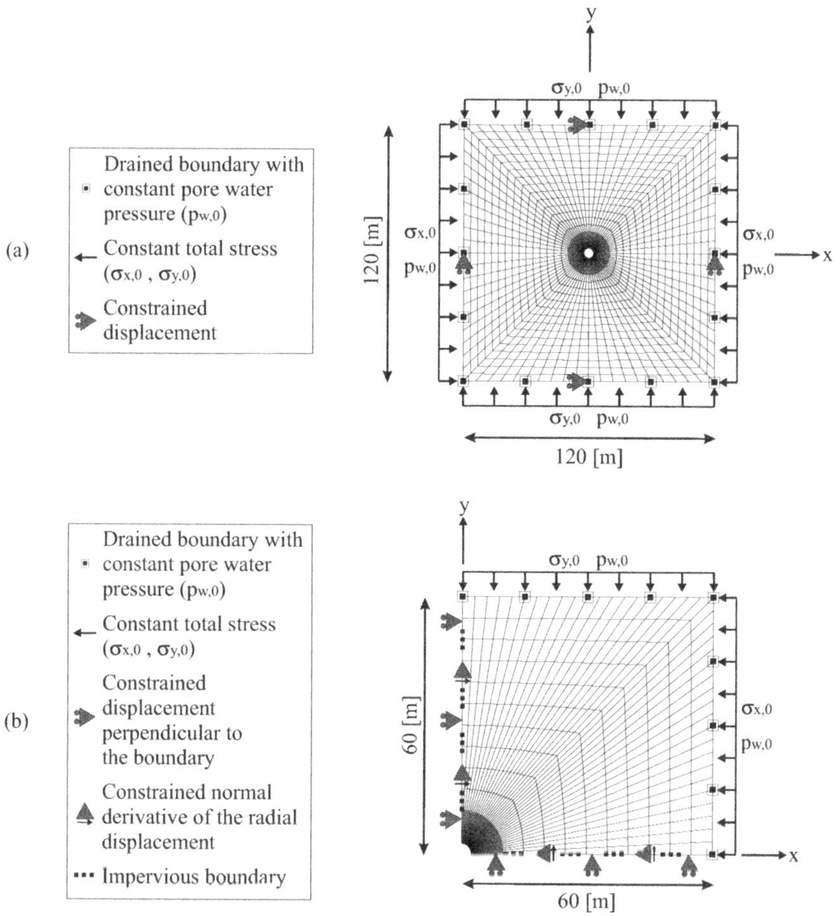

Figure 7.11. *Schematic representation of the models used for the modeling of a gallery excavation: (a) full gallery and (b) quarter of a gallery*

The gallery excavation can now be considered. It is modeled by decreasing during 5 days the total stresses and the pore water pressure at the gallery wall from their initial values to the atmospheric pressure of 100 kPa. After the excavation, the calculation is extended to 1,000 days under constant total radial stress to highlight possible long-term effects (Figure 7.12). This stress imposition is representative of unsupported galleries.

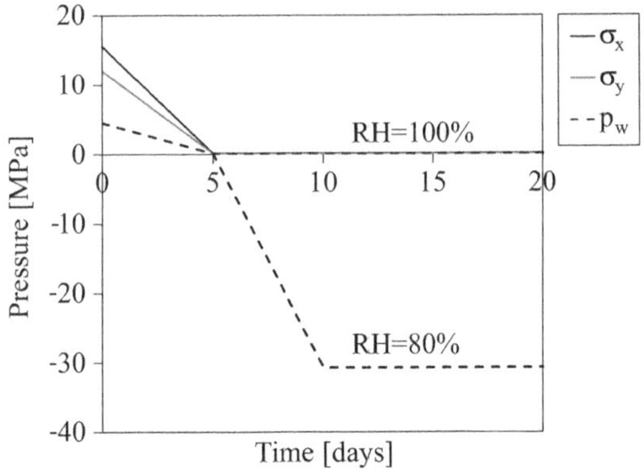

Figure 7.12. *Imposed total stresses and pore water pressure at the gallery wall for the modeling of a gallery excavation with and without air ventilation*

To model the air ventilation inside the gallery, a classical flow boundary condition is assumed and imposes the suction corresponding to the relative humidity of the cavity air at the tunnel wall. Two cases are considered for the air inside the gallery (Figure 7.12). In the first case, there is no ventilation inside the gallery; thus, the air is saturated with water vapor and this maximum concentration corresponds to $RH = 100\%$. According to Kelvin's law, the corresponding pore water pressure at the gallery wall is the atmospheric pressure $p_w = 100$ kPa. The pore water pressure is then maintained constant after the end of the excavation and the rock mass remains almost saturated. In the second case, air ventilation is taken into account, since ventilation is usually realized in the galleries composing underground structures. It may drain the water from the rock, desaturate it and modify the structure and the fracturing pattern, as well as the size of the fractured zone. Air ventilation can thus be modeled in order to observe its effects on the rock material. A theoretical ventilation, with constant air relative humidity, is envisaged to obtain a first outlook of the ventilation effect on shear banding. The air that is injected in the gallery is dryer than previously and a lower relative humidity of 80% with a temperature of 25°C ($T = 298.15$ K) is considered. Following Kelvin's law, this humidity corresponds to a pore water pressure at gallery wall of $p_w = -30.7$ MPa. To reach this value, the decrease in p_w is performed in two steps: first, it decreases from its initial value to the atmospheric pressure during the excavation (5 days), and then an initiation phase of ventilation is considered (5 days) to reach the final value. After this initiation phase, a constant ventilation is maintained.

The imposed boundary conditions at the gallery wall, for total stresses and pore water pressure evolutions, are presented in Figure 7.12 for the two considered cases. It is worth mentioning that the ventilation effect on the shear banding is therefore represented by the hydromechanical model. In fact, the ventilation influences the pore water pressures and the effective stresses, which then influence the shear strain localization structure and behavior.

The main purpose of this numerical modeling is to represent the fractures with shear strain localization and to reproduce, as well as possible, the *in situ* measurements and observations with an isotropic mechanical model.

7.6.2. *Influence of stress and permeability anisotropies*

Creation and evolution of the fractured zone can be observed through the evolution of shear strain localization. The latter is not *a priori* assured to be symmetric around the gallery and many solutions could emerge (Sieffert *et al.* 2009). To avoid any early symmetry assumption, the excavation of a full gallery is first modeled with incompressible solid grains $b = 1$ and no ventilation. With a circular gallery and an isotropic state, it is not possible to trigger the shear strain localization and the deformation remains diffuse. Strain localization can be triggered through the introduction of an imperfection in the material.

However, in case of anisotropic stress state of the rock with $\sigma_{x,0} = 15.6$ MPa and $\sigma_{y,0} = \sigma_{z,0} = 12$ MPa, the shear strain localization appears without adding an imperfection in the rock. Figure 7.13 illustrates the evolution of the strain localization around the gallery, during and after drilling. The numerical results presented are the total deviatoric strain, the plastic zone and the deviatoric strain increment, which represents the band activity:

$$\hat{\varepsilon}_{eq} = \sqrt{\frac{2}{3}\,\hat{\varepsilon}_{ij}\,\hat{\varepsilon}_{ij}} \qquad [7.93]$$

where $\hat{\varepsilon}_{ij}$ is the deviatoric total strain field calculated from the total strain tensor ε_{ij}:

$$\hat{\varepsilon}_{ij} = \varepsilon_{ij} - \frac{\varepsilon_{kk}}{3}\,\delta_{ij} \qquad [7.94]$$

$$\kappa_{eq} = \frac{\dot{\hat{\varepsilon}}_{eq}\,dt}{\int \dot{\hat{\varepsilon}}_{eq}\,dt} \qquad [7.95]$$

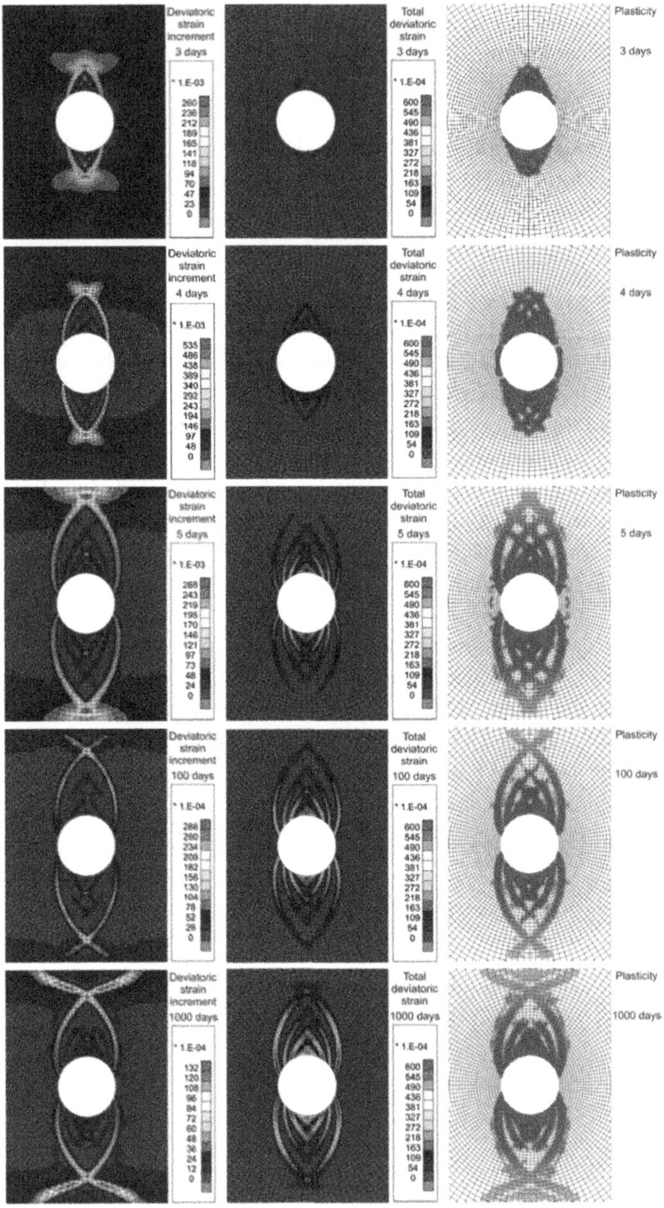

Figure 7.13. *Evolution of strain localization during and after the gallery excavation (5 days of excavation) for a full gallery and for a rock having anisotropic hydraulic permeability and anisotropic stress state. For a color version of this figure, see www.iste.co.uk/stefanou/instabilities.zip*

The modeling exhibits a symmetric chevron fracture pattern around the gallery similar to *in situ* observations for galleries parallel to σ_h. The chevron fractures appear during the excavation and are mainly concentrated above the gallery because of the material anisotropic stress state. On the contrary, introducing only the anisotropy of the intrinsic water permeability with $k_{w,h} = 4 \times 10^{-20}\ m^2$ and $k_{w,v} = 1.33 \times 10^{-20}\ m^2$ does not lead to strain localization unless an imperfection is introduced. This means that the appearance and shape of the strain localization are mainly due to mechanical effects linked to the anisotropic stress state. The shear banding zone develops preferentially in the direction of the minor principal stress in the gallery section. More recent modeling of the effect of material anisotropic behaviour on shear banding around underground galleries can be found in Pardoen *et al.* (2015b).

7.6.3. *Influence of second gradient boundary condition*

The previous modeling highlights that the anisotropic stress state is at the origin of a symmetry in the localization pattern around the gallery. So, it would be convenient, in the following, to consider only a quarter of a gallery. However, in the context of second gradient theory, a boundary condition of higher order should be considered in addition to the classical boundary condition of constrained displacement perpendicular to the boundary (Zervos *et al.* 2001a). This second kinematic condition specifies that the normal derivative of the radial displacement has to cancel on the symmetry axes.

To illustrate the necessity of this second gradient boundary condition, the strain localization pattern of Figure 7.13 is compared to the pattern obtained on a quarter of a gallery. The modeling on a quarter of a gallery is computed with the specific second gradient boundary condition, and with $b = 1$ and no ventilation as previously. In Figure 7.14, one can observe that using the second gradient boundary condition produces a shear strain localization pattern that is similar to the full gallery results. Thus, it is confirmed that, for calculation simplicity and symmetry reasons, a quarter of a gallery can be adopted for future modeling, provided that the specific second gradient boundary condition is used.

7.6.4. *Influence of Biot's coefficient*

Even if strain localization seems to be mainly controlled by mechanical effects, hydraulic conditions can also impact the shear banding pattern. Here, the focus is on the influence of Biot's coefficient for the case without ventilation. In the first calculation, it is assumed that the solid grains are incompressible, which implies $b = 1$ (Figure 7.15). In the second calculation, a value of $b = 0.6$ is used (Figure 7.16). A comparison of Figures 7.15 and 7.16 indicates that Biot's coefficient significantly influences the shear band pattern. With a value of 0.6, less bands appear and the shear strain localization is delayed. In fact, the strain remains diffuse until the fourth day of the excavation; nonetheless, the localization appears before the end of the excavation. This can be explained by examining the stresses close to the gallery. At the gallery wall, the total stresses and the pore water pressure are imposed. Consequently, following Biot's effective stress definition for unsaturated materials of equation [7.82], the lower the Biot coefficient, the higher the effective compressive stress at the gallery wall. This implies that the rock close to the gallery wall is more resistant and that the shear strain localization appears later.

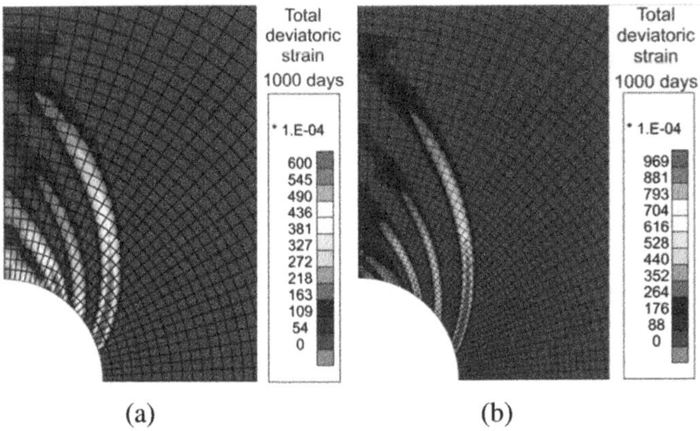

Figure 7.14. *Comparison of the strain localization pattern at the end of the calculation for the modeling of (a) a full gallery and (b) a quarter of a gallery with the second gradient boundary condition. For a color version of this figure, see www.iste.co.uk/stefanou/instabilities.zip*

7.6.5. *Influence of gallery ventilation*

The modeling presented hereafter includes the initial anisotropies, a Biot's coefficient value of 0.6 and the gallery ventilation. The drilling phase is not influenced by the ventilation and the same results as in Figure 7.16 are obtained until 5 days of computation. The results obtained after the excavation, displayed in Figure 7.17, indicate that the suction imposed at the wall strongly influences the results. Following the effective stress definition, the higher the suction, the higher the effective stress (Figure 7.20). As noted before, this involves the material being more resistant, and in this case, it becomes elastic again close to the gallery. This inhibits the shear strain localization around the gallery.

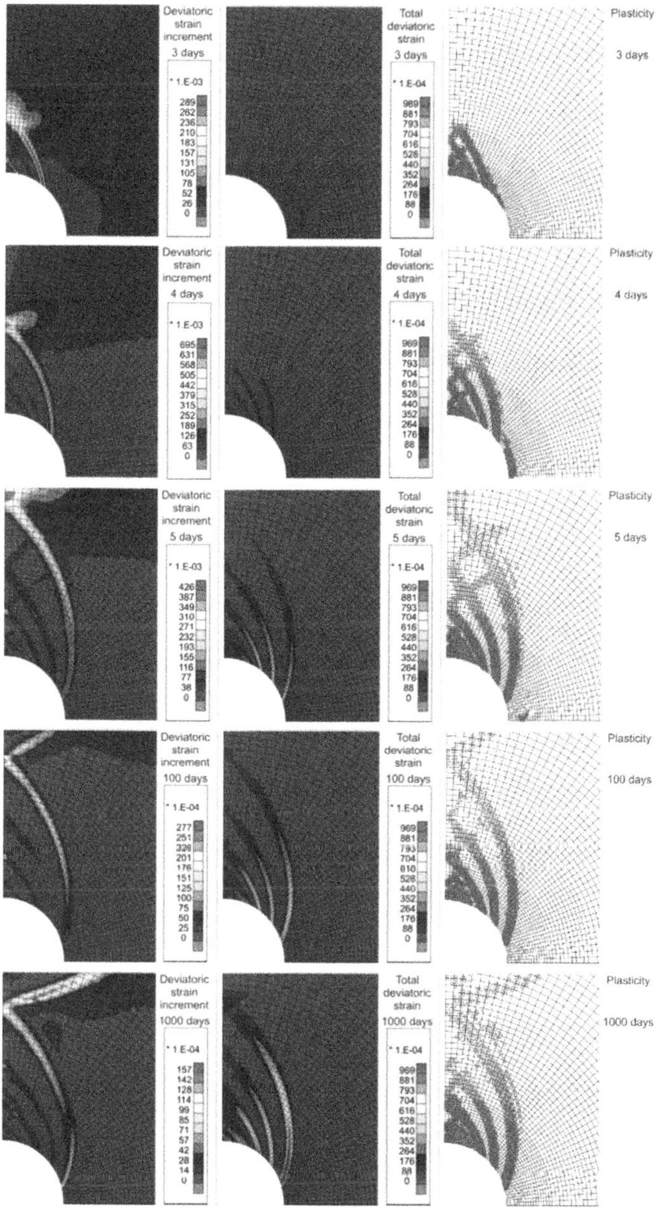

Figure 7.15. *Evolution of strain localization during and after the gallery excavation (5 days of excavation), without gallery ventilation and for a Biot coefficient value of 1. For a color version of this figure, see www.iste.co.uk/stefanou/instabilities.zip*

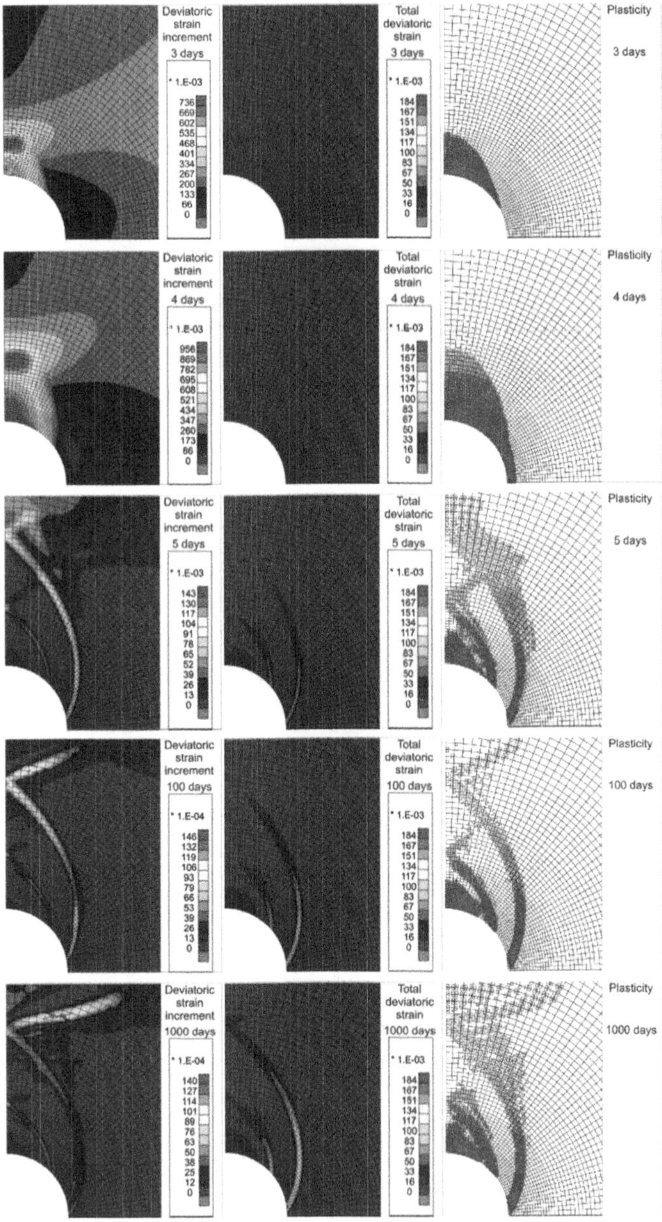

Figure 7.16. *Evolution of strain localization during and after the gallery excavation (5 days of excavation), without gallery ventilation and for a Biot coefficient value of 0.6. For a color version of this figure, see www.iste.co.uk/stefanou/instabilities.zip*

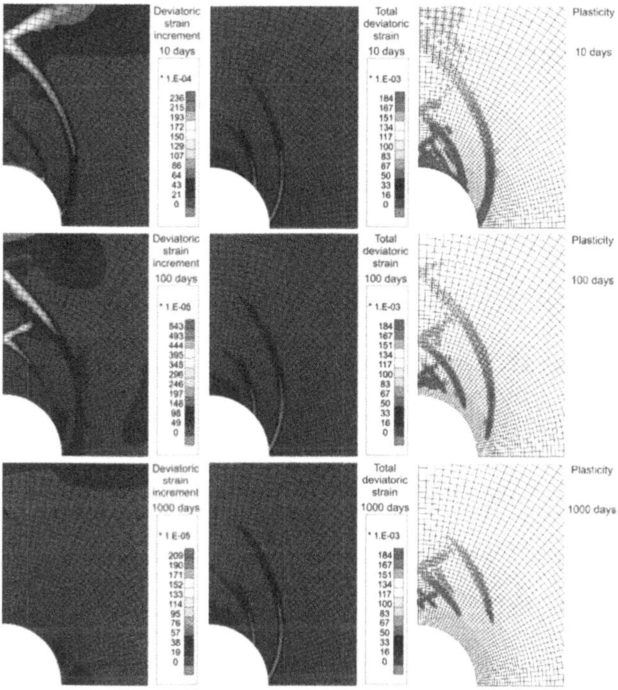

Figure 7.17. *Evolution of strain localization after the gallery excavation, with gallery ventilation and for a Biot coefficient value of 0.6. For a color version of this figure, see www.iste.co.uk/stefanou/instabilities.zip*

Various numerical results, coming from the gallery wall and the rock mass, are interpreted hereafter in order to emphasize the influence of the gallery air ventilation. The results come from the selected cross-sections and observation points on the gallery wall presented in Figure 7.18. The vertical cross-section goes through the shear bands and the results along it highlight the effects of strain localization, which is not the case for the horizontal cross-section. Furthermore, the results are compared for the cases considering ($RH = 80$ %) or not ($RH = 100$ %) the ventilation.

The evolution of pore water pressure for the vertical and horizontal cross-sections is detailed in Figure 7.19. In the rock mass, an increase of pressure is observed in the vertical direction and a decrease is observed in the horizontal direction up to a radial distance of about 30 m. These overpressures are related to the hydromechanical coupling induced by the anisotropy of the initial stress state. The influence of the strain localization bands is visible vertically but not horizontally. It is illustrated by the fluctuations of the pore water pressure in limited zones, with a decrease in the shear band. The influence of the shear band can mostly be observed during the first

50 days of calculation, then it tends to vanish. This is due to the strain increment inside the bands (band activity) and the hydromechanical coupling. As expected, the influence of the ventilation is marked close to the gallery wall, but tends to disappear deeper in the rock.

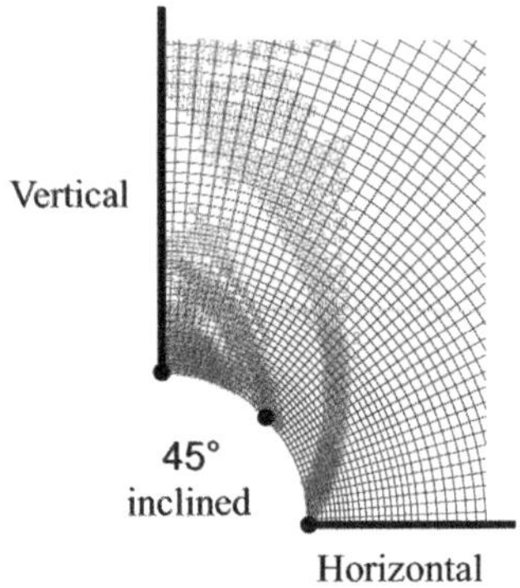

Figure 7.18. *Positions of cross-sections and gallery wall observation points. For a color version of this figure, see www.iste.co.uk/stefanou/instabilities.zip*

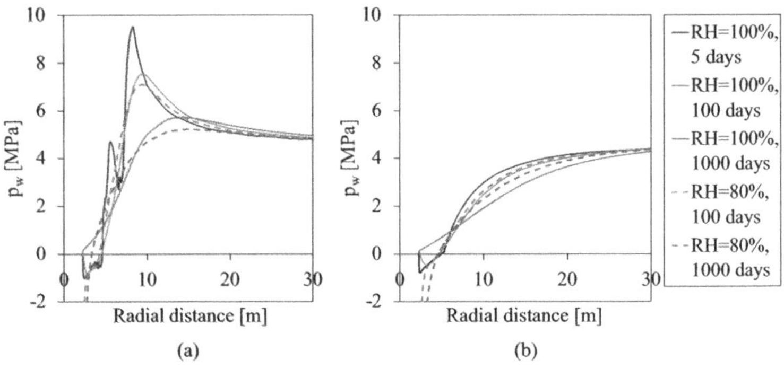

Figure 7.19. *Evolution of pore water pressure along (a) vertical and (b) horizontal cross-sections, after the gallery excavation. For a color version of this figure, see www.iste.co.uk/stefanou/instabilities.zip*

The stress paths at the gallery wall are detailed in Figure 7.20 where q is the deviatoric stress:

$$q = \sqrt{3}\, II_{\hat{\sigma}'} \qquad [7.96]$$

and p' is the mean effective stress. As mentioned before, in the case of ventilation, the effective stresses are much higher due to suction. This explains the difference between the stress paths of the modeling with and without ventilation, after the end of the drilling phase.

All these results are evidences that noticeable differences exist whether ventilation is applied or not. For the modeling with ventilation, p_w remains negative close to the gallery (Figure 7.19), the effective stresses increase after the excavation (Figure 7.20) and the material becomes elastic again. Consequently, the desaturation of the rock close to the gallery inhibits the shear strain localization (Figure 7.17), which has the effect of restricting further deformation. On the contrary, without ventilation, p_w close to the gallery wall increases after the excavation (Figure 7.19), the effective stresses reduce (Figure 7.20) and the material remains partly plastic close to the gallery (Figure 7.16). This increases the deformation and the gallery convergence.

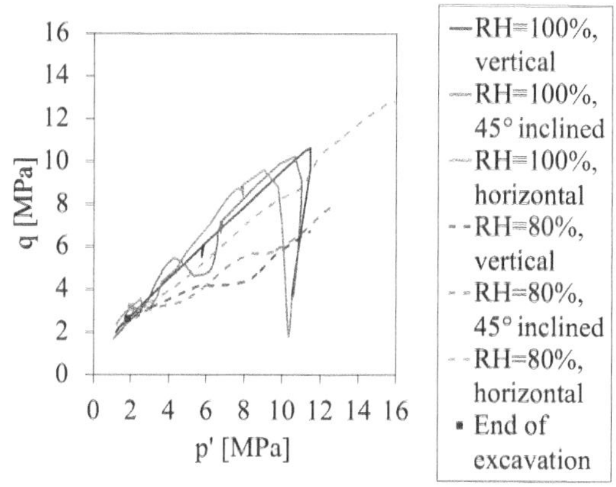

Figure 7.20. *Stress paths at the gallery wall, during and after gallery excavation. For a color version of this figure, see www.iste.co.uk/stefanou/instabilities.zip*

If the problem is studied with an isotropic model, without considering strain localization but modeling the gallery ventilation, then the horizontal and vertical convergences are more or less equal. Only the fracturing and strain localization processes allow us to produce the convergence anisotropy. In fact, neither the fracturing pattern nor the gallery convergence can be well reproduced with a classical approach (Pardoen and Collin 2017). The creation of fractures, globally above the gallery due to the material anisotropic stress state, increases both the vertical and the horizontal convergences. In that latter direction, the proximity of the shear bands induces excessive deformations. In the long term, the delayed deformations that are observed in saturated conditions are explained by consolidation. In contrast to this, when gallery ventilation is reproduced, the material close to the gallery wall becomes elastic again, which restricts the plastic deformation and convergence in the long term. The performed numerical modeling highlights the effect of gallery ventilation on the hydraulic transfer and progressive drainage of the surrounding rock. However, the considered air ventilation is theoretical and a real ventilation could be considered (Charlier *et al.* 2013; Pardoen *et al.* 2016). Moreover, the water transfer and its kinetics close to the gallery are mainly conditioned by the hydromechanical property changes inside the excavation damaged zone. In fact, the damaged zone developing around galleries due to the drilling process is composed of fractures having a significant irreversible impact on flow and transport characteristics (Tsang *et al.* 2005). For a shear banding approach, the impact of fracturing on the transport properties can be addressed by associating the intrinsic permeability increase with mechanical deformation (Pardoen *et al.* 2016), which is amplified in the strain localization discontinuities. Such a dependence allows us to reproduce a significant permeability increase in several orders of magnitude in the excavation damaged zone (Pardoen *et al.* 2016), in agreement with available experimental measurements (Armand *et al.* 2014).

7.7. Conclusion

Rupture in geomaterials is often preceded by a localization of the deformations within thin bands. The strain localization is thus an important process, which has been studied both experimentally and theoretically. The developments of geomechanics in the field of coupled multiphysic processes impose the study of strain localization to these new conditions. Interactions between the different processes can indeed occur. Furthermore, the numerical modeling of shear bands with classical finite element suffers from a mesh dependency problem. An internal length scale has to be introduced in the model. Among the different regularization techniques, we propose a second gradient coupled model for an application to gallery excavation. It has been shown that the model regularizes the solution but does not restore its uniqueness. The extension of such theories to other multiphysic context is more an experimental problem than a numerical one. Experiments still have to exhibit the influence of temperature, suction or chemical concentration on the occurrence and the thickness of the strain localization!

7.8. References

Aifantis, E.C. (1984). On the microstructural origin of certain inelastic models. *J. Eng. Mater. Technol.*, 106(4), 326–330.

Alshibli, K., Batiste, S., Sture, S. (2003). Strain localization in sand: Plane strain versus triaxial compression. *J. Geotech. Geoenviron. Eng.*, 129(6), 483–494.

Armand, G., Leveau, F., Nussbaum, C., de La Vaissiere, R., Noiret, A., Jaeggi, D., Landrein, P., Righini, C. (2014). Geometry and properties of the excavation-induced fractures at the Meuse/Haute-Marne URL drifts. *Rock Mech. Rock Eng.*, 47(1), 21–41.

Bazant, Z.P., Belytschko, T.B., Chang, T.P. (1984). Continuum theory for strain softening. *J. Eng. Mech.*, 110(12), 1666–1692.

Biot, M.A. (1941). General theory for three-dimensional consolidation. *J. Appl. Phys.*, 12(2), 155–164.

Biot, M.A. and Willis, D.G. (1957). The elastic coefficients of the theory of consolidation. *J. Appl. Mech.*, 24, 594–601.

Borja, R.I. and Alarcón, E. (1995). A mathematical framework for finite strain elastoplastic consolidation part 1: Balance laws, variational formulation, and linearization. *Comput. Methods Appl. Mech. Engrg.*, 122(1-2), 145–171.

de Borst, R. and Mühlhaus, H.B. (1992). Gradient-dependent plasticity: Formulation and algorithm aspects. *Int. J. Numer. Meth. Engng.*, 35(3), 521–539.

Bésuelle, P., Desrues, J., Raynaud, S. (2000). Experimental characterisation of the localisation phenomenon inside a vosges sandstone in a triaxial cell. *Int. J. Rock Mech. Min. Sci.*, 37(8), 1223–1237.

Bésuelle, P., Chambon, R., Collin, F. (2006). Switching deformation modes in post-localization solutions with a quasibrittle material. *J. Mech. Mater. Struct.*, 1(7), 1115–1134.

Chambon, R. and Moullet, J.C. (2004). Uniqueness studies in boundary value problems involving some second gradient models. *Comput. Methods Appl. Mech. Engrg.*, 193(27–29), 2771–2796.

Chambon, R., Caillerie, D., Hassan, N.E. (1998). One-dimensional localisation studied with a second grade model. *Eur. J. Mech. A-Solid*, 17(4), 637–656.

Chambon, R., Caillerie, D., Matsushima, T. (2001a). Plastic continuum with microstructure, local second gradient theories for geomaterials: Localization studies. *Int. J. Solids Struct.*, 38(46–47), 8503–8527.

Chambon, R., Crochepeyre, S., Charlier, R. (2001b). An algorithm and a method to search bifurcation points in non-linear problems. *Int. J. Numer. Meth. Engng.*, 51(3), 315–332.

Charlier, R., Radu, J.P., Barnichon, J.D. (1997). Water movement effect on the strain localisation during a biaxial compression. In Pande, G., Pietruszczak, S. (eds). *Numerical Models in Geomechanics, NUMOG VI*, Balkema, Rotterdam, 219–224.

Charlier, R., Collin, F., Pardoen, B., Talandier, J., Radu, J.P., Gerard, P. (2013). An unsaturated hydro-mechanical modelling of two in-situ experiments in Callovo-Oxfordian argillite. *Eng. Geol.*, 165, 46–63.

Collin, F., Chambon, R., Charlier, R. (2006). A finite element method for poro mechanical modelling of geotechnical problems using local second gradient models. *Int. J. Numer. Meth. Engng.*, 65(11), 1749–1772.

Collin, F., Caillerie, D., Chambon, R. (2009a). Analytical solutions for the thick-walled cylinder problem modeled with an isotropic elastic second gradient constitutive equation. *Int. J. Solids Struct.*, 46(22–23), 3927–3937.

Collin, F., Levasseur, S., Chambon, R. (2009b). Numerical post failure methods in multiphysical problems. *Eur. J. Environ. Civ. Eng.*, 13(7–8), 983–1004.

Cosserat, E. and Cosserat, F. (1909). *Théorie des Corps Déformables*. Éditions Hermann, Paris.

Coulomb, C.A. (1773). Essai sur une application des règles de maximis et minimis à quelques problèmes de statique, relatifs à l'architecture. *Académie Royale des Sciences*, 7, 343–382.

Coussy, O. (1995). *Mechanics of Porous Continua*. John Wiley & Sons, Chichester.

Coussy, O. (2004). *Poromechanics*. John Wiley & Sons, Chichester.

Desrues, J. (1984). La localisation de la déformation dans les matériaux granulaires. PhD thesis, Université Joseph Fourier, Institut National Polytechnique, Grenoble.

Desrues, J. (2005). Hydro-mechanical coupling and strain localization in saturated porous media. *Rev. Eur. Génie Civ.*, 9(5–6), 619–634.

Desrues, J. and Viggiani, G. (2004). Strain localization in sand: An overview of the experimental results obtained in Grenoble using stereophotogrammetry. *Int. J. Numer. Anal. Meth. Geomech.*, 28(4), 279–321.

Detournay, E. and Cheng, A.H.D. (1993). Fundamentals of poroelasticity. In *Comprehensive Rock Engineering: Principles, Practice and Projects 2: Analysis and Design Method*, J.A. Hudson (ed.). Pergamon Press, Oxford.

Diederichs, M.S. (2003). Rock fracture and collapse under low confinement conditions. *Rock Mech. Rock Eng.*, 36(5), 339–381.

Ehlers, W. and Volk, W. (1998). On theoretical and numerical methods in the theory of porous media based on polar and non-polar elasto-plastic solid materials. *Int. J. Solids Struct.*, 35(34–35), 4597–4617.

Fernandes, R. (2009). Modélisation numérique objective des problèmes hydromécaniques couplés dans le cas des géomatériaux. PhD thesis, Université Joseph Fourier, Grenoble.

Finno, R., Harris, W., Mooney, M., Viggiani, G. (1996). Strain localization and undrained steady state of sands. *J. Geotech. Engrg.*, 122(6), 462–473.

Finno, R., Harris, W., Mooney, M., Viggiani, G. (1997). Shear bands in plane strain compression of loose sand. *Géotechnique*, 47(1), 149–165.

van Genuchten, M.T. (1980). A closed-form equation for predicting the hydraulic conductivity of unsaturated soils. *Soil. Sci. Soc. Am. J.*, 44(5), 892–898.

Germain, P. (1973). The method of virtual power in continuum mechanics. Part 2: Microstructure. *SIAM J. Appl. Math.*, 25(3), 556–575.

Guy, N., Seyedi, D.M., Hild, F. (2012). A probabilistic nonlocal model for crack initiation and propagation in heterogeneous brittle materials. *Int. J. Numer. Meth. Engng.*, 90(8), 1053–1072.

Han, C. and Drescher, A. (1993). Shear bands in biaxial tests on dry coarse sand. *Soils Found*, 33(1), 118–132.

Jenq, Y.S. and Shah, S.P. (1988). Mixed-mode fracture of concrete. *Int. J. Fracture*, 38(2), 123–142.

Jouan, G., Kotronis, P., Collin, F. (2014). Using a second gradient model to simulate the behaviour of concrete structural elements. *Finite Elements in Analysis and Design*, 90, 50–60 [Online]. Available at: http://www.sciencedirect.com/science/article/pii/S0168874X14001085.

Khoa, H.D.V., Georgopoulos, I.O., Darve, F., Laouafa, F. (2006). Diffuse failure in geomaterials: Experiments and modelling. *Comput. Geotech.*, 33(1), 1–14.

Kotronis, P. (2008). Stratégies de modélisation de structures en béton soumises à des chargements sévères. Thesis, Université Joseph-Fourier-Grenoble I. [Online]. Available at: http://tel.archives-ouvertes.fr/tel-00350461.

Kotronis, P., Chambon, R., Mazars, J., Collin, F. (2005). Paper 5712: Local second gradient models and damage mechanics: Application to concrete. In *11th International Conference on Fracture*, Curran Associates, Inc., Turin.

Kotronis, P., Collin, F., Bésuelle, P., Chambon, R., Mazars, J. (2007). Local second gradient models and damage mechanics: 1D post-localization studies in concrete specimens. In *Bifurcation, Instabilities and Degradation in Geomechanics*, Exadaktylos, G., Vardoulakis, I. (eds). Springer, 127–142.

Kotronis, P., Holo, S.A., Bésuelle, P., Chambon, R. (2008). Shear softening and localization: Modelling the evolution of the width of the shear zone. *Acta. Geotechnica.*, 3(2), 85–97.

Lasry, D. and Belytschko, T. (1988). Localization limiters in transient problems. *Int. J. Solids Struct.*, 24(6), 581–597.

Lenoir, N., Bornert, M., Desrues, J., Bésuelle, P., Viggiani, G. (2007). Volumetric digital image correlation applied to X-ray microtomography images from triaxial compression tests on argillaceous rock. *Strain*, 43(3), 193–205.

Marinelli, F., Sieffert, Y., Chambon, R. (2014). Hydromechanical modelling of an initial boundary value problem: Studies of non-uniqueness with a second gradient continuum. *Int. J. Solids Struct.*, 54, 238–257 [Online]. Available at: https://www.sciencedirect.com/science/article/pii/S0020768314003837.

Matsushima, T., Chambon, R., Caillerie, D. (2002). Large strain finite element analysis of a local second gradient model: Application to localization. *Int. J. Numer. Meth. Engng.*, 54(4), 499–521.

Mindlin, R.D. (1964). Micro-structure in linear elasticity. *Arch. Ration. Mech. An.*, 16(1), 51–78.

Mindlin, R.D. (1965). Second gradient of strain and surface-tension in linear elasticity. *Int. J. Solids Struct.*, 1, 417–438.

Mokni, M., Desrues, J. (1999). Strain localisation measurements in undrained plane-strain biaxial test on Hostun RF sand. *Mech. Cohes-Frict. Mat.*, 4(4), 419–441.

Mualem, Y. (1976). A new model for predicting the hydraulic conductivity of unsaturated porous media. *Water Resour. Res.*, 12(3), 513–522.

Nuth, M., Laloui, L. (2008). Effective stress concept in unsaturated soils: Clarification and validation of a unified framework. *Int. J. Numer. Anal. Meth. Geomech.*, 32(7), 771–801.

Papanastasiou, P.C. and Vardoulakis, I.G. (1992). Numerical treatment of progressive localization in relation to borehole stability. *Int. J. Numer. Anal. Meth. Geomech.*, 16(6), 389–424.

Pardoen, B. and Collin, F. (2017). Modelling the influence of strain localisation and viscosity on the behaviour of underground drifts drilled in claystone. *Comput. Geotech.*, 85, 351–367.

Pardoen, B., Levasseur, S., Collin, F. (2015a). Using local second gradient model and shear strain localisation to model the excavation damaged zone in unsaturated claystone. *Rock Mech. Rock Eng.*, 48(2), 691–714.

Pardoen, B., Seyedi, D.M., and Collin, F. (2015b). Shear banding modelling in cross-anisotropic rocks. *International Journal of Solids and Structures*, 72, 63–87. Doi: 10.1016/j.ijsolstr.2015.07.012.

Pardoen, B., Talandier, J., Collin, F. (2016). Permeability evolution and water transfer in the excavation damaged zone of a ventilated gallery. *Int. J. Rock Mech. Min. Sci.*, 85, 192–208.

Peerlings, R.H.J., de Borst, R., Brekelmans, W.A.M., de Vree, J.H.P. (1996a). Gradient enhanced damage for quasi-brittle materials. *Int. J. Numer. Meth. Engng.*, 39(19), 3391–3403.

Peerlings, R.H.J., de Borst, R., Brekelmans, W.A.M., de Vree, J.H.P., Spee, I. (1996b). Some observations on localisation in non-local and gradient damage models. *Eur. J. Mech. A/Solids*, 15(6), 937–953.

Peerlings, R.H.J., Geers, M.G.D., de Borst, R., Brekelmans, W.A.M. (2001). A critical comparison of nonlocal and gradient-enhanced softening continua. *Int. J. Solids Struct.*, 38(44–45), 7723–7746.

Pietruszczak, S.T. and Mróz, Z. (1981). Finite element analysis of deformation of strain-softening materials. *Int. J. Numer. Meth. Engng.*, 17(3), 327–334.

Pijaudier-Cabot, G. and Bazant, Z.P. (1987). Nonlocal damage theory. *J. Eng. Mech.*, 113(10), 1512–1533.

Rice, J.R. (1976). The localization of plastic deformation. In *Theoretical and Applied Mechanics 1*, Koiter, W.T. (ed.). North-Holland Publishing Company, Delft, 207–220.

Rudnicki, J.W. and Rice, J.R. (1975). Conditions for the localisation of deformation in pressure sensitive dilatant materials. *J. Mech. Phys. Solids*, 23, 371–394.

Sieffert, Y., al Holo, S., Chambon, R. (2009). Loss of uniqueness of numerical solutions of the borehole problem modelled with enhanced media. *Int. J. Solids Struct.*, 46(17), 3173–3197.

Skempton, A.W. (1960). Effective stress in soils, concrete and rocks. In *Pore Pressure and Suction in Soils Conference*, Butterworths, London, 4–16.

Soufflet, M., Jouan, G., Kotronis, P., Collin, F. (2018). Using a penalty term to deal with spurious oscillations in second gradient finite elements. *International Journal of Damage Mechanics*, 28(3) [Online]. Available at: https://doi.org/10.1177/1056789518769341.

Toupin, R. (1962). Elastic materials with couple-stresses. *Arch. Ration. Mech. An.*, 11(1), 385–414.

Tsang, C.F., Bernier, F., Davies, C. (2005). Geohydromechanical processes in the Excavation Damaged Zone in crystalline rock, rock salt, and indurated and plastic clays – in the context of radioactive waste disposal. *Int. J. Rock Mech. Min. Sci.*, 42(1), 109–125.

Vardoulakis, I. and Sulem, J. (1995). *Bifurcation Analysis in Geomechanics*. Blackie Academic and Professional, London.

Vardoulakis, I., Goldscheider, M., Gudehus, Q. (1978). Formation of shear bands in sand bodies as a bifurcation problem. *Int. J. Numer. Anal. Meth. Geomech.*, 2(2), 99–128.

Wu, S. and Wang, X. (2010). Mesh dependence and nonlocal regularization of one-dimensional strain softening plasticity. *J. Eng. Mech.*, 136(11), 1354–1365.

Zauderer, E. (1989). *Partial Differential Equations of Applied Mathematics*, 2nd edition. Wiley, Chichester.

Zervos, A., Papanastasiou, P., Vardoulakis, I. (2001a). Modelling of localisation and scale effect in thick-walled cylinders with gradient elastoplasticity. *Int. J. Solids Struct.*, 38(30–31), 5081–5095.

Zervos, A., Papanastasious, P., Vardoulakis, I. (2001b). A finite element displacement formulation for gradient elastoplasticity. *Int. J. Numer. Meth. Engng.*, 50(6), 1369–1388.

Zhang, H.W., Sanavia, L., Schrefler, B.A. (2001). Numerical analysis of dynamic strain localisation in initially water saturated dense sand with a modified generalised plasticity model. *Comput. Struct.*, 79(4), 441–459.

Zienkiewicz, O.C., Taylor, R.L. (2000). *The Finite Element Method*, 5th edition. Butterworth-Heinemann, Stonchem.

8

Multiphysics Couplings and Strain Localization in Geomaterials

Jean SULEM[1] and Ioannis STEFANOU[2]

[1] *Navier, Ecole des Ponts ParisTech, Université Paris-Est, Marne-la-Vallée, France*

[2] *GeM, Ecole Centrale de Nantes, France*

The purpose of this chapter is to provide a review of recent research regarding the effects of temperature, pore pressure, chemical reactions and microstructure on strain localization in geomaterials. Examples have been taken in relation with seismic slip and with compaction banding. Strain localization is treated as instability from a homogeneous deformation state.

8.1. Introduction

Although strain localization in the form of shear band formation can occur with a negative or positive rate of strain hardening, the latter being possible for deformation states close to plane strain (Rudnicki and Rice 1975), softening behavior definitely favors shear banding. This softening behavior may correspond to a mechanical degradation of the rock properties (microcracking, grain crushing and grain size reduction, etc.) (Das *et al.* 2011), but various other physical processes can be responsible for it (Regenauer-Lieb *et al.* 2013). The effect of an infiltrated pore fluid that interacts with a rock mass can lead to hardening or softening behavior depending on the volumetric response of the rock (dilatant or contractant). The effect of rapid heating of saturated geomaterial leads to pore-fluid pressurization due to the discrepancy between the thermal expansion of water and solid grains. Thermal pressurization is a softening mechanism as it results in a decrease in the effective mean stress and thus in the shear strength. Chemical reactions, such as

Instabilities Modeling in Geomechanics,

dissolution/precipitation and mineral transformation at high temperatures (dehydration of minerals, decomposition of carbonates, etc.), affect the solid phase of the rock, sometimes release a new fluid phase in the system (dehydration reactions) and can induce a positive feedback in the progressive mechanical degradation. On the other hand, mechanical damage increases the reaction surface between the reactive fluid and the solid and enhances dissolution and further material weakening (Hu and Hueckel 2007a, 2007b).

A key parameter when studying multiphysics effects on the formation and evolution of deformation bands is the actual width of the localized zone. Obviously, this parameter plays a major role in the energy budget of the system as it controls the feedback of the dissipative terms in the energy balance equation. As emphasized by Rice *et al.* (2014), narrow deforming zones concentrate on the frictional heating, which leads to large temperature rises and thus to more rapid weakening. The width of the deforming zone is determined by the various physical processes involved in the weakening mechanisms but it also controls the multiphysics couplings, which occur during dynamic slip.

It is well known that strain localization analyses performed for rate-independent materials within the frame of classical continuum theories lead to an infinitesimally narrow localized zone. This reflects the ill-posedness of the underlying mathematical problem and can be traced to the absence of a material length in the constitutive equations. Viscous regularization by considering strain rate hardening is commonly considered to overcome this problem. Another approach is to resort to continuum models with microstructure to describe the localization phenomena on a more physical basis. These generalized continua usually contain additional kinematical degrees of freedom (Cosserat continuum) and/or higher deformation gradients (higher grade continuum). They introduce material internal lengths and also characteristic time scales. The internal length and the microinertia introduced from generalized continua permits us to describe localization phenomena in zones of finite thickness and lead to a finite evolution rate of a deformation band like in strain rate dependent constitutive laws (Sulem *et al.* 2011; Veveakis *et al.* 2012, 2013).

In this chapter, we review some multiphysics couplings which enhance strain localization in geomaterials. The first part of the paper focuses on shear banding, emphasizing thermal and chemical effects in relation with shear heating. As mentioned above, the localization zone thickness can be captured either by considering rate dependency of the constitutive law or by resorting to higher order continua that possess an internal length. In order to explore the link between the two different modeling approaches, a comparison between (1) a rate-dependent Cauchy continuum and (2) a rate-independent Cosserat continuum is shown for the scaling of the localized zone thickness. The comparison is made on the basis of a simplified example in order to illustrate the main differences and aspects of each modeling strategy. The last part of the chapter is focused on compaction band formation in porous materials triggered by dissolution as an example of another type of chemomechanically induced strain localization.

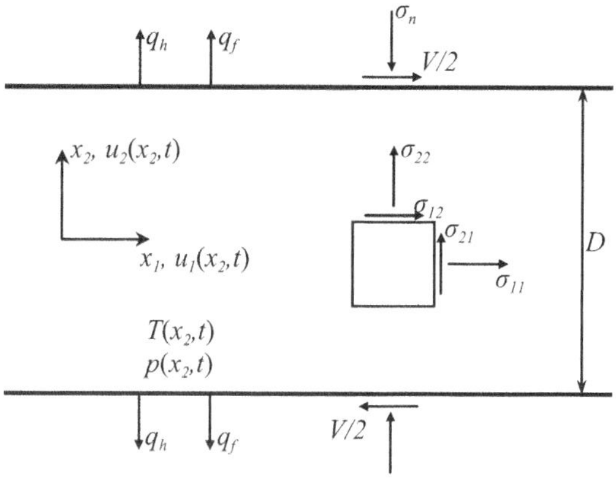

Figure 8.1. *Model of a fault zone as an infinite layer under uniform shear strain rate* $\dot{\gamma} = \frac{V}{D}$

8.2. Thermo-chemo-chemical couplings and stability of shear zones

8.2.1. *Problem statement*

We consider a layer of saturated rock with thickness D deformed in shear at a slip rate V (Figure 8.1). Several investigations have shown that the ultracataclastic gouge zones forming the fault core have a much lower permeability ($<10^{-19}$ m^2) than that in the surrounding damage zone (e.g. Wibberley and Shimamoto 2002; Sulem *et al.* 2004). Therefore, as fluids and heat are trapped inside the slip zone during an earthquake, it is interesting to investigate the stability of undrained adiabatic shearing of such a gouge layer. This is done by assuming that drainage and heat flux are prohibited at the boundaries of the layer. It is also assumed that the normal stress σ_n acting on the layer is constant.

In this 1D model, the velocity components, $u_1(x_2, t)$ and $u_2(x_2, t)$, in the direction parallel and normal to the fault, respectively, depend only on the time since the onset of slip and on the position x_2 in the direction normal to the band. Inside such a shear-band, the pore pressure p_f and the temperature T are assumed to be functions only of time t and the position x_2 in the direction normal to the band.

The governing equations of the system are the balance laws of linear momentum, mass and energy. Neglecting inertia effects (Rice 2006), mechanical equilibrium leads to shear and normal stress that do not depend on x_2

$$\frac{\partial \sigma_{12}}{\partial x_2} = 0, \quad \frac{\partial \sigma_{22}}{\partial x_2} = 0 \qquad [8.1]$$

The fluid mass balance equation (equation [8.2]) is the sum of three terms: the diffusion term, the thermal pressurization term and the term corresponding to the effect on pore pressure of inelastic porosity change (for example dilatancy). Chemical effects and their impact on the localization zone will be discussed later in the chapter.

$$\frac{\partial p_f}{\partial t} = c_{hy} \frac{\partial^2 p_f}{\partial x_2{}^2} + \Lambda \frac{\partial T}{\partial t} - \frac{1}{\beta^*} \frac{\partial n^p}{\partial t} \qquad [8.2]$$

In equation [8.2], c_{hy} is the hydraulic diffusivity, β^* is the storage capacity, $\partial n^p / \partial t$ is the rate of inelastic porosity change and $\Lambda = n(\lambda_f - \lambda_n) / \beta^*$ is the undrained thermal pressurization coefficient, where λ_f is the pore fluid thermal expansion coefficient and λ_n is the thermal expansion coefficient of the pore volume, and n is the porosity of the rock. Typical values for Λ range from 0.1 to 1 MPa/°C (Ghabezloo and Sulem 2008).

The energy balance equation (equation [8.3]) is the sum of two terms: the diffusion term and the source term corresponding to the frictional heat.

$$\frac{\partial T}{\partial t} = c_{th} \frac{\partial^2 T}{\partial x_2{}^2} + \frac{1}{\rho C} \tau \dot{\gamma}_0^p \qquad [8.3]$$

In equation [8.3], c_{th} is the thermal conductivity, ρC is the specific heat and $\tau \dot{\gamma}_0^p$ is the plastic work that is assumed to be entirely converted into heat. The shear stress τ is proportional to the effective Terzaghi stress: $\tau = f(\sigma_n - p_0)$, where f is the friction coefficient.

Considering that the layer is homogeneous and uniformly sheared at a constant strain rate $\dot{\gamma}_0 = V / D$, where V is the imposed slip rate, and pore pressure and temperature are independent of x_2, a closed-form solution for their evolution in time has been given by Lachenbruch (1980), Rice (2006) and Sulem *et al.* (2007):

$$p_f = p_{f0} + (\sigma_n - p_0)\left(1 - \exp\left(-\frac{Vt}{L}\right)\right)$$
$$T = T_0 + \frac{(\sigma_n - p_0)}{\Lambda}\left(1 - \exp\left(-\frac{Vt}{L}\right)\right), \text{ with } L = \frac{\rho C h}{f\Lambda} \quad [8.4]$$

The pore pressure increases toward the imposed normal stress σ_n and the temperature increases toward $T = T_0 + (\sigma_n - p_{f0})/\Lambda$. In due course of the shear heating and fluid pressurization process, the shear strength is reduced toward zero. We observe that the length scale L, which controls the shear stress evolution, is the same as the one that controls the temperature evolution.

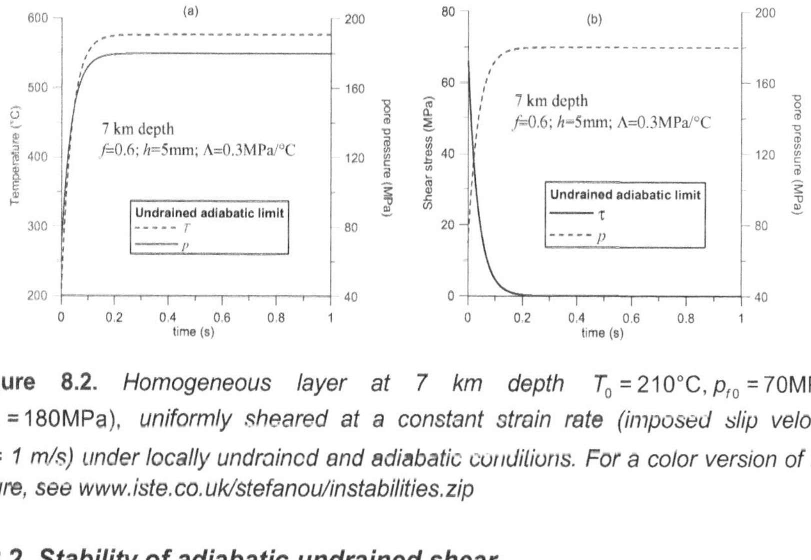

Figure 8.2. *Homogeneous layer at 7 km depth* $T_0 = 210°C, p_{f0} = 70MPa$, ($\sigma_n = 180MPa$), *uniformly sheared at a constant strain rate (imposed slip velocity V = 1 m/s) under locally undrained and adiabatic conditions. For a color version of this figure, see www.iste.co.uk/stefanou/instabilities.zip*

8.2.2. *Stability of adiabatic undrained shear*

The stability of the above undrained adiabatic (uniform) solution can be studied by performing a linear perturbation analysis. Details of such an analysis are found in the paper of Rice *et al.* (2014) (see also Sulem *et al.* 2011). Note that when performing this stability analysis, spatially dependent perturbations are considered inside the layer so that heat and fluid diffusion is allowed inside the layer, whereas zero heat and fluid fluxes are imposed at the boundaries of the layer. If we assume constant friction coefficient, this solution is unstable for all wavelengths of the perturbation. Therefore, strain will localize into a zone of zero thickness, which corresponds to the solution of slip on a plane given by Rice (2006) for which the temperature rise is $\Delta T = \left(1 + \sqrt{c_{hy}/c_{th}}\right)(\sigma_n - p_{f0})/\Lambda$. Considering the very low permeability of the fault gouge, c_{hy} and c_{th} are of the same order of magnitude so that the temperature is in that case about twice bigger than the undrained adiabatic uniform solution. As mentioned in the introduction to this chapter, the infinitesimally thin localized zone is reflecting the lack of a material length in the model (for example, the grain size) that will act as a localization limiter. Rice *et al.* (2014) have considered a rate-dependent friction coefficient derived from rate-and-state friction (RSF) laws, which are commonly used in seismology (Dieterich 1979), $f = f_0 + a\log(\dot{\gamma}/\dot{\gamma}_0)$, where a is a strain rate-hardening (or softening) parameter and it depends on temperature. Under temperature increase, a is not monotonous and experimental evidence shows that either thermal softening or thermal hardening takes place (Chester and Higgs 1992; Scholz 1998). This coupling is important as it plays a direct role in the stability of faults and of course in localization (Alevizos *et al.* 2014; Poulet *et al.* 2014; Veveakis *et al.* 2014). The choice of the RSF law depends on the available experimental data, and its mathematical form can lead to differences regarding temperature increase (Veveakis *et al.* 2010). As will be shown in the following sections, temperature increase and strain localization are key factors as they concern the interplay between mechanics, heat and pore pressure diffusion and the activation of chemical reactions induced by frictional heating.

Neglecting the direct effect of temperature increase on the RSF law, Rice *et al.* (2014) have obtained that for strain-rate softening ($a < 0$), the uniform solution of undrained adiabatic shearing is unstable for all wavelengths of the perturbation. For strain rate hardening ($a > 0$), only shear zones with a thickness h smaller than a critical value $h_{cr} = \pi\sqrt{\dfrac{a\,\rho C}{f_0\Lambda}\dfrac{c_{th}+c_{hy}}{(f_0+2a)\dot{\gamma}_0}}$ can support stable homogeneous deformation. This critical thickness is interpreted as the thickness of the localized shear zone and this statement was corroborated by numerical simulation in the postlocalization regime performed by Platt *et al.* (2014). Interestingly, this expression of the critical shear zone thickness exhibits two competing processes: fluid and thermal diffusion and rate-dependent frictional strengthening tend to expand the localized zone, while thermal pressurization tends to narrow it. As emphasized above, strain localization in a narrow zone leads to faster and stronger temperature rise than that predicted if localization is ignored.

For representative values of the material parameters, stress and temperature conditions of a fault at a seismogenic depth of 7 km, Rice *et al.* (2014) have obtained typical values of few micrometers to few tens of micrometers for the thickness of the localized zone. This value is comparable to the gouge grain size and therefore it might be important to take into account the granular microstructure of the material. This has been proposed by Sulem *et al.* (2011), who studied localization in a fluid-saturated material accounting for the motion of individual grains using a Cosserat microstructure. In a Cosserat continuum, each material point possesses additional rotational degrees of freedom so that a rolling resistance is attributed to the grains. Note that an internal length R related to the grain size of the material is introduced in the formulation of constitutive laws for a Cosserat material. A stability analysis of undrained adiabatic shearing of an elastoplastic layer with strain hardening and dilatancy has been proposed. It was shown that instability can occur even in the hardening regime of the underlying drained stress–strain response if dilatant hardening cannot compensate the thermal pressurization of the pore fluid. A remarkable result is that if we do not take into account the effect of microstructure and the associated microinertia, the underlying mathematical problem is ill-posed, i.e. for a hardening modulus lower than the critical hardening modulus at instability, the growth coefficient in time of the instability is infinite. The complete dynamic analysis for a Cosserat continuum shows that the growth coefficient of the instability is always finite and that it depends on the wavelength of the instability mode. The wavelength of the instability mode for which the growth coefficient is maximum will evolve faster and will dominate the others. This selected wavelength is depicted in Figure 8.3. For representative values of a fault zone at 7 km depth, it was found that the selected wavelength is about $200R$. The localized zone thickness is half of this value, so that for typical values of a grain size of few micrometers for the fault gouge, this corresponds to a shear band with thickness of few hundreds of micrometers, which is compatible with field observations in fault zones.

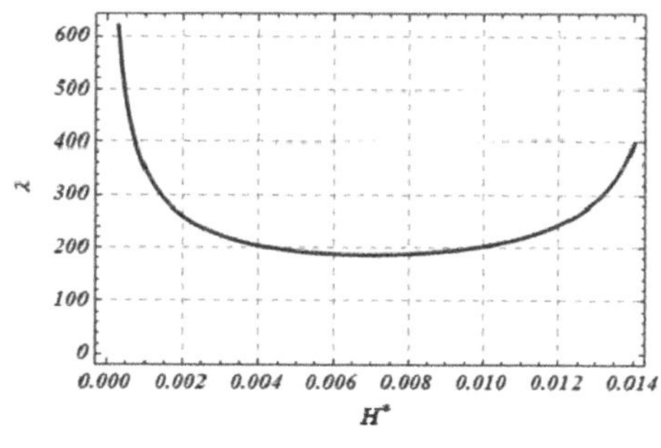

Figure 8.3. *Wavelength selection: λ is the wavelength of the perturbation normalized by the Cosserat internal length R, and H^* is the strain hardening modulus of the elastoplastic model. For a color version of this figure, see www.iste.co.uk/stefanou/instabilities.zip*

8.2.3. *Chemical weakening and earthquake nucleation*

An interesting situation corresponds to shear localization in a chemically weakening material. This case is relevant for understanding the nucleation of intermediate and deep earthquakes within subduction zones. As discussed by Green (2007), deep earthquakes have been a paradox since their discovery in the 1920s. The combined increase of pressure and temperature with depth precludes brittle failure or frictional sliding beyond a few tens of kilometers. Nevertheless, earthquakes still occur in subduction zones to 700 km. Growing evidence suggests that the great majority of subduction zone earthquakes shallower than 400 km are initiated by breakdown of hydrous phases and that deeper ones probably initiate as a shearing instability associated with breakdown of metastable olivine to its higher pressure polymorphs. Reaction weakening behavior has been extensively documented in the case of rocks containing dehydrating minerals such as gypsum and serpentinite. In the case of serpentinite dehydration, the reaction products may be weaker. This is the case for the dehydration of lizardite for which the produced ultra-fine grained olivine is weaker than the serpentinite aggregates (Rutter and Brodie 1988):

$$\underset{\text{lizardite}}{5\,Mg_3Si_2O_5(OH)_4} \rightarrow \underset{\text{talc}}{Mg_3Si_4O_{10}(OH)_2} + \underset{\text{olivine}}{6\,Mg_2SiO_4} + 9\,H_2O$$

This leads to a self-lubrication of the fault by the newly formed material. The intrinsic reaction-weakening process assumed here is thus an interesting possibility for dehydration-induced earthquakes at intermediate depths. Brantut and Sulem (2012) have assumed a simple evolution law for the friction coefficient f in order to describe chemical weakening during dehydration:

$$f = f_0 + a\ln\left(\dot{\gamma} / \dot{\gamma}_0\right) - b\mu \qquad [8.5]$$

f_0 is a reference friction coefficient for a reference strain rate $\dot{\gamma}_0$ and a and b are positive constitutive parameters. This friction law is similar to the commonly used RSF laws where the "state" is identified as the reaction extent μ.

The reaction rate is assumed to be of the first order, which accounts for temperature dependency following an Arrhenius law and depletion:

$$\frac{\partial \mu}{\partial t} = A(1-\mu)\exp\left(-\frac{E_a}{RT}\right), \qquad [8.6]$$

where A is a pre-exponential factor, E_a is the activation energy of the reaction and R is the gas constant. It is shown that due to the endothermic character of the mineral decomposition reaction, the temperature is buffered during the reaction to a value close to the critical temperature T_c at which the reaction is triggered (Sulem and Famin 2009; Brantut *et al.* 2011). Therefore, the reaction rate can be linearized above T_c:

$$\frac{\partial \mu}{\partial t} \approx \begin{cases} 0 & \text{if } T < T_c \\ c_T \left(T - T_c\right) - c_\mu \mu & \text{if } T \geq T_c \end{cases} \qquad [8.7]$$

The pore pressure evolution is given by the fluid mass balance (Sulem and Famin 2009; Brantut *et al.* 2011):

$$\frac{\partial p_f}{\partial t} = c_{hy} \frac{\partial^2 p_f}{\partial y^2} + \Lambda \frac{\partial T}{\partial t} + m_w \frac{1-\xi}{\rho_f \beta^*} \frac{\partial \mu}{\partial t}, \qquad [8.8]$$

where m_w is the mass of water that is released due to the reaction per unit of rock volume and ξ is the ratio of pore volume creation to fluid volume release due to the dehydration reaction.

The temperature evolution is given by the energy balance equation. Heat is generated by dissipation of the frictional energy (shear heating) and is partitioned into temperature change and diffusion and reaction enthalpy, which is a heat sink for endothermic reactions. Denoting m_0 the mass of reacting mineral per unit of total rock volume and ΔH the reaction enthalpy, we obtain (Sulem and Famin 2009; Brantut *et al.* 2011):

$$\frac{\partial T}{\partial t} = c_{th} \frac{\partial^2 T}{\partial y^2} + \frac{\tau \dot{\gamma}}{\rho C} - m_0 \frac{\Delta H}{\rho C} \frac{\partial \mu}{\partial t}, \qquad [8.9]$$

where τ is the applied shear stress on the fault and $\dot{\gamma}$ is the strain rate.

The linear stability analysis of the above system of equations has been performed by Brantut and Sulem (2012) and they have obtained the following expression for the critical wavelength (smallest wavelength for which the growth coefficient of the instability is positive), below which all perturbations vanish in time:

$$\lambda_{cr}^{ch} = 2\pi \sqrt{\frac{a c_{th}}{\dot{\gamma}_0} \frac{\rho C}{b \tau_0} \frac{c_\mu}{c_T}} \qquad [8.10]$$

This value corresponds to the smallest wavelength for which the growth coefficient of the instability is positive. We observe that it depends only on the thermochemical parameters and that it is not influenced by the pore pressure effect.

In order to illustrate this analysis, we use the parameter values given in Table 8.1. They are taken from Brantut and Sulem (2012) for lizardite dehydration of a layer at 30 km depth, which is sheared at a very low strain rate of 10^{-6} s^{-1}.

It is worth investigating the evolution of the localization zone in a simple case of a 5 m thick gouge layer. An initial small perturbation of the shear strain field is imposed with a wavelength equal to the gouge thickness. According to [8.10], this wavelength is unstable. In particular, the critical wavelength given by [8.10] is 0.12 m.

Figure 8.5(a) shows the evolution of the shear stress and of the shear strain rate in the middle of the localization zone. Figure 8.5(b) presents the profile of the shear strain at various time instants showing the localization and the delocalization process when the reaction is depleted. Note that in this example, a very low strain rate corresponding to a subducting slab was assumed. For this strain rate and due to partial dehydration (the material in the middle is depleted but not in the adjacent zones), chemical traveling waves are triggered and travel toward the boundaries of the gouge layer. Figure 8.5(c) depicts the temperature increase due to shearing in the middle of the shear band. It is worth mentioning that after a point, the (endothermic) chemical reaction buffers the temperature rise. Figure 8.5(d) shows the evolution of the chemical reaction and the pore pressure increases in the middle of the shear band.

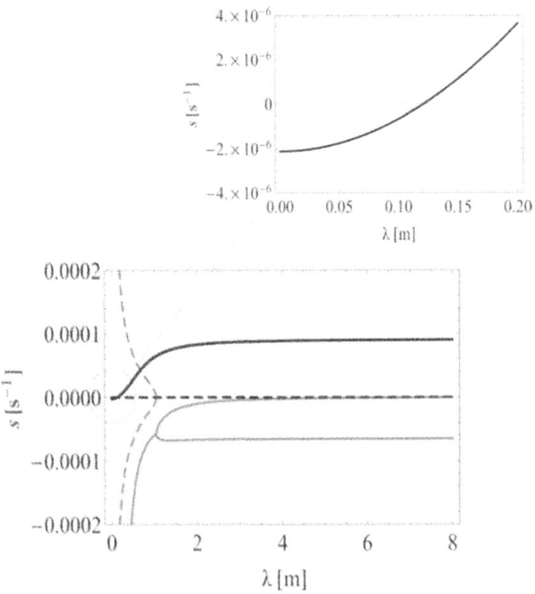

Figure 8.4. *Growth coefficient (Lyapunov exponent) in terms of perturbation wavelength. The real part is plotted in solid line and the imaginary part in dotted line. Black signifies the mode related to chemical instability and blue denotes the mode related to thermal pressurization (Re[s] < 0 for this mode). The smallest wavelength for which s becomes positive is given by equation [8.10]*

It is worth investigating the evolution of the localization zone in a simple case of a 5 m thick gouge layer. An initial small perturbation of the shear strain field is imposed with a wavelength equal to the gouge thickness. According to Figure 8.4, this wavelength is unstable. In particular, the critical wavelength given by [8.10] is 0.12 m (see Figure 8.4).

Quantity	Value
Friction coefficient, f_0	0.6
Rate strengthening parameter, a	0.002
Reaction weakening parameter b	0.5
Specific heat capacity, ρC	2.7 MPa °C^{-1}
Thermal dependency of the chemical kinetics, c_T	2.58×10^{-7} °C^{-1} s^{-1}
Depletion dependency of the chemical kinetics, c_μ	2.12×10^{-6} s^{-1}
Initial shear stress, τ_0	240 MPa
Nominal strain rate, $\dot{\gamma}_0$	10^{-6} s^{-1}
Thermal pressurization coefficient, Λ	0.5 MPa °C^{-1}
Thermal diffusivity, c_{th}	10^{-6} m^2 s^{-1}
Hydraulic diffusivity, c_{hy}	10^{-6} m^2 s^{-1}

Table 8.1. *Parameter values for lizardite dehydration at a depth of around 30 km (Brantut and Sulem 2012)*

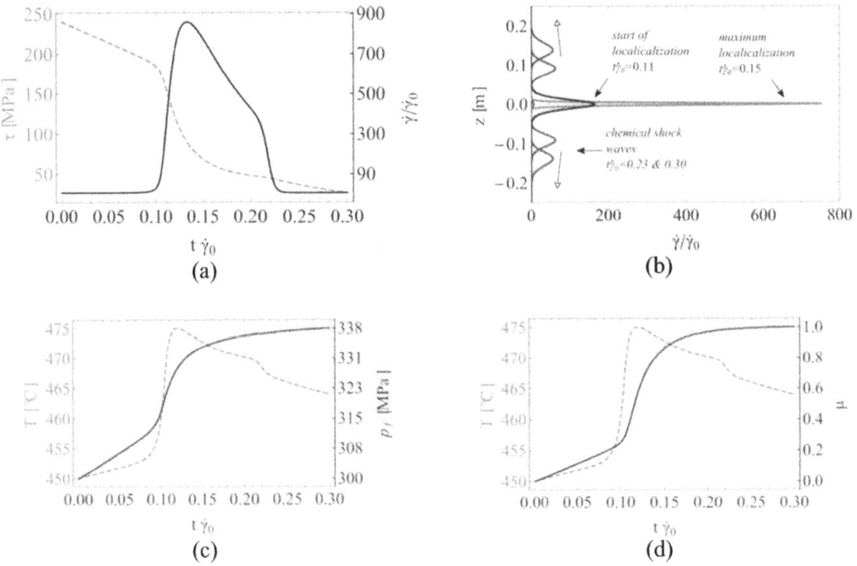

Figure 8.5. *Strain localization due to chemical softening in a 5 m thick dehydrating gouge layer: (a) shear stress drop in the middle of the shear band and shear strain evolution. The system localizes to a narrow band because of dehydration, but then it delocalizes due to depletion; (b) evolution of shear band localization – profile of shear strain rate; (c) temperature and pore pressure increase in the middle of the shear band. The chemical reaction buffers temperature rise; (d) evolution of the chemical reaction in the middle of the shear band. At* $t\dot{\gamma}_0 \simeq 0.1$*, the reaction effects become important and the material is rapidly depleted in the middle of the gouge (*$\mu \rightarrow 1$*) until* $t\dot{\gamma}_0 \simeq 0.3$*. Due to partial depletion, chemical shock waves are triggered and travel toward the boundaries of the gouge. For a color version of this figure, see www.iste.co.uk/stefanou/instabilities.zip*

8.3. Dissolution weakening and compaction banding

8.3.1. *Multiscale modeling of strong chemo-poro-mechanical coupling*

In a recent paper (Stefanou and Sulem 2014), instabilities in the form of compaction bands as triggered by chemical degradation of the solid skeleton have been studied. Chemical dissolution and grain breakage have been considered. The interest of the approach is in the strong chemo-poro-mechanical coupling, which was considered. As the stresses and the deformations evolve, the grains of the material break, leading to an increase in their specific surface. As the dissolution rate depends upon the area of contact between the reactive fluid and the minerals, dissolution is accelerated by grain fracturing and grain breakage and chemical softening is further enhanced.

The effect of chemical dissolution is important in both field and reservoir applications. For instance, the experimental results of Xie *et al.* (2011) showed that the chemical dissolution of a limestone leads to a significant increase in the porosity (from 23% for the intact rock to 27% for the degraded one). According to the same authors, the plastic pore collapse threshold is also reduced from about 30 to 20 MPa and the chemically degraded materials become more collapsible and more ductile due to the increase in porosity and the degradation of the intergranular cementation. This evidence is corroborated by other authors (e.g. Nova *et al.* 2003; Hu and Hueckel 2007a, 2007b; Buscarnera 2012; Ciantia *et al.* 2015) for a class of geomaterials and results in a contraction of the elastic domain only due to chemical reasons (chemical softening). On the other hand, in a saturated porous geomaterial, the progressive mechanical damage of the solid skeleton during compaction has as a result the increase of the interface area of the reactants (i.e. of the solution with the solid) and consequently the acceleration of the dissolution rate of the solid phase (Rimstidt and Barnes 1980). Thus, the solid skeleton is degraded more rapidly (mass removal because of dissolution), the overall mechanical properties of the system diminish (contraction of the elastic domain – chemical softening), deformations increase and the solid skeleton is further damaged (intergranular fractures, debonding, breakage of the porous network, etc.). Figure 8.6 schematically shows this positive feedback process, whose stability is not guaranteed. Note that chemical softening is central for compaction banding in the absence of other softening mechanisms such as mechanical softening.

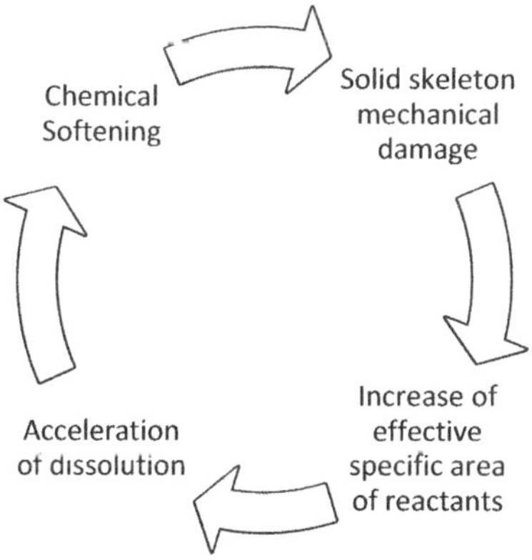

Figure 8.6. *Positive feedback process due to dissolution and solid skeleton damage (e.g. intergranular fracturing, breakage of the porous network, matrix cracking and grain–matrix debonding)*

A two-scale approach was proposed as the reaction kinetics is considered at the microscale (grain level), whereas the balance and constitutive equations are written at the macroscale, i.e. the Representative Elementary Volume (REV). Due to the existing heterogeneity of the microstructure (e.g. different grain sizes and constituents in the REV) the dissolution rate may not be homogeneous over the REV. The size of the REV is a finite statistical quantity that depends upon the geomaterial at hand. The chemical softening rate of the yield surface is therefore related to the average, over the REV, of the reaction rate at the grain level. This average procedure naturally introduces a characteristic length (size of the REV). This approach is directly inspired from the development of non-local continuum theories. For heterogeneous materials, the constitutive law at a point of the continuum should involve weighted averages of a state variable over a certain neighborhood of that point. This leads to an integral type of constitutive equations. Along the same lines of thinking, gradient-type constitutive models take into account the field in the vicinity of the considered point by enriching the local constitutive equations with higher order gradients of the deformation field (Germain 1973; Vardoulakis and Sulem 1995). As emphasized by Bažant and Jirásek (2002), resorting to a non-local continuum appears to be an effective means for regularizing boundary value problems with strain softening. In our problem, the introduction of this "chemical" characteristic length gives a length scale to the problem and naturally leads to a selection of a particular wavelength when performing a linear stability analysis of the system. Details of the analysis can be found in the paper of Stefanou and Sulem (2014). In the following we just recall the main assumptions and the principal results.

Microscale

We consider a dissolution process of the form: $\underset{(3)}{\text{solid}}+\underset{(1)}{\text{solvent}} \rightleftharpoons \underset{(2)}{\text{solution}}$. For example, the dissolution/precipitation of quartz in water is described by the following chemical equation:

$$SiO_2\,(\text{solid}) + 2H_2O(\text{liquid}) \rightleftharpoons H_4SiO_4\,(\text{aqueous solution})$$

Another example is the dissolution of calcite with water that is saturated with carbon dioxide:

$$CaCO_3\,(\text{solid}) + H_2CO_3\,(\text{aqueous solution}) \rightleftharpoons Ca\left(HCO_3\right)_2\,(\text{aqueous solution})$$

This equation represents a set of consecutive reactions that take place and are responsible for the dissolution of carbonate rocks (Grgic 2011). The reaction kinetics is written in a simplified manner as:

$$\frac{\partial w_2}{\partial t} = k^* \frac{S}{e}\left(1 - \frac{w_2}{w_2^{eq}}\right), \qquad [8.11]$$

where w_2 is the mass fraction of the dissolution product in the fluid, k^* is a reaction rate coefficient, e is the void ratio, S is the specific surface of a single grain (which is inversely proportional to the grain size) and w_2^{eq} is the mass fraction of dissolution product to the fluid mass at chemical equilibrium. It should be emphasized that equation [8.11] is written at the microscale and that w_2, S and e represent local quantities, which are not necessarily homogeneous over the RVE. Grain crushing is also described at the microscale. We use here a simple empirical expression for the evolution of the grain size in terms of the mechanical work input as proposed by Lade *et al.* (1996):

$$D = D_0\left(\frac{a}{a + E_T}\right), \qquad [8.12]$$

where D_0 is the effective grain size of the initial gradation, E_T is the total energy input density and a is a fitting parameter, which represents the grain crushability. Considering that the specific effective surface of a grain S is inversely proportional to the grain diameter, it is natural to assume the same type of relationship:

$$S = S_0\left(1 + \frac{E_T}{a}\right) \qquad [8.13]$$

Macroscale

At the macroscale, the constitutive equations are derived from a modified Cam-Clay elastoplastic model for which an associative flow rule is assumed for sake of simplicity:

$$F = q^2 + M^2 p'(p' - p'_c) = 0, \qquad [8.14]$$

where p' and q are, respectively, the Terzaghi effective mean stress and the shearing stress intensity (square root of the second invariant of the deviatoric part of the stress tensor). p'_c is a material parameter (namely the yield stress under isotropic loading), which is assumed to decrease from p'_0 (initial reference state) to p'_R (residual yield stress once the chemical reaction is completed) according to the following law $p'_c \equiv p'_R - (p'_R - p'_0)\zeta^\kappa$. κ is an exponent that can be experimentally determined for the material and the chemical process at hand, and $\zeta = M_s / M_0$ is the ratio of the current solid mass over its initial value, which is taken here as a chemical softening parameter.

This chemical softening parameter ζ can be expressed in terms of the *average* mass fraction of the dissolution product over the RVE $w_2^M = \frac{1}{V_T}\int_{V_T} w_2 dV$. Assuming that $w_2 = w_2(z,t)$ (oedometric conditions) is a function that can be expanded into Taylor series up to the second order in z:

$$w_2^M \approx w_2 + \ell_c^2 \frac{\partial^2 w_2}{\partial z^2}, \qquad [8.15]$$

where $\ell_c = \ell_{REV}\sqrt{\frac{1}{24}} \simeq \frac{\ell_{REV}}{5}$ appears as a characteristic internal length and ℓ_{REV} is the size of the RVE in z-direction.

8.3.2. *Compaction banding in oedometric compression*

In the following, we refer to the main results obtained in Stefanou and Sulem (2014) for oedometric compaction in a carbonate reservoir at 1.8 km depth because of CO_2 injection. In Table 8.2, we present some typical values for the chemomechanical parameters of a porous carbonate rock. At this depth, we assume that the water pressure is $p_f \simeq 18\text{MPa}$ and the total vertical stress is $\sigma_n \simeq 45\text{MPa}$.

Stefanou and Sulem (2014) have derived the instability conditions for which compaction bands formation is possible from a linear stability analysis. The details of the analysis can be found in the aforementioned paper. In particular, there is a region in the $q - p'$ plane where compaction band formation is possible. In other words, homogeneous deformations are unstable and the system bifurcates to non-homogeneous solutions. Figure 8.7 shows the instability zone for the chemo-poro-mechanical parameters of the carbonate grainstone considered in this example.

Inside the instability region, there exists a minimum critical wavelength above which perturbations are unstable (positive growth coefficient). This critical wavelength is related to the characteristic internal length, ℓ_c, which was introduced in the previous section. If $\ell_c = 0$, then the system is unstable for any perturbation wavelength.

Quantity	Value
Hydraulic diffusivity, c_{hy}	$10^{-3}\,m^2\,s^{-1}$
Grain diameter, D_0^{50}	0.2 mm
Initial effective grain surface area to volume ratio, S	25 mm^{-1}
Bulk modulus, K	5 GPa
Shear modulus, G	5 GPa
Porosity, n	0.25
CSL slope, M	0.9
Initial yield stress, p_0'	35 MPa
Residual yield stress, p_R'	30% p_0'
Chemical softening exponent, κ	2
Grain crushing parameter, a	1 MPa
Fluid density, ρ_f	1 g cm^{-3}
Solid density, ρ_s	2.65 g cm^{-3}
Dissolution rate, k^*	1.6×10^{-10} m/s

Table 8.2. *Indicative material properties of a carbonate grainstone*

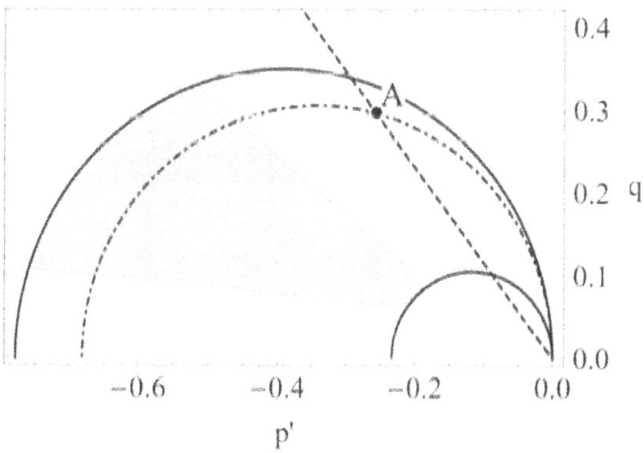

Figure 8.7. *Instability region (shaded) for compaction bands under oedometric conditions for a carbonate grainstone. The outer ellipse (solid line) represents the initial strength of the material. The straight dashed line depicts the linear elastic oedometric path. Point A corresponds to the initial stress state at 1.8 km depth. The inner ellipse (solid line) represents the residual strength of the grainstone after complete dissolution of the rock. The stresses are normalized by the applied vertical stress at 1.8 km depth, which is constant*

Initially, we assume that the material is in a state of elastic deformation (point A) under the applied total vertical stress of 45 MPa. At time $t=0$, the injection of the CO_2 solution starts. It is assumed that the CO_2 solution is continuously renewed in such a way that practically open flow conditions hold ($w_2(t>0)=0$). In the field, CO_2 injection open flow conditions would correspond to a zone outside the gas plume, where the formation fluid is saturated with CO_2, but is not in chemical equilibrium with the rock, so that carbonate dissolution occurs continuously ($w_2 < w_2^{eq}$).

As a result of CO_2 injection, the system is not in chemical equilibrium and dissolution occurs. Consequently, the material is progressively degraded due to chemical softening and the chemical softening parameter decreases from its initial value $\zeta_0 = 1$. When $\zeta = \zeta_A \simeq 0.9$ the material yields, plastic strains are accumulated and solid skeleton damage occurs (equation [8.13]). This phase of deformation under constant applied loading (i.e. the overburden) corresponds to the creep behavior that is observed due to CO_2 injection (Le Guen *et al.* 2007; Liteanu and Spiers 2009; Rutqvist 2012).

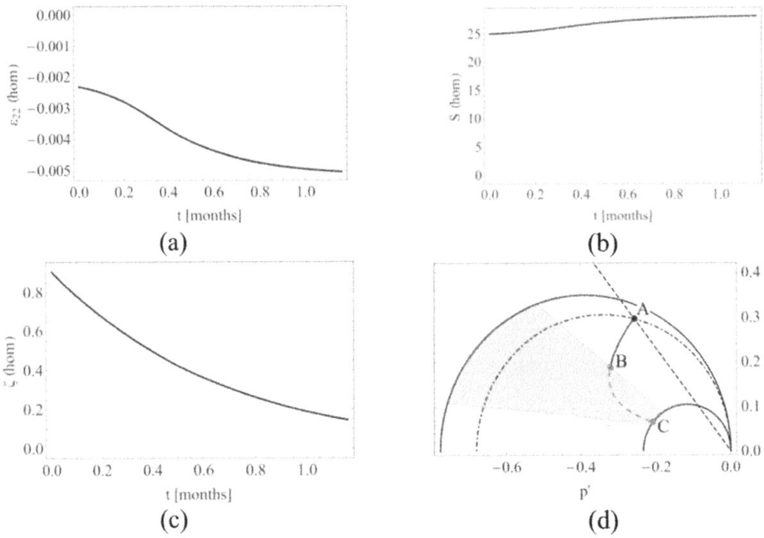

Figure 8.8. *Evolution in time of (a) deformation, (b) the specific surface, (c) the chemical softening parameter and (d) the stress path during imposed homogeneous dissolution of a specimen under oedometric conditions*

We emphasize that linear stability analysis only gives the conditions for compaction band triggering. However, in order to assess the evolution of the system and the gradual strain localization inside the band, one has to study the postbifurcation (Sulem and Stefanou 2016). A numerical simulation is first performed by imposing homogeneous deformation. The conditions are oedometric and the numerical analysis starts from point A ($\zeta_A = 0.9$). Figure 8.8 shows the evolution of (a) the total (homogeneous) deformation, (b) the specific surface, (c) the chemical softening parameter ζ and (d) the stress path followed during combined dissolution and chemical softening. Due to dissolution, we observe a slow increase in the compressive deformations from 2×10^{-3} to 5×10^{-3} (creep) and a limited increase in the specific surface of the grains. At the end of the simulation, the grainstone is depleted ($\zeta = 0$) and no more dissolution is possible. The situation is different when the unphysical constraint of homogeneous deformation is removed.

The system is now free to develop non-homogeneous deformation and strain localization in the form of compaction banding. An initial perturbation with a wavelength equal to the height of the sample and amplitude 10^{-5} is imposed (200 mm). Compaction band formation is possible for this perturbation wavelength only when the stress state falls into the unstable zone (Figure 8.8(d)). At the beginning of CO_2 injection, the system creeps from point A to point B. No compaction band is triggered and the system behaves as in the previous case of homogeneous deformation. Once the stress path crosses the shaded area, non-homogeneous deformations start to grow and a compaction band forms. Figure 8.9 shows the profile of the vertical deformation at various times. The deformations localize into a narrow band whose thickness depends on the characteristic internal length, ℓ_c. Here, we chose $\ell_c = 4$ mm (~20 grains for a typical grain size of 200 μm). In Figure 8.10, we present the average vertical deformation over the entire sample, the vertical deformation at the peak of the compaction band and the vertical deformation of a point far from the localization zone. The vertical deformation at the peak of the compaction band is 14 times larger than the vertical deformation far from it. Figure 8.11 shows the stress path of the point at the peak of the compaction band and the stress path of a point far from the localization zone. In the beginning, the stress paths coincide, but after entering the instability zone they start to diverge due to compaction band formation.

It is worth emphasizing that compaction band thickness depends on the chosen characteristic length. This is shown by linear stability analysis (Stefanou and Sulem 2014) and is corroborated numerically by choosing a different characteristic length, $\ell_c = 16$ mm (~80 grains). Figure 8.12 shows how a larger characteristic length leads to larger compaction band thickness. On the contrary, in the absence of internal length ($\ell_c = 0$), the compaction band thickness is zero.

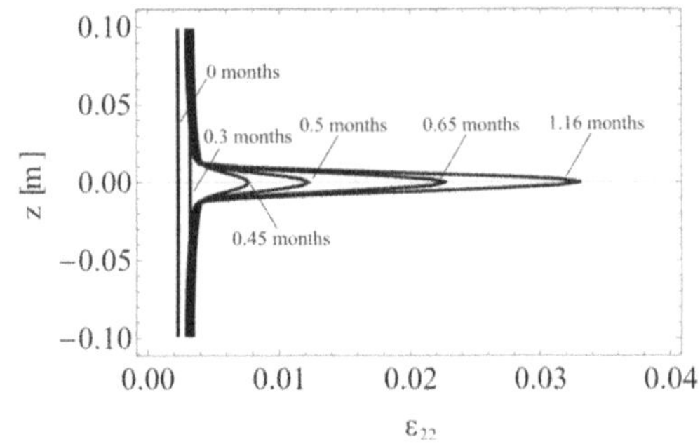

Figure 8.9. *Profile of the vertical deformation at various times and for* $\ell_c = 4$ mm *(20 grains). The deformations localize into a narrow band, i.e. compaction band*

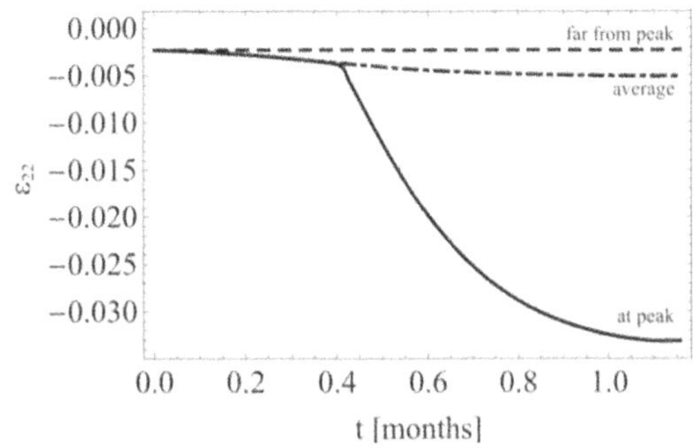

Figure 8.10. *Vertical deformation in function of time: at the peak of the compaction band, average over the entire sample and at a point far from the localization zone*

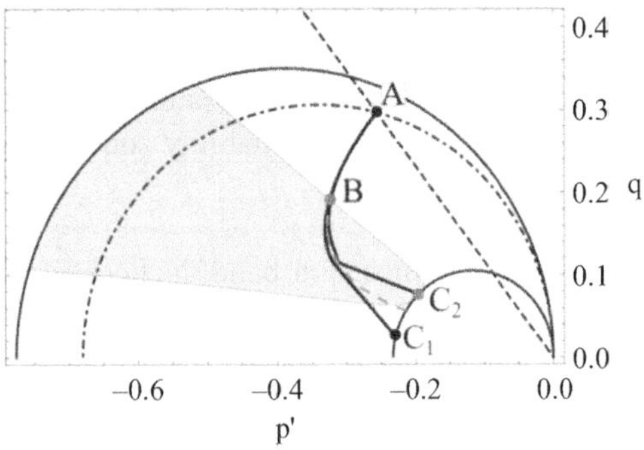

Figure 8.11. *Stress paths of the point at the peak of the compaction band (ABC_2) and a point far from the localization zone (ABC_1). The dashed line corresponds to the stress path of the homogeneous deformation (see Figure 8.8(d))*

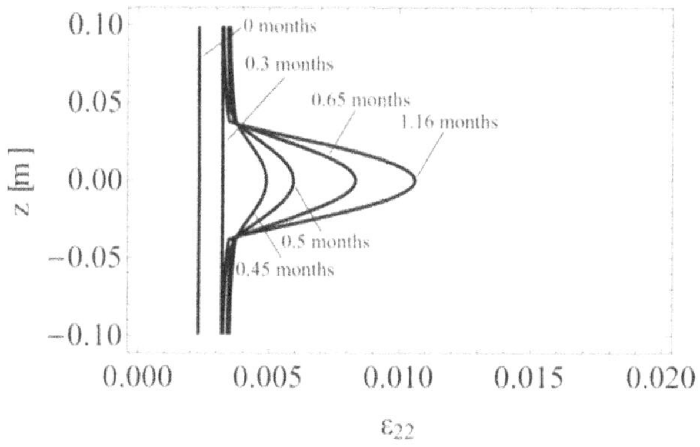

Figure 8.12. *Profile of the vertical deformation at various times for* $\ell_c = 16\ \mathrm{mm}$ *(80 grains). The deformations localize into a band that is thicker than in the case of* $\ell_c = 4\ \mathrm{mm}$ *(20 grains) (see Figure 8.9)*

8.4. Conclusion

In this chapter we have reviewed some recent work on the effects of temperature, pore pressure and chemical reactions on strain localization in geomaterials. Examples have been taken in relation to seismic slip and with compaction banding. Strain localization is triggered by the softening of the material strength. Weakening mechanisms can be of various origins: mechanical (e.g. microcracking), hydraulic (e.g. pore fluid pressurization), thermal (e.g. shear heating) and chemical (e.g. dissolution).

We have presented the framework of localized failure in earthquake rupture as the result of instability of the deformation process. It was shown that it is necessary to introduce a localization limiter in order to obtain a finite thickness for the localized shear zone. Rate-independent constitutive models for a classical (Cauchy) continuum lead to an infinitesimally thin zone. On the contrary, a rate-dependent friction law or a Cosserat framework lead to a finite value for the critical wavelength of the perturbation, above which homogeneous deformation is unstable.

We have also analyzed compaction banding induced by dissolution. A strong chemomechanical coupling is considered: the material softens in due course of dissolution, whereas the dissolution process accelerates with increasing damage (because of the increase in the specific surface of the grain and thus of the area of contact between the reactive fluid and the minerals). A two-scale approach has been proposed in order to account for the heterogeneity of the dissolution process over the REV. This naturally introduces a “chemical” material length related to the non-local character of the relation between the softening rate of the yield surface (at the macroscale) and the dissolution process (at the grain scale). Here again, the introduction of this material length is crucial to limit the compaction band thickness to a finite value. Postlocalization robust computations have been performed to simulate the progressive evolution of compaction band under open flow conditions.

Some challenging open questions remain such as the modeling of coupled thermo-chemo-hydro-mechanical phenomena with evolution of the microstructure of the material through various mechanical and chemical processes. This would require methods for cross scale couplings, advanced homogenization techniques and enriched continua models.

8.5. References

Alevizos, S., Poulet, T., Veveakis, M. (2014). Thermo-poro-mechanics of chemically active creeping faults. 1: Theory and steady state considerations. *J. Geophys. Res. Solid Earth*, 119(6), 4558–4582.

Bažant, Z.P. and Jirásek, M. (2002). Nonlocal integral formulations of plasticity and damage: Survey of progress. *J. Eng. Mech.*, 128(11), 1119–1149.

Brantut, N. and Sulem, J. (2012). Strain localization and slip instability in a strain-rate hardening, chemically weakening material. *J. Appl. Mech.*, 79(3), 031004.

Brantut, N., Sulem, J., Schubnel, A. (2011). Effect of dehydration reactions on earthquake nucleation: Stable sliding, slow transients, and unstable slip. *J. Geophys. Res. Solid Earth*, 116, 1–16.

Buscarnera, G. (2012). A conceptual model for the chemo-mechanical degradation of granular geomaterials. *Géotechnique Lett.*, 2(July–September), 149–154.

Chester, F.M. and Higgs, N.G. (1992). Multimechanism friction constitutive model for ultrafine quartz gouge at hypocentral conditions. *J. Geophys. Res.*, 97(B2), 1859.

Ciantia, M.O., Castellanza, R., di Prisco, C. (2015). Experimental study on the water-induced weakening of calcarenites. *Rock Mech. Rock Eng.*, 48(2), 441–461.

Das, A., Nguyen, G.D., Einav, I. (2011). Compaction bands due to grain crushing in porous rocks: A theoretical approach based on breakage mechanics. *J. Geophys. Res.*, 116(B8), 1–14.

Dieterich, J.H. (1979). Modeling of rock friction: 1. Experimental results and constitutive equations. *J. Geophys. Res.*, 84(B5), 2161.

Germain, P. (1973). The method of virtual power in continuum mechanics. Part 2: Microstructure. *SIAM J. Appl. Math.*, 25(3), 556–575.

Ghabezloo, S. and Sulem, J. (2008). Stress dependent thermal pressurization of a fluid-saturated rock. *Rock Mech. Rock Eng.*, 42(1), 1–24.

Green, H.W. (2007). Shearing instabilities accompanying high-pressure phase transformations and the mechanics of deep earthquakes. *Proc. Natl. Acad. Sci. USA*, 104(22), 9133–9138.

Grgic, D. (2011). Influence of CO_2 on the long-term chemomechanical behavior of an oolitic limestone. *J. Geophys. Res.*, 116(B7), B07201.

Le Guen, Y., Renard, F., Hellmann, R., Brosse, E., Collombet, M., Tisserand, D., Gratier, J.-P. (2007). Enhanced deformation of limestone and sandstone in the presence of high PCO_2 fluids. *J. Geophys. Res.*, 112(B5), B05421.

Hu, L.-B. and Hueckel, T. (2007a). Creep of saturated materials as a chemically enhanced rate-dependent damage process. *Int. J. Numer. Anal. Methods Geomech.*, 31(14), 1537–1565.

Hu, L.-B. and Hueckel, T. (2007b). Coupled chemo-mechanics of intergranular contact: Toward a three-scale model. *Comput. Geotech.*, 34(4), 306–327.

Lachenbruch, A.H. (1980). Frictional heating, fluid pressure, and the resistance to fault motion. *J. Geophys. Res.*, 85(B11), 6097.

Lade, P.V., Yamamuro, J.A., Bopp, P.A. (1996). Significance of particle crushing in granular materials. *J. Geotech. Eng.*, 122(4), 309–316.

Liteanu, E. and Spiers, C.J. (2009). Influence of pore fluid salt content on compaction creep of calcite aggregates in the presence of supercritical CO_2. *Chem. Geol.*, 265(1–2), 134–147.

Nova, R., Castellanza, R., Tamagnini, C. (2003). A constitutive model for bonded geomaterials subject to mechanical and/or chemical degradation. *Int. J. Numer. Anal. Methods Geomech.*, 27(9), 705–732.

Platt, J.D., Rudnicki, J.W., Rice, J.R. (2014). Stability and localization of rapid shear in fluid-saturated fault gouge: 2. Localized zone width and strength evolution. *J. Geophys. Res. Solid Earth*, 119(5), 4334–4359.

Poulet, T., Veveakis, M., Regenauer-Lieb, K., Yuen, D.A. (2014). Thermo-poro-mechanics of chemically active creeping faults: 3. The role of serpentinite in episodic tremor and slip sequences, and transition to chaos. *J. Geophys. Res. Solid Earth*, 119, 4606–4625.

Regenauer-Lieb, K., Veveakis, M., Poulet, T., Wellmann, F., Karrech, A., Liu, J., Hauser, J., Schrank, C.E., Gaede, O., Trefry, M.G. (2013). Multiscale coupling and multiphysics approaches in earth sciences: Theory. *J. Coupled Syst. Multiscale Dyn.*, 1(1), 49–73.

Rice, J.R. (2006). Heating and weakening of faults during earthquake slip. *J. Geophys. Res. Solid Earth*, 111(B5), B05311.

Rice, J.R., Rudnicki, J.W., Platt, J.D. (2014). Stability and localization of rapid shear in fluid-saturated fault gouge: 1. Linearized stability analysis. *J. Geophys. Res. Solid Earth*, 119(5), 4311–4333.

Rimstidt, J.D. and Barnes, H.L. (1980). The kinetics of silica-water reactions. *Geochim. Cosmochim. Acta*, 44(11), 1683–1699.

Rudnicki, J.W. and Rice, J.R. (1975). Conditions for the localization of deformation in pressure-sensitive dilatant materials. *J. Mech. Phys. Solids*, 23(6), 371–394.

Rutqvist, J. (2012). The geomechanics of CO_2 storage in deep sedimentary formations. *Geotech. Geol. Eng.*, 30(3), 525–551.

Rutter, E.H. and Brodie, K.H. (1988). Experimental "sytectonic" dehydration of serpentinite under conditions of controlled pore water pressure. *J. Geophys. Res.*, 93(B5), 4907.

Scholz, C.H. (1998). Earthquakes and friction laws. *Nature*, 391, 37–42.

Stefanou, I. and Sulem, J. (2014). Chemically induced compaction bands: Triggering conditions and band thickness. *J. Geophys. Res. Solid Earth*, 119(2), 880–899.

Sulem, J. and Famin, V. (2009). Thermal decomposition of carbonates in fault zones: Slip-weakening and temperature-limiting effects. *J. Geophys. Res.*, 114(B3), B03309.

Sulem, J. and Stefanou, I. (2016). Thermal and chemical effects in shear and compaction bands. *Geomech. Energy Environ.*, 6, 4–21.

Sulem, J., Lazar, P., Vardoulakis, I. (2007). Thermo-poro-mechanical properties of clayey gouge and application to rapid fault shearing. *Int. J. Numer. Anal. Methods Geomech.*, 31(3), 523–540.

Sulem, J., Stefanou, I., Veveakis, M. (2011). Stability analysis of undrained adiabatic shearing of a rock layer with Cosserat microstructure. *Granul. Matter*, 13(3), 261–268.

Sulem, J., Vardoulakis, I., Ouffroukh, H., Boulon, M., Hans, J. (2004). Experimental characterization of the thermo-poro-mechanical properties of the Aegion Fault gouge. *Comptes Rendus – Geosci.*, 336, 455–466.

Vardoulakis, I. and Sulem, J. (1995). *Bifurcation Analysis in Geomechanics*. Blackie Academic and Professional, Glascow.

Veveakis, M., Alevizos, S., Vardoulakis, I. (2010). Chemical reaction capping of thermal instabilities during shear of frictional faults. *J. Mech. Phys. Solids*, 58(9), 1175–1194.

Veveakis, E., Sulem, J., Stefanou, I. (2012). Modeling of fault gouges with cosserat continuum mechanics: Influence of thermal pressurization and chemical decomposition as coseismic weakening mechanisms. *J. Struct. Geol.*, 38, 254–264.

Veveakis, M., Stefanou, I., Sulem, J. (2013). Failure in shear bands for granular materials: Thermo-hydro-chemo-mechanical effects. *Géotechnique Lett.*, 3(April-June), 31–36.

Veveakis, M., Poulet, T., Alevizos, S. (2014). Thermo-poro-mechanics of chemically active creeping faults: 2. Transient considerations. *J. Geophys. Res. Solid Earth*, 119(6), 4583–4605.

Wibberley, C.A.J., Shimamoto, T. (2002). Internal structure and permeability of major strike-slip fault zones: The median tectonic line in Mie Prefecture, Southwest Japan. *J. Struct. Geol.*, 25, 59–78.

Xie, S.Y., Shao, J.F., Xu, W.Y. (2011). Influences of chemical degradation on mechanical behaviour of a limestone. *Int. J. Rock Mech. Min. Sci.*, 48(5), 741–747.

9

On the Thermo-poro-mechanics of Chemically Active Faults

Manolis VEVEAKIS
Duke University, Durham, USA

Shear zones in outcrops and core drillings on active faults commonly reveal two scales of localization, with centimeter to tens of meters thick deformation zones embedding much narrower zones of millimeter to centimeter scale. The narrow zones are often attributed to some form of fast instability such as earthquakes or slow slip events. Surprisingly, the double localization phenomenon seem to be independent of the mode of failure, as it is observed in brittle cataclastic fault zones, as well as ductile mylonitic shear zones. In both, a very thin layer of chemically altered, ultrafine-grained ultracataclasite or ultramylonite is noted. We present an extension to the classical solid mechanical theory, where both length scales emerge as part of the same evolutionary process of shearing the host rock. We highlight the important role of any type of solid-fluid phase transitions that govern the second-degree localization process in the core of the shear zone. In both, brittle and ductile shear zones chemistry stops the localization process caused by a multiphysics feedback loop leading to an unstable slip. The microstructural evolutionary processes govern the time scale of the transition between slow background shear and fast, intermittent instabilities in the fault zone core. The fast cataclastic fragmentation processes are limiting the rates of forming the ultracataclasites in the brittle domain, while the slow dynamic recrystallization prolongs the transition to ultramylonites into a slow slip instability in the ductile realm.

Instabilities Modeling in Geomechanics,
coordinated by Ioannis STEFANOU and Jean SULEM.

9.1. Introduction

The geologist in the field is often confronted with two scales of localization when investigating shear or fault zones (Ben-Zion and Sammis 2003; Chester and Chester 1998). Brittle fault zones show striking examples of extremely localized slip events, occurring within a thin shear zone, $< 1 - 5$ mm thick, called the principal slipping zone (PSZ) Sibson (2003). This localized PSZ lies within a finely granulated fault zone of typically tens to hundreds millimeter thickness. These fault zones are either cataclastic in the brittle regime or mylonitic in the ductile regime.

Exhumed field examples for the morphology of these brittle fault zones can be found in the North Branch San Gabriel fault (Chester *et al.* 1993), the Punchbowl fault of the San Andreas system in southern California (Chester and Chester 1998), the Median Tectonic Line fault in Japan (Wibberley and Shimamoto 2003) and in the Hanaore fault in southwest Japan (Noda and Shimamoto 2005). Active faults have been intersected in wells in the Aigion system, central Greece, where the fault zone of clay size particles consists of finely crushed radiolarites, extended to about 1 m. This localized zone of deformation was found to be intercepted by a "freshdistinct slip surface of sub-millimeter size (Cornet *et al.* 2004; Sulem *et al.* 2007). Other wells that have been specifically designed to intersect active faults are the completed San Andreas Fault project (Holdsworth *et al.* 2011) and the ongoing Alpine Fault drilling project (Townend 2009). The San Andreas Fault project revealed a fault zone of 1–2.5 m width at 3 km depth, with several ultralocalized PSZs.

Similarly, in the ductile field, the dynamic recrystallization of the matrix minerals is telltale of crystal plastic or diffusion creep, forming shear zones of vast thickness interrupted by ultralocalized anastomosing patterns of extreme grain size reduction in the ultramylonite. Exhumed thrusts in creeping carbonates around the world, such as the Naukluft thrust in Namibia (Rowe *et al.* 2012), the McConnell thrust in Alberta Canada (Kennedy and Logan 1997) or the Glarus Thrust in Switzerland (Herwegh *et al.* 2008), all present a common structure of a meter-wide (1–5 meters) zone, which accommodates several thin (mm–cm at most) veins of ultralocalized deformation and chemical alteration.

The brittle field features micromechanisms, governed by grain breakage, rolling, cleavage and brittle fragmentation processes. The ductile field features dislocation, diffusion, and dissolution mechanism. Although these mechanisms are completely different, they surprisingly exhibit the same dual pattern of localization: a broad meter-wide shear zone with ultralocalized PSZs in its middle.

Each of the different phenomena can be explained independently but to date no comprehensive theory has been suggested that can investigate the reason for the double degree localization process and the potential commonalities between brittle and ductile localization. This is because the dynamic process of the extreme localization and the quasi-static response of the background flow of the shear zone are treated as two different fundamental theories. The former instability is described in solid mechanics and the latter flow in fluid dynamics.

While solid mechanics is capable of accurately describing the conditions for failure, it does not emphasize the long-term postfailure behavior. This is due to the nature of the solid materials that maintain a high degree of structural integrity after failure and are not significantly affected by the elapsed time. The consequence of this is that a quasi-static approach is preferred, thus making it possible to cast the weak rate effects into approximate laboratory determined hardening laws.

Fluid dynamics on the opposite does not have the concept of failure; it knows nothing about the processes that happen prior to failure and is totally concerned with the description of the rate of flow of the material under an applied stress. Since most mantle convection codes are based on the fluid dynamical approach, the geophysical community over the recent years has expended a lot of effort into incorporating the theory of solid mechanics.

The fact that both theories apply to geology has been universally acknowledged and the after-dinner contribution of Markus Reiner (1964) on the "Deborah Number" is well worth reading. The number is defined as the ratio of the stress relaxation time over the time of observation. Reiner's quote "The mountains flowed before the Lord" implies that given enough time, one may smoothly transition from one to the other theory. This can be easily conceptionalized in a Newtonian (linear viscous) viscoelastic framework. However, the inclusion of nonlinear temperature dependence of the creep processes gave rise to the time–temperature superposition (TTS) in polymer sciences (Tobolsky and Andrews 1945). The TTS principle implies that the temperature-dependent behavior of materials can be substituted by appropriate experiments at different time scales for a reference temperature. It is thereby assumed that the time evolution described in the temperature equation is the only time scale relevant for the long-term mechanical behavior of solids. This simple concept allows us to directly extend it for the nonlinear viscous behavior of solids within a single theoretical framework, incorporating elasticity, plastic failure and viscous postfailure evolution.

In this chapter, we summarize a theory that couples the solid and fluid-like behavior. This theory must be able to provide the framework for modeling both the formation of faults (refer, for more in depth reading, to Regenauer-Lieb *et al.* (2013a,b)) and their postfailure evolution. We show that the formation and postfailure evolution of faults depend strongly on the coupled multiphysical effects affecting pressure, temperature, and chemical conditions. We thus aim at providing in this chapter a comprehensive guide through the energetics of faults and explain how they apply to the different types of failure.

9.2. Time-independent formation of shear zones from solid mechanics

The theory describing the onset of localized failure from uniform deformation has been a direct extension of the classic Mohr's theory of the strength of materials and is applied to geomaterials using the so-called Thomas–Hill–Mandel shear-band model (Hill 1962; Mandel 1966), which was introduced in the early 1960s. A seminal paper was contributed by (Rudnicki and James 1975). More recently, the mathematical formulation of bifurcation and postbifurcation phenomena and related instabilities were summarized in Vardoulakis and Sulem (1995) to form the basis of an improved continuum theory of failure of geomaterials.

This theory defines failure as stationary elastic wave and does not consider time evolution as a degree of freedom. While velocity-dependent solutions are sought, they are independent of their evolution in true time. This is known as quasi-static deformation. The fundamental stress and strain solutions therefore degenerate into a geometric problem where failure lines appear with a finite width d.

These concepts of shear banding as material bifurcation of an elastoplastic skeleton have led to the identification of a material length scale defining the width of shear bands. The critical stress as well as the orientation and the thickness of the localized shear failure planes (shear zones) are calculated through the eigenvalues of the elastoplastic stiffness modulus C^{ep}_{ijkl} of the material (Rudnicki and James 1975) obeying a rate-independent rheology, $\dot{\sigma}'_{ij} = C^{ep}_{ijkl}\dot{\epsilon}_{kl}$ (the prime denoting effective stress). Within the framework of the solid mechanical instabilities, the shear (fault) zone thickness emerges as a solution, depending on the microstructure (Mühlhaus and Vardoulakis 1987; Papanicolopulos and Veveakis 2011).

This rate-independent regime is typically used as an adequate description of brittle processes, thus placing the formation of the shear zones (onset of localization) near the maximum deviatoric stress (Rudnicki and James 1975). For example, in linear elastic fracture mechanics, brittle fracture occurs without thermal activation when a critical stress level is reached to split the bonds. At a critical energy threshold, an elastodynamic fast time scale instability ensues where a variety of dissipative processes kick in such as grain/particle rotations, which release heat in extremely fast time scales. The fast time scale instability simplifies the processes, as temperature and fluids do not have time to diffuse. This regime is called undrained-adiabatic and is characterizing extremely fast co-seismic slip (Rice 2006; Sulem *et al.* 2011) of brittle failure events.

Within this brittle, solid mechanical framework recent studies (Sulem *et al.* 2011; Veveakis *et al.* 2012, 2013) derived different levels of localization corresponding to different energy (temperature) regimes. This result is a direct generalization of the extension of the viscoelastic Deborah number concept to different temperatures using the time temperature superposition principle Tobolsky and Andrews (1945) for elastoplastic media. This approach is generalizing the concept of viscoelastic relaxation experiment at different temperatures to that of nonlinear visco-elasto-plastic solutions where quasi-static solutions are sought for different temperatures.

In order to so we first freeze time by adopting classical, rate-independent, elasto-plastic theory for a fault zone material incorporating chemical, hydraulic and thermal sensitivity. In the brittle field, these fault zones are characterized by a damage zone surrounding the fault zone with an embedded PSZ, as shown in Figure 9.1(a). In the brittle crust, the width of this hierarchical damage zone can be up to 3–5 km wide (Allam and Ben-Zion 2012; Ben-Zion and Sammis 2003), containing in principle multiple localization zones (Faulkner *et al.* 2003). In order to keep the mathematical model tractable, we restrict the present work to the study of a single fault zone.

We therefore calculate the extent (thickness) of the brittle fault zone at different temperature regimes, marked by (1) the temperature inferred from geothermal consideration for the boundary of the fault zone, (2) the activation temperature of the dominant chemical reaction observed in the PSZ. The details of the approach can be found in Sulem *et al.* (2011); Veveakis *et al.* (2012, 2013) and the results are summarized in Figures 9.1(b) and (c).

9.2.1. *Shear zone thickness at boundary temperature conditions*

The general solution for a quasi-static shear zone considers the average of the background particle size d_{50}, the fluid pressure and the dominant chemical reaction at a given temperature. Both fluid pressure and reaction rate are strongly affected by the temperature. At ambient geothermal temperatures of the shear zone, the chemical reaction is inactive and the thickness of the shear zone is predominantly controlled by the thermal-elastic pressurization of the poro-elastic skeleton (Figure 9.1(b)), characterized by the pressurization coefficient $\Lambda = \frac{\alpha_f - \alpha_s}{K_s + K_f}$ (MPa/K). In this expression, α is the thermal expansion coefficient and K is the compressibility, with subscripts s, f denoting solid and fluid phases, respectively.

The solid mechanical thickness of the shear zone can be calculated directly as a function of the average particle size d_{50}, and the thermal pressurization Λ; as well as a multiplier C representing the elasto-plastic material:

$$\frac{d}{d_{50}} = \sqrt{\frac{C}{\Lambda}} \qquad [9.1]$$

for a Drucker–Prager Cosserat material, it was shown by (Veveakis *et al.* 2013; Rattez *et al.* 2017a,b,c) that $C = 17.5 \times 10^3 \mathrm{KMPa}^{-1}$. For typical values of Λ between

10^{-4} (MPa/K) (Cecinato *et al.* 2011) and 1 (MPa/K) Rice (2006), with d_{50} of the order of 1 mm, the shear zone width ranges between 0.1 and 13 m.

9.2.2. *Shear zone thickness at elevated temperature*

The effect of chemical reaction can be incorporated by considering the reaction rate to follow Arrhenius kinetics. In this framework, the reaction rate depends on the activation enthalpy Q_F and the reaction rate frequency k_F. The reaction is activated at elevated temperatures, near the activation temperature of the reaction at given pressure conditions. At these temperature conditions, and considering calcite decomposition as a typical reaction for a carbonate host rock (Famin *et al.* 2008; Sulem and Famin 2009), the thickness of the PSZ is several orders of magnitude smaller than the low temperature shear zone as shown in Figure 9.1(c).

This result is reconciling field observations from two different temperature regimes, but has no predictive power on how these temperatures may have been achieved. The traditional concept of TTS is not extendable to obtain extended solution for nonlinear solids. We therefore need a nonlinear superposition that allows fault zone evolution through time. The nonlinear concept in this work is introduced by a Taylor expansion of the plastic strain rate, accounting for higher order terms describing the nonlinearities.

In geomechanics, several studies have been devoted to the thickness of shear bands in soils. Under purely mechanical effects, the thickness of the initial shear band (with thickness D in Figure 9.1) in a granular medium is around $16d_{50}$ with d_{50} the average (50 weight) grain diameter (Mühlhaus and Vardoulakis 1987). This zone appears to be too wide to be considered as the PSZ (Rice 2006). In principle, the PSZ represents a very fine shear localization, consisting of ultrafine particles that have usually undergone mechanochemical degradation (gelification, decarbonation and dehydration reactions, melting, as thoroughly discussed by Toro *et al.* (2011)). Therefore, the initial mechanism such as the shear band formation from homogenous deformation (Mühlhaus and Vardoulakis 1987) is a consequence of the solid mechanical failure, while the formation of PSZs may be considered an effect of the fluid-like deformation following the onset of the initial structure.

Such postfailure behavior may, for instance, be significantly influenced by the presence of fluids interacting with the rock, hence inducing coupled effects including shear heating and pore fluid pressurization. These mechanisms are therefore expected to control the strength during fault slip (Lachenbruch 1980; Sibson 1973; Wibberley and Shimamoto 2005) and the weakening effect displayed in catastrophic landslides (Blasio and Elverhøi 2008; Goren and Aharonov 2007; Goren *et al.* 2010; Vardoulakis 2002a; Veveakis *et al.* 2007). Thermo-poro-mechanical couplings due to shear heating (Rice 2006; Sulem and Perdikatsis 2005) can be also associated with chemical effects such as dehydration of minerals or decomposition of carbonates, theoretically studied recently by (Brantut *et al.* 2010; Brantut *et al.* 2011b; Sulem and Famin 2009; Veveakis *et al.* 2010) and reported to take place in real faults (Hirono *et al.* 2007) and experiments at laboratory conditions (Brantut *et al.* 2011a; Colletini *et al.* 2013; Paola *et al.* 2011; Ferri *et al.* 2010; Han *et al.* 2007).

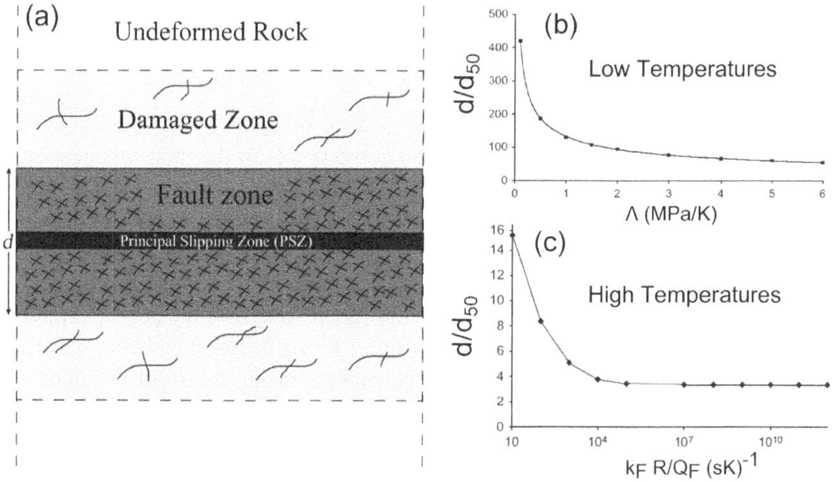

Figure 9.1. *(a) Conceptual model of the internal structure of a fault described by two scales: the PSZ in the center surrounded by the fault zone (b) using the geothermal temperature as a thermal condition for the boundary of the fault zone, we obtain the thickness of the fault zone as a function of its internal structure (average particle size) and the thermal pressurization. (c) Using the higher temperature required for chemical reactions as recorded in the PSZ, a much length scale down to three times the average grain/particle size is obtained*

9.3. Time-dependent evolution of shear zones

To this end, we assume a smooth function of the effective stress, temperature T and additional internal variables ξ, expressing the constitutive behavior of the material

$$\dot{\epsilon}^{p}_{ij} = f(\sigma'_{ij}, T, \xi) \tag{9.2}$$

the temperature and ξ obey their own evolution laws, namely the energy balance for T and experimentally deduced evolution laws for ξ. By expanding equation [9.2] around the effective yield stress σ'_Y, we obtain

$$\dot{\epsilon}^p_{ij} = f'\left(\frac{\bar{\sigma}_{ij}}{\sigma'_n}\right) + \sum_{m\geq 2} f^{(m)}\left(\frac{\bar{\sigma}_{ij}}{\sigma'_n}\right)^m \qquad [9.3]$$

where $\bar{\sigma}_{ij} = \sigma'_{ij} - \sigma_Y$, σ'_n a reference stress and $f^{(m)} = \frac{1}{m!}\left|\frac{d^m f(\sigma'_{ij},T,\xi)}{d\sigma'^m_{ij}}\right|_{\sigma'_{ij}=\sigma'_Y}$.

The first order of the Taylor expansion corresponds to the time-independent solid mechanical regime presented earlier and describes the formation of the fault near initial yield. Once the fault has been formed after initial yield, it continues to deform plastically (irreversible deformation) upon continuation of loading. This restricts our modeling efforts to the behavior of the material inside the solid mechanical shear zone obtained from the first-order expansion. Therefore, by considering explicitly the rate sensitivity emerging from the higher order terms of the expansion, we study the nonlinear (fluid-like) evolution of the mechanisms inside the solid mechanical fault zone solution.

The higher order terms of equation [9.3] give a natural extension of the classical TTS in the nonlinear viscoplastic space. If the solid mechanical yield is considered as a large-scale reference state of our model, we can describe the higher order terms in the nonlinear fluid dynamic space by the so-called overstress $\bar{\sigma}_{ij} = \sigma'_{ij} - \sigma'_Y$, defined as the stress that follows the evolution post the initial yield point (Perzyna 1966), hence $\bar{\sigma}_{ij} > 0$ is always satisfied. The rate of plastic strain is therefore expressed as

$$\dot{\epsilon}^p_{ij} = \sum_m f^{(m)}\left(\frac{\bar{\sigma}'_{ij}}{\sigma'_n}\right)^m \quad when\ \bar{\sigma}_{ij} > 0, \qquad [9.4]$$

where $f^{(m)}$ has now dimensions of $(s)^{-1}$, being thus a reference strain rate.

In this sense, the nonlinear TTS expresses a straightforward extension of the linear TTS with the difference that the time scale now plays an explicit role. In the classical TTS, the time scale is simply derived from relaxation experiments returning the system near to its equilibrium, whereas in the nonlinear TTS the system is always driven far from equilibrium. In the nonlinear framework, the relaxation experiments therefore must be replaced by far from equilibrium energy considerations, which define the time scales over which the rate of processes exchange energy inside the fault zone (Regenauer-Lieb *et al.* 2009).

9.3.1. *Energy considerations*

We introduce the Helmholtz free energy ψ, being a function of the elastic strain, temperature and the internal variables ξ. The energy balance equation, together with the second law of thermodynamics and Fourier's law of diffusion, provides the local form of the entropy production equation (Regenauer-Lieb *et al.* 2009; Rosakis *et al.* 2000):

$$\frac{\partial T}{\partial t} + v_k^m \frac{\partial T}{\partial x_k} = c_{th} \frac{\partial^2 T}{\partial x_k^2} + \frac{q_e}{(\rho C)} + \frac{\Phi_m}{(\rho C)} + \frac{L}{(\rho C)}, \quad [9.5]$$

where v_k^m is the barycentric velocity of the system, $c_{th} = \frac{k_{th}}{(\rho C)}$ is the thermal diffusion coefficient of the mixture (usually depending weakly on temperature and strongly on the microstructure, here considered constant), $C = -T\frac{\partial^2 \psi}{\partial T^2}$ is the specific heat capacity of the mixture under constant volume and $q_e = \rho_m T \frac{\partial^2 \psi}{\partial T \partial \epsilon_{ij}^e} \dot{\epsilon}_{ij}^e$ is the thermoelastic rate of heating. Note that in the framework of solid mechanics the advective terms are neglected, thus $v_k^m = 0$.

The term $L = \rho_m T \frac{\partial^2 \psi}{\partial T \partial \xi} \dot{\xi} = \Delta h\ r$ represents the latent energy produced or absorbed during a higher order energy transition (Δh is the enthalpy of the energy transition and r its rate). This includes any microstructural changes such as grain size reduction and damage in the case that the state variable is a grain size or a damage parameter (Lyakhovsky and Ben-Zion 2014b,a), respectively. In the case where the state variable refers to the chemical constituents, it corresponds to the well-known latent heat of a phase change. Finally, $\Phi_m$is the local dissipation, i.e. the mechanical power that is converted into heat in the system. In generalized thermodynamics, the dissipation is defined by the product of a thermodynamic force times a thermodynamic flux. In mechanics, the thermodynamics force is the stress σ'_{ij} and the thermodynamic flux is the plastic strain rate $\dot{\epsilon}_{ij}^p$. In the generalized framework, when additional state variables/processes are considered, the force becomes a generalized force such as a damage force E_ξ and the thermodynamic flux is the rate of change of the state variable $\dot{\xi}$.

$$\Phi_m = \sigma'_{ij}\dot{\epsilon}_{ij}^p - E_\xi \dot{\xi} = \chi \sigma'_{ij}\dot{\epsilon}_{ij}^p. \quad [9.6]$$

with $E_\xi = \rho_m \frac{\partial \psi}{\partial \xi}$ the energy dual of the internal variable ξ. The Taylor–Quinney ratio χ expresses the amount of the mechanical energy converted into heat (Taylor and Quinney 1934), and is in principle a history-dependent quantity rather than a constant ($0 < \chi < 1$). Its importance is discussed in the following section.

9.3.2. *The Taylor–Quinney coefficient*

The Taylor–Quinney coefficient is a function of the elastic strain, the temperature and the internal variables:

$$\chi = \chi(\epsilon^{e}_{ij}, \xi, T) = 1 - \frac{E_{\xi}\dot{\xi}}{\sigma'_{ij}\dot{\epsilon}^{p}_{ij}}. \qquad [9.7]$$

The Taylor–Quinney coefficient is unity if all deformation work is converted into heat. For values smaller than unity and larger than zero, it describes the portion of deformation work that is stored in microstructure. The coefficient can be derived through thermographic deformation experiments (Rosakis *et al.* 2000). The Taylor-Quinney coefficient as derived from thermography often starts at initial yield being close to zero and progressively evolves in the transient deformation regime toward a steady-state value of 0.9 for most materials (Chrysochoos and Belmahjoub 1992). Its evolution between these two values is therefore prescribed by the development of the internal state variables. As such, it has a fundamental meaning for discriminating deformation mechanisms.

If the dissipation process is, for instance, a rapid fracture, the evolution time of the Taylor–Quinney is extremely fast and obtained from the bond energy between the covalent or electronic bonds of the atoms. This equates macroscopically to an elastic interaction potential. In linear elastic fracture mechanics, brittle fracture occurs at explosive time scales (of the order of seconds). The Taylor–Quinney coefficient prior to fracture is zero as all mechanical work is stored elastic deformation. At the critical threshold, an elastodynamic fast time scale instability ensues where a variety of dissipative processes kick in such as grain/particle rotations, which release heat, therefore increasing the Taylor–Quinney coefficient. This is the typical mechanism for brittle fracture.

Similarly in ductile deformation processes, the Taylor–Quinney coefficient is zero at initial yield; however, in contrast to brittle deformation, the failure is thermally activated. A series of micromechanical processes such as dislocation and diffusion mechanisms kick in that feature a time scale that is relatively slow compared to the elastodynamic time scale. The thermal activation is often described by an Arrhenius temperature activation. This rate-dependent failure mechanism is known as ductile (or creep) fracture.

$$\Phi_m = \Phi_0\, \chi(\xi) \left(\frac{\bar{\sigma}_{ij}}{\sigma'_n}\right)^{m+1} e^{-\frac{T_0}{T}}, \qquad [9.8]$$

where $\Phi_0 = \sigma'_n \dot{\epsilon}_0$ is the reference dissipation, σ'_n is a reference stress value and $T_0 = Q_d/R$ the thermal sensitivity (activation temperature) of dissipation.

9.3.3. *Chemical reactions*

From the analysis of the brittle regime (section 9.2), we obtained that chemical reactions have a critical role in the system, defining the ultimate width of localization (the PSZ thickness). In this work, we will emphasize on fluid release reactions due to their fundamental importance in earth systems. Fluid-release reactions occur when either a hydrous mineral such as a clay mineral, serpentinite, mica, gypsum, etc., loses its water at a critical activation enthalpy. They also occur or when a mineral that is made of fluid phase and a solid constituent breaks down such as carbonate breaking to lime plus carbon dioxide upon critical activation enthalpy (Famin *et al.* 2008; Sulem and Famin 2009).

We treat these shear zone minerals as generalized solids characterized by the bonded chemical species A and B forming the solid composite AB. We assume B to represent fluid species filling the porous matrix of AB at initial conditions. At high temperatures, the solid AB breaks down, producing excess B fluid, and increasing the fluid pore pressure through a general fluid-release reaction of the form $\nu_1 AB_{(s)} \rightleftharpoons \nu_2 A_{(s)} + \nu_3 B_{(f)}$.

The kinetics of this reaction are assumed to obey the standard Arrhenius dependency on temperature (see Appendix 9.8). Following these considerations, the rates of the forward (r_F) and reverse (r_R) first-order reactions (and for $\nu_1 = \nu_2 = \nu_3 = 1$) can be calculated as

$$r_F = \frac{\rho_{AB}}{M_{AB}}(1-\phi)(1-s)k_F e^{-Q_F/RT} \qquad [9.9]$$

$$r_R = \frac{\rho_A \rho_B}{M_A M_B}(1-\phi)s\Delta\phi_{chem} k_R e^{-Q_R/RT}$$

The total reaction rate is subsequently

$$r = \left[(1-s) - s\Delta\phi_{chem}\frac{\rho_A\rho_B}{\rho_{AB}^2}\frac{M_{AB}^2}{M_A M_B}K_c^{-1}e^{\Delta h/RT}\right] \times (1-\phi)\rho_{AB}k_F e^{-Q_F/RT}, \qquad [9.10]$$

where $K_c = k_F/k_R$ and $\Delta h = Q_R - Q_F$. In these expressions, k_F, k_R, Q_F, Q_R are the pre-exponential factors and activation enthalpies of the forward and reverse reaction, respectively, and M_i and ρ_i are the molar mass and density of the ith constituent. The porosity ϕ consists of an initial value ϕ_0 and the new interconnected pore volume created from the reaction $\Delta\phi_{chem}$, $\phi = \phi_0 + \Delta\phi_{chem}$. The partial solid ratio s is the volume ratio of the produced solid A in the solid matrix. The expressions for the dependency of ϕ and s on the reaction kinetics are given in Alevizos *et al.* (2014) and are summarized here in Appendix 9.8. This formulation is essentially a damage mechanics formulation, with porosity being the damage parameter being controlled by the physics of the chemical reactions.

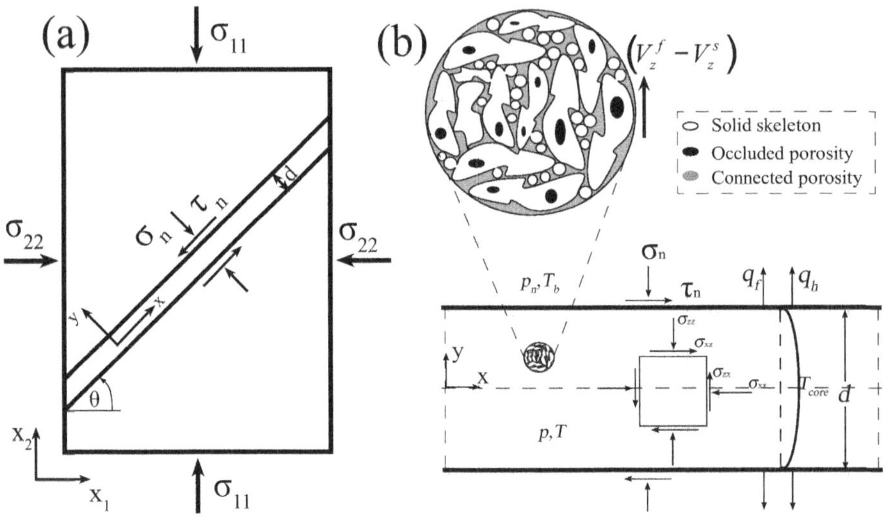

Figure 9.2. *(a) Sketch of the continuum mechanical concept of shear failure as a material bifurcation (Rudnicki and James 1975). At a critical stress ratio of the original stresses (in the coordinate system $x_1 - x_2$), a conjugate pair of shear zones emerges having a thickness d and dipping at angles $\pm\theta$, thereby defining the rotated coordinate system of failure, $x - y$. For visual purposes, only the shear zone dipping at θ is shown here. (b) Processes inside the shear zone of (a). The loading conditions and the filter velocity $v_z^f - v_z^s$, along the content of a saturated rock, are depicted. We assume that any chemical reaction is taking place at the solid–pore interface. Hence, the produced fluid contributes to the interconnected pore volume and is concentrated on the grain boundaries. We also assume that the produced solid is added to the skeleton, establishing a common velocity field v_z^s with the reactant solid*

9.4. Postfailure evolution of a shear zone

We focus on the postfailure evolution of a shear zone with thickness d formed by material bifurcation (Figure 9.2). Due to the small thickness of the shear zone compared to the thickness of the overburden, its momentum balance prescribes a constant stress profile across the shear zone, thus $\sigma_{yx} = \tau_n(t)$ and $\sigma_{yy} = \sigma_n(t)$ (Rice 2006; Veveakis *et al.* 2010). In addition, in the presence of a fluid, the stress can be decomposed according to Terzaghi's principle (Vardoulakis and Sulem 1995) to $\sigma'_{ij} = \sigma_{ij} + p_f\delta_{ij}$, where δ_{ij} is the Kronecker's delta and $p_f = p_n + \Delta p$ is the pore fluid pressure consisting of a hydrostatic part p_n at the boundary and the excess pore pressure Δp. The final system of equations is obtained once the energy equation is coupled with the mass balance equation (Alevizos *et al.* 2014; Veveakis *et al.* 2010;

Veveakis and Regenauer-Lieb 2014; Veveakis *et al.* 2014a):

$$\frac{\partial \Delta p^\star}{\partial t^\star} = \frac{\partial}{\partial y^\star}\left[\frac{1}{Le}\frac{\partial \Delta p^\star}{\partial y^\star}\right] + \Lambda \frac{T_c}{\bar{\sigma}'_n}\frac{\partial \theta}{\partial t^\star} + (1-\phi)(1-s)\mu_r e^{\frac{Ar\theta}{1+\theta}} \qquad [9.11]$$

$$\frac{\partial \theta}{\partial t^\star} = \frac{\partial^2 \theta}{\partial y^{\star 2}} + \delta \left[Gr\left(1-\Delta p^\star\right)^m e^{\frac{aAr}{1+\theta}} - (1-\phi)(1-s) + (1-\phi)s\Delta\phi_{chem}\eta K_c^{-1} e^{\frac{xAr}{1+\theta}}\right] e^{\frac{Ar\theta}{1+\theta}}$$

where the fields were normalized with the help of a reference temperature T_c:

$$t^\star = \frac{c_{th}}{(d/2)^2}t,\ y^\star = \frac{y}{d/2},\ \theta = \frac{T-T_c}{T_c},\ \Delta p^\star = \frac{\Delta p}{\bar{\sigma}'_n} \qquad [9.12]$$

The kinetics of the reactions are normalized using the Arrhenius scaling

$$Ar = \frac{Q_F}{RT_c},\ a = 1 - \frac{Q_d}{Q_F},\ x = 1 - \frac{Q_R}{Q_F} \qquad [9.13]$$

The remaining dimensionless quantities appearing in the system [9.11] are defined as

$$\delta = \frac{|\Delta h|(d/2)^2}{k_{th}T_c}k_F \rho_{AB} e^{-Ar} \qquad [9.14]$$

$$\eta = \frac{\rho_A \rho_B}{\rho_{AB}^2}\frac{M_{AB}^2}{M_A M_B}$$

$$\mu_r = \left(\frac{\rho_{AB}}{\rho_B}\frac{M_B}{M_{AB}}\right)\frac{(d/2)^2}{c_{th}\bar{\sigma}'_n}\frac{k_F}{(\beta_f+\beta_s)}e^{-Ar}$$

$$Le = \frac{c_{th}\mu_f(\beta_f+\beta_s)}{k}$$

$$Gr = \frac{\chi(\xi)\Phi_0}{k_F|\Delta h|(\rho_{AB}/M_{AB})}$$

where $k_{th} = c_{th}(\rho C)$ is the thermal conductivity of the system. The system of equations [9.11] is fully described when eight dimensionless numbers (a, x, μ_r, Le, δ, Gr, K_c and the boundary temperature θ_b since Ar is implicitly considered in these groups) are determined. However, the system has a much lower dimensionality, which can be revealed through a comprehensive analysis of its steady-state and transient responses. These analyses has been presented in (Alevizos *et al.* 2014; Poulet *et al.* 2014a; Veveakis *et al.* 2014b), identifying the dominant parameters of the system. Of critical importance is the Gruntfest number Gr, as it incorporates the energetics of the microstructure through the Taylor–Quinney coefficient, as well as the mechanical loading and the characteristics of the chemical reaction. High values of Gr correspond to a regime where the mechanical input is sufficient to trigger the chemical reaction. Due to its fundamental nature, Gr will be treated as a bifurcation parameter.

The postfailure evolution of the shear zone is summarized in Figure 9.3. The first panel (a) illustrates the fundamental three-phase stability S-curve typical to all systems of the generalized type discussed above (equation [9.11]) (Yuen and Schubert 1977). The first phase is illustrated in details in panel (b) showing a steady-state creep solution acting as a global stable material response without instabilities. The panel (c) is the solution for the phase of elevated Gruntfest numbers, but below the critical point B, as shown in Figure 9.3(a). For any initial conditions below line BC of (a), the system relaxes back to the stable branch AB. For any initial condition above line BC of (a), the system features a solitary oscillation. Figure 9.3(d) illustrates the behavior in domain III where a stable oscillator emerges. This stable oscillator is self-sustained and has a fundamental role in the postfailure evolution of shear zones. This area, where $Gr > Gr_B$, is the domain where localization of dissipation occurs owing to critical energy transitions. This means that the latent energy term L of equation [9.5] is triggered and energy is released. This domain does not depend on the exact micro mechanism, as shown in Veveakis *et al.* (2010), but describes a generalized material instability where the changes in the state variable form the localization phenomenon. The state variable can be, for instance, a grain size, a damage or porosity change as discussed in this study.

Alevizos *et al.* (2014) provided asymptotic criteria for the area of oscillations to be admitted. This occurs when the following inequalities simultaneously hold:

$$Gr > \left(1 + \frac{e^{-Ar\theta_b}}{Ar\delta}\right) e^{-aAr}\ ,\ \mu_r Le < Ar^{-3/2}\ ,\ \log(K_c) \gg 1 \qquad [9.15]$$

These inequalities correlate most of the dimensionless groups of the system, implying that in environments where strong (irreversible) endothermic reactions are favored (hence $\log(K_c) \gg 1$ is always satisfied), any two of the remaining groups can always be selected as control over oscillatory instabilities.

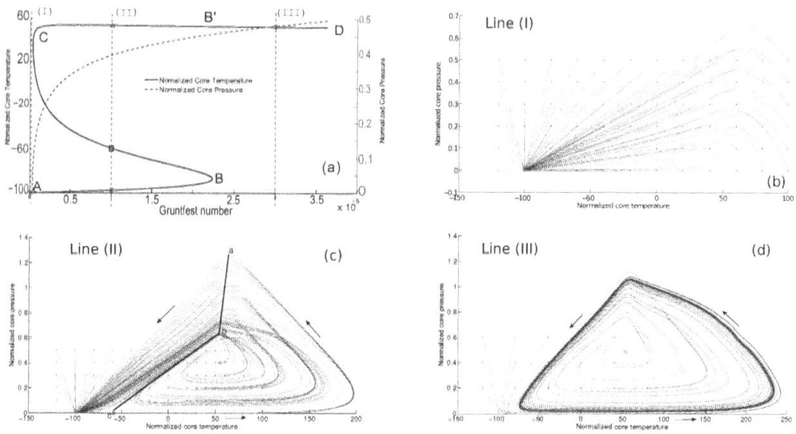

Figure 9.3. *Summary of the system's response when equations [9.15] are satisfied. (a) Steady-state response of the dimensionless core temperature (solid line) and the dimensionless core excess pore pressure (dashed line). The three lines annotated correspond to the three areas of interest, providing the phase diagrams of (b) a stable, aseismic creep behavior (line I), where all the initial conditions in the $\Delta p - T$ phase plane end up at the stable node depicted as a circle, (c) non-periodic events (line II), where the line a-b-c separates the linear paths toward the lower stable node (circle) from the homoclinic orbits tending to the circle via the unstable spiral orbit of the upper solution (triangle), and (d) periodic instabilities (line III), appearing in the $\Delta p - T$ phase plane as stable limit cycles around the unstable upper steady-state solution (circle). The magnitude of the cycle increases with decreasing Gr, obtaining its maximum amplitude at point B, where a homoclinic bifurcation takes place, with the periodic orbit colliding with the saddle point B. Note that values of Δp larger than 1 indicate possible hydraulic fracturing in the vicinity of the fault's core, limiting the present model. For a color version of this figure, see www.iste.co.uk/stefanou/instabilities.zip*

9.4.1. *Analysis of the system's response*

The solution discussed in Figure 9.3 is obtained under the assumption of a constant loading stress and chemistry, as well as internal microstructure, thus giving a constant Gruntfest number. In the more general case, either of these parameters can evolve in time, thus allowing the system to cross all three areas of the phase diagram. This leads potentially to a multiplicity of chaotic responses marked by phases of self-organized chaos, transition from aseismic creep to seismic slip and back. For example, if the system follows the hysteretic loop ABB'CA as shown in Figure 9.3, we find all three different system responses in one and the same shear zone.

Of particular interest for plate tectonic loading is the particular attractor depicted in Figure 9.3(d), where a constant plate velocity would lead to a constant shear stress over time, allowing for a steady-state evolution of the microstructure and chemistry. The periodic instability that emerges in this regime corresponds to a stick-slip type of instability, as illustrated in Figure 9.4 where the limit cycle is plotted in the logarithmic strain rate versus normalized time space. The limit cycle is characterized by two distinctly different time scale phenomena, the first being a long time scale of slow creep interrupted by a short time scale fast response owing to fluid the release reaction.

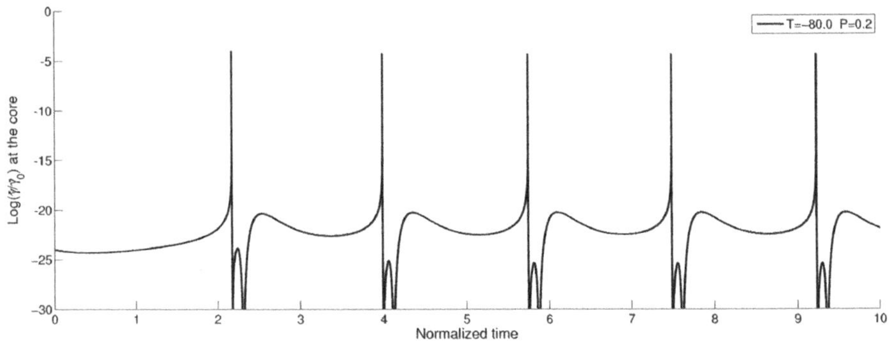

Figure 9.4. *Evolution of the core strain rate on line III in Figure 9.3(a). Irrespectively of the initial conditions, the system in this area undergoes cycles of abrupt acceleration followed by prolonged relaxation*

In addition to the two time scales characterizing the system's time evolution, its spatial manifestation also comprises two length scales. Figure 9.5 shows profiles of strain rate (red line), porosity (blue line) and solid product (green line) inside the solid mechanical shear zone at their maximum points in the limit cycle. Starting from a flat initial profile when the solid mechanical shear zone is established, the strain rate localizes in an ultrathin core zone during the fast time scale. The extreme localization of strain rate in the center of the solid mechanical shear zone is a robust outcome for any chemical reaction. The thickness of this ultralocalized shear band inside the global mechanical shear zone depends on the activation enthalpy of the forward reaction. For large enthalpy reactions, the localized zone is ultra-sharp and broadens with reducing enthalpy. This localization is then followed by a chemical–mechanical localization wave propagating toward the boundaries of the shear zone (Veveakis *et al.* 2014b, 2015; Stefanou and Sulem 2014).

We expect from this result two important outcomes. The first is that the shear zone will show at least two different scales of localization, linked to the corresponding time scales. The broad solid mechanical zone acts as a vessel for the long creeping time scale, whereas the sharp fluid-dynamic length scale accompanies the fast pressurization time scale. The second outcome is that the width of the fluid dynamic localization can be used to identify the activation energy of the chemical reaction involved. The spatial extent of the fluid release reaction is shown in the blue porosity curve in Figure 9.5 to broaden with reducing enthalpy of the reaction. The same trend applies to the solid constituent of the reaction. This additional information may be used in the field as added constraint for identification of activation enthalpies.

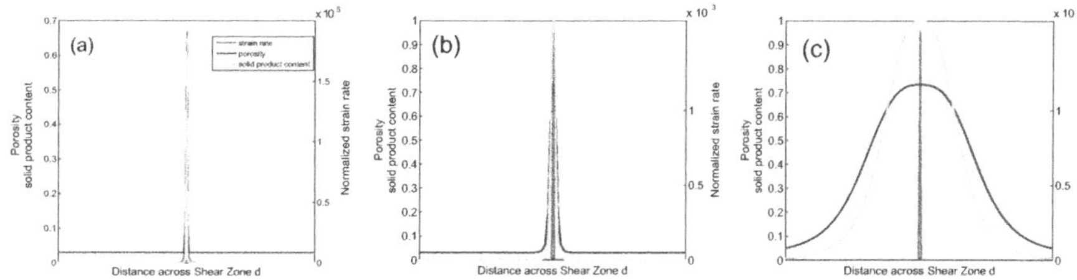

Figure 9.5. *Influence of the characteristics of the reaction. Snapshots of the profiles of porosity ϕ, solid product content and strain rate near the time of their maximum values (the point of maximum temperature in the limit cycle in Figure 9.3(d). (a) For $\Delta h = Q_F$, (b) $\Delta h = 3/4Q_F$ and (c) $\Delta h = Q_F/2$. We note that as Δh decreases, the reaction broadens its extent inside the shear zone d. For $\Delta h \approx Q_F$, porosity and s are weakly produced at the core of the shear zone, and the strain rate is smaller in magnitude (a). Once s and ϕ reach their maximum values (in this case $s = 1$ and $\phi = 0.75$), the reaction takes place in an increasingly broader zone as Q_R increases (b–c). For a color version of this figure, see www.iste.co.uk/stefanou/instabilities.zip*

9.4.2. *Time scales of the system*

We expand the temperature field with respect to $\epsilon = 1/Ar$ (Fowler 1997). At leading order the chemical reaction is inactive, thus reducing the system to a single equation, known as the Frank–Kamenetskii limit:

$$\frac{\partial \theta}{\partial t^\star} = \frac{\partial^2 \theta}{\partial y^{\star 2}} + \lambda e^\theta \qquad [9.16]$$

where $\lambda = \delta\left(Gre^{aAr} - 1\right)$. Equation [9.16] is correspondingly the Frank–Kamenetskii equation (Fujita 1969), known to have the semianalytic solution (Veveakis *et al.* 2007):

$$\theta = \theta_{core} - ln\left[\lambda(t_I - t^\star) + \frac{y^{\star 2}}{4(c_1 - ln(t_I - t^\star))}\right], \qquad [9.17]$$

where θ_{core} is the initial temperature at the core (maximum), $t_I = 1/\lambda$ is the time that temperature presents a singularity (blow-up time) and the constant c_1 is determined by the boundary conditions (Veveakis *et al.* 2007). We note that the initial condition should be the temperature at which the fault is initially formed. For flat, isothermal profile inside the shear zone, $\theta_{core} = \theta_b$.

As discussed in Veveakis *et al.* (2007), past a critical strain rate achieved at $t \approx 0.88t_I$, the analytical solution of equation [9.17] indicates that dissipation (strain rate) localizes toward the center of the shear band, while abruptly increasing. When reaching the critical value of temperature to trigger chemical pressurization, then the system enters the pressurization regime and excess pore pressure is being generated from the reaction in an undrained adiabatic setting. The time at which the temperature at the center of the fault zone ($y = 0$) reaches the activation temperature of the reaction, θ_{cr}, is approximately the time scale of the frictional (stable creep under zero overpressures) process,

$$t^{\star}_{cr} = \frac{1}{\lambda}\left[1 - e^{(\theta_{core} - \theta_{cr})}\right] \qquad [9.18]$$

Thus, the time at which chemical pressurization will set is a function of the initial configuration of the system θ_{core}, the pressurization temperature θ_{cr} and all the materials, chemical and loading parameters of the problem, incorporated into λ. Recalling that Gr (hence λ) incorporates the Taylor–Quinney coefficient, $t^{\star}_{cr}$ is directly influenced by the evolution of the internal variables, thus of the microstructure.

Note that when $\theta_{cr} = \theta_{core}$ then $t^{\star}_{cr} = 0$, meaning that the fault will enter the pressurization regime directly, without admitting any period of creep. On the other hand, when $\theta_{cr} \gg \theta_{core}$, then $t^{\star}_{cr} = 1/\lambda$ and the fault will admit all its creeping capability. Once the chemical reaction is triggered, it evolves in a fast time scale, estimated by the higher order of the expansion, where undrained-adiabatic conditions establish. In this regime, the time scale is inversely proportional to μ_r (Veveakis *et al.* 2010).

$$t^{\star}_{p} \sim \frac{1}{\mu_r} \qquad [9.19]$$

9.5. Comparison to field observations

We have identified in the above theoretical considerations two fundamentally different processes (rate/temperature independent and rate/temperature dependent) with similar outcomes. Both processes lead to two scales of localization. The large scale is associated with the solid mechanical solution and the small scale with the fluidized fault zone material inside the master shear zone.

In the rate/temperature-independent solution, the ultralocalized PSZ appears without a creeping phase. From equation [9.18], we conclude that the system is already in a critical state and does not require shear heating to be brought to criticality. In physical terms, this could be seen as a transition that is equivalent to allowing phase changes, or grain size reduction, or damage to start at a negligible input of energy. This leads to the identification of the fast time scale elastodynamic instability, where the energy released in the PSZ causes extreme localization with an internal time scale governed by the energy change process. In our example, we postulated a fluid release reaction and obtained the pressurization time scale of equation [9.19]. Other mechanisms, such as damage or grain size, impose different time scales for instability based on their energetics. This dual localization is therefore the hallmark of brittle shear zones as illustrated in Figure 9.6(a).

In the case where the system is not close to criticality and is creeping, the equivalent ductile localization mechanism can emerge. Upon a finite time after release of deformational work into heat inside the creeping shear zone, the system can reach the critical point for a fast energy transition. At this point, the creeping zone forms a PSZ upon which the micromechanical or chemical changes of the fast energy process occur. We have discussed the example of a chemical breakdown reaction, which results into an accelerated slip instability forming the PSZ. A number of other microstructural processes are often activated in the course of this instability leading to a rich microstructure inside the PSZ. This could be dynamic recrystallization, fluid release, dissolution-precipitation, etc. The complex nature of these instabilities leads to complex geometries and time series such as slow slip/earthquake signatures. The dual feature of creeping zone and ultralocalized PSZ therefore is also a hallmark of ductile shear zones as illustrated in Figure 9.6(b).

To juxtapose the outcomes of the theoretical approach with field evidence, we compare our solutions with observation from brittle and ductile shear zones. Figure 9.6 shows in (a) the famous Punchbowl cataclastic fault described by Chester and Chester (1998). The fault shows a cataclastic fault zone of the order of meter thickness with an ultralocalized and pulverized PSZ. The mineralogy of the host rock and PSZ indicate a transition from quartz and feldspar dominated to clay mineralogy in the cataclastic fault zone and ultimately higher ordered smectite and quartz in the PSZ (Chester and Chester 1998). This indicates a series of mineral dissolution-precipitation reactions that are typical for the ingression of water under lower temperature environments. The addition of water can decrease the activation temperature and thereby render the system critical at ambient conditions.

The same style of deformation is also recorded in the UNESCO world heritage Glarus fault that shows a meter thick, chemically altered tectonite (known as Lochsite tectonite), blending the hanging wall and footwall minerals (Poulet *et al.* 2014b). The tectonite is deformed as folded by ductile deformation and has an ultralocalized PSZ (Figure 9.6(b)) in its middle. Field evidence include a multiplicity of PSZs inside the tectonite, implying repetitive fast events interrupting ductile creep. The key chemical reaction controlling both the tectonite formation and the thickness of the PSZ was identified to be carbonate decomposition and precipitation (Herwegh *et al.* 2008). For the ductile localization instability, additional heat is required as the carbonate decomposition reaction happens at much larger temperature than the boundary of the thrust zone. The addition of temperature is readily available from the long-term dissipation of the creep process, which becomes a prerequisite for the ductile instability. Because the shear zone has a characteristic background temperature, the chemical reaction has a critical temperature to be triggered and the microstructure has equilibrated over time, to quasi-steady state, the ductile instability is inferred to have periodic signature in time. This implies a regular stick-slip type of behavior for the ductile PSZ, whereas the brittle PSZ is expected to be chaotic in time. The reason is that brittle instabilities are always at criticality, while ductile instabilities require time to be brought to criticality.

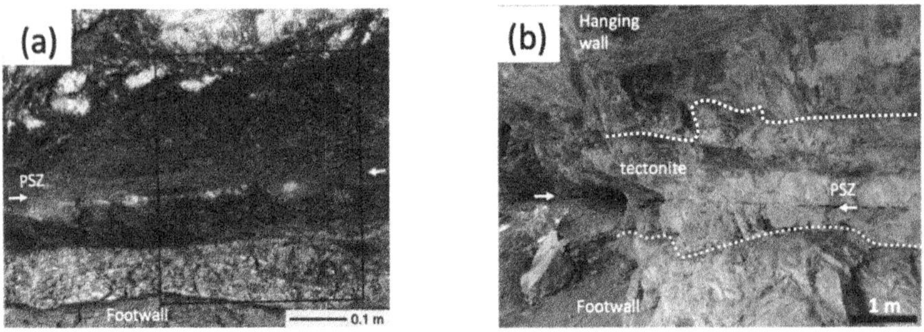

Figure 9.6. *(a) Punchball fault from Chester and Chester (1998) featuring two scales of localization, the cataclastic fault zone and the ultracatlastic PSZ annotated by arrows. (b) The Glarus Thrust at the Lochsite, featuring two scales of localzation, the creeping zone forming the carbonate-rich tectonite and the ultralocalized PSZ in its center. For a color version of this figure, see www.iste.co.uk/stefanou/instabilities.zip*

9.6. Application to ETS sequences

The temporal predictions of the model presented can be tested against real measurements from episodic tremor and slip (ETS) events at the plate interface during the subduction of oceanic plates below the continental ones. In these subduction environments, key minerals controlling both the rheology and the fluid-release reaction are the minerals of the serpentinite family (Poulet *et al.* 2014a). We may therefore specify the fluid release reaction to be the dehydration of serpentinite minerals. For the example of antigorite dehydration, AB could correspond to antigorite ($Mg_{48}Si_{34}O_{85}(OH)_{62}$), where A would represent all the solid products such as olivine ($MgSiO_4$) and enstatite ($MgSiO_3$) and B is the fluid H_2O. In this section, we use the values presented by (Poulet *et al.* 2014b) for the material properties of the system and study regular and irregular GPS signals from two different subduction environments.

9.6.1. *Regular sequences – Cascadia ETS sequence*

We begin with the periodically consistent case of the Cascadia ETS sequence in Canada (for a more detailed analysis, refer to (Alevizos *et al.* 2014; Veveakis *et al.* 2014a; Poulet *et al.* 2014a)). Episodic transient movements are continuously recorded from a network of GPS sensors revealing consistent periodicity between slip events over 20 years. The magnitude of the slip events is about 10 mm, being accumulated over approximately two weeks. We model the detailed displacement data available from the GPS stations in Vancouver Island, exhibiting a period of 14 months between the slow slip events. Because of the nearly perfect periodic sequence, it is expected that the model will be able to predict the evolution with a single fundamental mode of oscillation. All the GPS data used in this section were de-trended from the background to obtain a zero average velocity. No further geophysical techniques were used to remove annual effects, so that the model can be juxtaposed against the raw signals.

The model replicates the temporal sequence of the last 20 years, as shown in Figure 9.7, highlighting the validity of the suggested interplay between dehydration reactions and the mechanical response. The inverted values of the parameters used (Poulet *et al.* 2014b) show that the serpentinite family minerals creep in an interface shear zone of about 6 m thickness in which the ETS sequence takes place.

The GPS data show some irregularities and offsets between the events, thus suggesting perturbations to the system due to seasonal or secondary transient effects. However, within the 20 years interval these perturbations are recovered by the system back to its fundamental mode, the stable multiphysical oscillator of Figure 9.7(b). This in turn could suggest that there is small variation in the parameters and only one eigenmode is preferred. Thus, the system can absorb these perturbations due to the stable nature of the fundamental limit cycle and cannot switch to a different one.

In the case of Cascadia ETS sequence, the loading conditions used (τ_n and σ_n) were constant in time. Therefore, as suggested above, the 20 years of GPS recordings were matched with a single limit cycle of the system (Figure 9.7(b)). This result reveals that Cascadia is a self-sustained process, equilibrating near the minimum dissipation capacity of the subduction system.

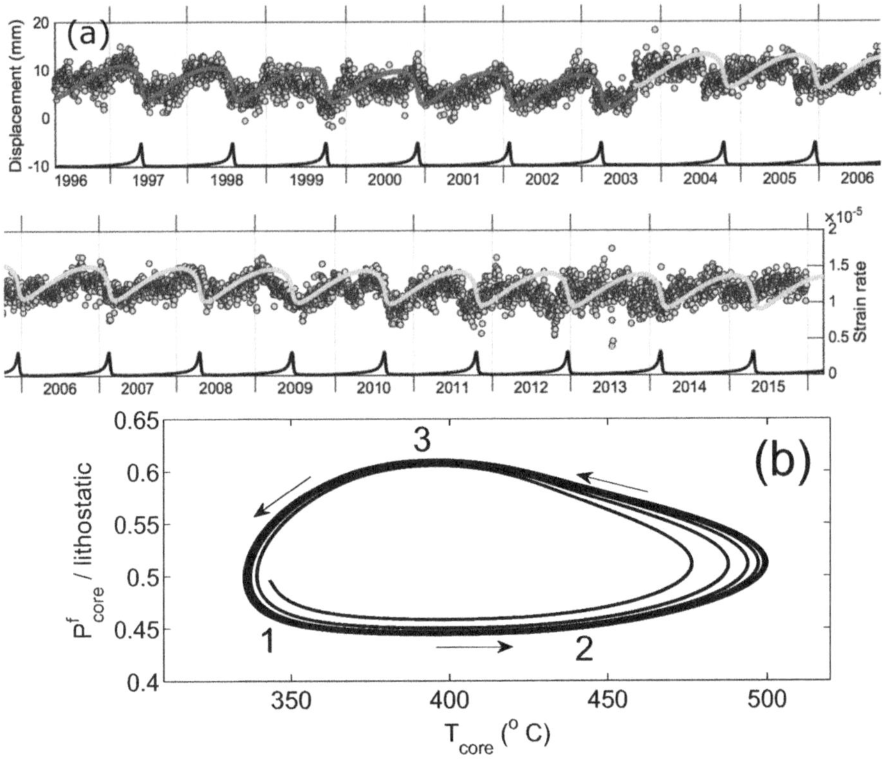

Figure 9.7. *Regular ETS sequence in South Vancouver Island, Cascadia, station ALBH. (a) The GPS data (blue dots) represent raw displacement data with its linear trend removed. The continuous sequence is shown split into the period from mid-1996 to mid-2006 (upper row) and beginning 2006 to end-2015 (lower row). The 20-year ETS sequence is matched (red and green lines) by a single stable limit cycle (b) and the strain rate curve (black line) incorporates the long slow creep periods (path 1–2) as well as the slow earthquake events due to the fluid release during the reaction (path 2–3). Any perturbations are recovered bringing the system back to the same cycle, as reflected in the GPS data of (a). The perturbation of mid-2000 was recovered by the system in 2001 and the perturbation, which affected the system in 2008, was recovered in 2011. Note that in October 2003 the sequence was interrupted due to a jump of the GPS sensor. The numerical fit is shifted (green line) to account for this jump. For a color version of this figure, see www.iste.co.uk/stefanou/instabilities.zip*

A question that naturally arises is whether the suggested formulation can also be used for less regular GPS signals. The application of the present model to an irregular case study was performed by (Poulet *et al.* 2014b). The basic assumption is that, unlike the Cascadia case, the existence of irregular ETS sequences can be attributed to non-constant parameter values, which forces the system to switch between different material oscillators. Such a case could be obtained if, for instance, the dimensionless groups are allowed to evolve in time (as shown in (Poulet *et al.* 2014a)) or in space, accounting for the different geologies and ambient conditions intersected by the fault. A natural consequence of the temporal and spatial variation of the parameters is that the system would turn into a multiperiod, multiscale chaotic attractor and the subduction system will depart from its minimum dissipation limit and driven toward the maximum dissipation.

The Grunfest and the Lewis numbers are bearing the information of the loading conditions, and are therefore the parameters that can vary in time. Having already pointed out the role of the critical Gruntfest number Gr_B as a global attractor of the oscillator, we assume here that its evolution is the most probable cause for the presence of alternating oscillatory modes in the system. Following the discussion in Poulet *et al.* (2014b), the time variation of the Gruntfest and associated Lewis number can be affected by short-term seismic stress perturbations in the vicinity of the subduction zone. As pointed out in the case of Cascadia, the fast time scale of the process is approximately two weeks. This means that a seismic event happens on an even faster time scale, which will provoke an almost instantaneous jump in the stress applied on the shear zone boundaries. The latter can be equated to step-like perturbations on the Gruntfest number in the framework of the present chemical oscillator model.

From the fitting process of the GPS sequence in New Zealand (Figure 9.8), it was deduced that the oscillators comprise two modes. The first mode (mode 1 in Figure 9.8) is exactly identical to the one derived from the Cascadia fit. The second mode is shown as mode 2 in Figure 9.8 and has a significantly larger temperature variation of around 270 °C, from 330 °C to approximately 600 °C. During both cycles, the fluid pressure rises from hydrostatic values (40% of the lithostatic) to about 60% of the lithostatic when the forward reaction is the dominant mechanism. A major outcome of this fitting process is that, similar to Cascadia, the width of the shear zone where all the mechanisms presented operate is calculated around 6 m wide. In addition, as suggested in (Poulet *et al.* 2014b), the common oscillator is a fundamental mode (or eigenmode in a mathematical context) of serpentinite deformation.

As far as material constants are concerned, these were considered to be the same for Cascadia and Hikurangi, suggesting the common serpentinite nature of the lubricating shear zones. The only difference between the two cases is the emergence of a second oscillator in New Zealand that was attributed to an overstress transient of the order of 60 kPa in shear (Poulet *et al.* 2014b). This derivation suggests that the elastodynamic transients of the surrounding environment cannot be neglected in the Hikurangi trench. Since mode 2 displays the higher amplitude oscillations, it is identified as the one closer to the global attractor (the homoclinic point) and mode 1 is interpreted as a higher Gruntfest number mode.

An interpretation of the sudden perturbation of the system could be the propagation of a brittle fracture into the ductile domain. The result of this would be a stress drop that would act as a perturbation to the system and could shift the oscillator from mode 1 to mode 2. Once the stress transients are dissipated, the elastic loading of the subducting plate would naturally lead the system back to an increased applied shear stress, therefore imposing the transition from mode 2 to mode 1 (Figure 9.8(b)).

This means that in Hikurangi the dynamics of the serpentinite oscillator between the two modes depicted in Figure 9.8 are driven by evolving (in time) boundary conditions. Unlike Cascadia, where constant stress boundary conditions could reproduce 20 years of ETS activity, in New Zealand the loading stress exhibits time variations thus not placing the subduction environment in the minimum dissipation limit. However, since the system is not admitting damped oscillations, Hikurangi is also not driven at its maximum dissipation limit. It is therefore in an intermediate loading regime, oscillating continuously between two different modes and driving both the slow earthquake activity and the fast seismic events in the vicinity of Gisborne (see also (Poulet *et al.* 2014b)).

The New Zealand case, on the other hand, shows frequent seismic events and the subducting activity seems to be strongly influenced by the elastic feedback of the surroundings, which drives the dynamics of the process and alters its eigenmode. Of particular interest for the above hypothesis is the recorded December 20, 2007 Gisborne earthquake (magnitude 6.8). Such an event could indeed be sufficient to cause the GPS perturbation from the base mode 1 (black) to the excited mode 2 (red), which coincides exactly (see the green box in Figure 9.8) with the Gisborne event. The same effect could be attributed to the magnitude 6.8 earthquake of 2001 (October 21), perturbing the system to the excited mode 2 recorded in the GPS signal shown in Figure 9.8 between 2002 to 2004. However, the lack of recordings before 2002 does not allow for further analysis of this period.

9.7. Discussion

In this contribution, we link the observed double localization patterns in fault zones with a discussion on their possible driving energetics and time scales of formation. We have shown that the thick fault zone is a result of a solid mechanical instability, acting as a vessel for the ultralocalized instability. In the case of brittle failure, negligible additional energy is required as the system behaves in a near temperature independent, close to criticality regime. In the case of ductile failure, the shear zone is not necessarily at criticality through time, but the fluid-like creep motions raise the local dissipation to eventually trigger an energy transformation in an ultralocalized PSZ. Under the assumption of microstructural steady state of the shear zone, the latter instability is periodic.

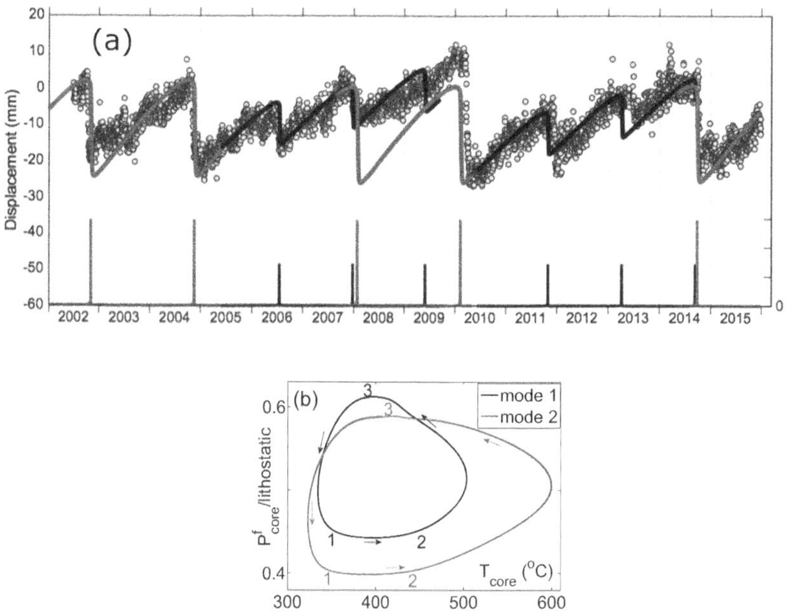

Figure 9.8. *Irregular ETS sequence of Gisborne, New Zealand, station GISB. (a) The GPS data (blue dots) represent raw displacement data with its linear trend removed. The signal is nonperiodic and the suggested fit consists of two modes. The first mode (red line) has higher displacement and strain rates (shown at the bottom) per event and a period of 2 years, and the second one (black line) has a period of about 14 months. Figure until end of 2013 is as published in June 2014 (Poulet et al. 2014b). (b) Limit cycles of the two modes identified. Note that this bimodal chaotic oscillator is driven from the elastic surrounding, with the transition from mode 1 to mode 2 being the natural mode of the elastic loading. The reverse transition requires additional excitation that would relax the applied shear stress, e.g., from nearby seismic events in the vicinity of the system. Such an event was recorded at the end of 2007, causing the perturbation recorded by the GPS station, and shifting the sequence back to mode 1 of the oscillator. For a color version of this figure, see www.iste.co.uk/stefanou/instabilities.zip*

We have presented in Figure 9.6 a direct comparison of brittle and ductile/creep fractures in shear zones. In the brittle case, it is known from field and laboratory observations that the PSZ represents an ultralocalized process zone, consisting of ultrafine particles that have usually undergone mechanochemical degradation (gelification, decarbonation and dehydration reactions, melting, as thoroughly discussed in Toro *et al.* (2011)). Therefore, although the mechanism of the fault zone formation is the material bifurcation from homogenous deformation (Mühlhaus and Vardoulakis 1987), the formation of PSZs should be considered the effect of the chemical–mechanical deformation following the onset of the initial structure.

The ductile shear zone in Figure 9.6 features a chemical transition from host rock to hanging rock and a strong grain size reduction toward the PSZ (Herwegh *et al.* 2008). The mechanism of localization, although different from the brittle case, has the same style of energetic transition where microstructural changes are captured by their latent heat effects inside the PSZ. The main difference in terms of energetics is the time scale of deformation, which we have shown to be governed by the energy equation and lastly the thermal activation.

In the brittle zone, thermal activation is not considered important. However, a similar energetic transition may be taking place. The activation process can equally be triggered by a lowering of the critical threshold to near ambient condition through the access of fluids interacting with the host rock. It can be argued that in the brittle case, the thermal time scale control is replaced by the mechanism of fluid transfer during fault slip. These mechanisms can be low temperature pressurization effects (Lachenbruch 1980; Rice 2006; Sibson 1973; Sulem and Perdikatsis 2005; Wibberley and Shimamoto 2005) or high temperature chemical effects such as dehydration of minerals or decomposition of carbonates, theoretically studied recently (Brantut *et al.* 2010; Brantut *et al.* 2011b; Sulem and Famin 2009; Veveakis *et al.* 2010) and reported to take place in real faults (Famin *et al.* 2008; Hirono *et al.* 2007) and experiments at laboratory conditions (Brantut *et al.* 2011a; Colletini *et al.* 2013; Paola *et al.* 2011; Ferri *et al.* 2010; Han *et al.* 2007). Additional mechanisms that include dry phase transitions such as breakage can be formulated in a similar thermodynamic framework (Einav 2007a,b; Lyakhovsky and Ben-Zion 2014b,a). Therefore, the above-described framework illustrates an energy-based formulation for instabilities of solid matter with emergent length scales of localization.

This fundamental outcome is illustrated in the present approach by two scales of localization naturally emerging and corresponding to two different energy regimes. These are the traditional solid mechanical localization during failure leading to a finite width shear band that can be described by internal variables and the fluid dynamical postfailure localization related to a phase change that counterintuitively leads to ultra thin shear zones within the master solid mechanical containment. We identify these fundamental outcomes to be a plausible explanation for the two-scale localization phenomena described in this work.

9.8. Appendix: poro-chemical model

At high temperatures, the solid AB breaks down, producing excess B fluid and increasing the fluid pore pressure through a general fluid-release reaction of the form:

$$\nu_1 AB_s \rightleftharpoons \nu_2 A_s + \nu_3 B_f \qquad [9.20]$$

We assume the following relations for the partial molar reaction rates for the species

$$r_{AB} = -\left[\frac{\rho_{AB}}{M_{AB}}(1-\phi)(1-s)\right]^{\nu_1} k_F exp(-Q_F/RT) \qquad [9.21]$$
$$r_A = \left[\frac{\rho_A}{M_A}(1-\phi)s\right]^{\nu_2} k_R exp(-Q_R/RT)$$
$$r_B = \left[\Delta\phi_{chem}\frac{\rho_B}{M_B}\right]^{\nu_3} k_R exp(-Q_R/RT)$$

From the stoichiometry of the considered reaction (equation [9.20]), it should hold that:

$$-\frac{r_{AB}}{\nu_1} = \frac{r_A}{\nu_2} = \frac{r_B}{\nu_3} \qquad [9.22]$$

From equations [9.21]–[9.22], and for $\nu_1 = \nu_2 - \nu_3 = 1$, we derive the poro-chemical model

$$\Delta\phi_{chem} = A_\phi \frac{1-\phi_0}{1+\frac{\rho_B}{\rho_A}\frac{M_A}{M_B}\frac{1}{s}}, \qquad [9.23]$$
$$s = \frac{\omega_{rel}}{1+\omega_{rel}}, \quad and$$
$$r_{rel} = \frac{\rho_{AB}}{\rho_A}\frac{M_A}{M_{AB}} K_c exp\left(\frac{\Delta h}{RT}\right).$$

In equation [9.23], $K_c = k_F/k_R$ is the ratio of the pre-exponential factors of the Arrhenius reaction rates and $\Delta h = Q_R - Q_F$ the difference of the forward and reverse activation energies. The parameter A_ϕ is a coefficient that determines the amount of the interconnected pore volume (porosity) created due to the reaction. We assume that all the fluid generated contributes to the interconnected pore volume, and thus set $A_\phi = 1$.

Following these considerations, the rates of the forward (ω_F) and reverse (ω_R) first-order reactions can be calculated to be

$$r_F = r_{AB} = \frac{\rho_{AB}}{M_{AB}}(1-\phi)(1-s)k_F e^{-Q_F/RT} \qquad [9.24]$$

$$r_R = r_A r_B = \frac{\rho_A \rho_B}{M_A M_B}(1-\phi)s\Delta\phi_{chem} k_R e^{-Q_R/RT}. \qquad [9.25]$$

Note that, for simplicity we have assumed in equations [9.21] that the two products are produced with the same pre-exponential factor and activation energies. If this is not the case, the above-described model should be modified accordingly. The net reaction rate would then be $r = r_F - r_R \frac{M_{AB}}{\rho_{AB}}$ (the reverse reaction rate was normalized with the reference concentration $\frac{\rho_{AB}}{M_{AB}}$ for dimensional purposes), which however would be essentially irreversible ($r_F \gg r_R$) in the case $K_c = k_F/k_R \gg 1$.

9.9. References

Alevizos, S., Poulet, T., Veveakis, E. (2014). Thermo-poro-mechanics of chemically active creeping faults 1: Theory and steady state considerations. *Journal of Geophysical Research: Solid Earth*, 119(6), 4558–4582.

Allam, A.A. and Ben-Zion, Y. (2012). Seismic velocity structures in the southern California plate-boundary environment from double-difference tomography. *Geophys. J. Int.*, 190, 1181–1196.

Ben-Zion, Y. and Sammis, C.G. (2003). Characterization of fault zones. *Pure Appl. Geophys.*, 160, 677–715.

Blasio, F.D. and Elverhøi, A. (2008). A model for frictional melt production beneath large rock avalanches. *J. Geophys. Res.*, 113, F02014.

Brantut, N., Schubnel, A., Corvisier, J., Sarout, J. (2010). Thermochemical pressurization of faults during coseismic slip. *J. Geophys Res.*, 115, B05314.

Brantut, N., Han, R., Shimamoto, T., Findling, N., Schubnel, A. (2011a). Fast slip with inhibited temperature rise due to mineral dehydration: Evidence from experiments on gypsum. *Geology*, 39(1), 59–62.

Brantut, N., Sulem, J., Schubnel, J. (2011b). Effect of dehydration reactions on earthquake nucleation: Stable sliding, slow transients and unstable slip. *J. Geophys Res.*, 116, B05304.

Cecinato, F., Zervos, A., Veveakis, E. (2011). A thermo-mechanical model for the catastrophic collapse of large landslides. *International Journal for Numerical and Analytical Methods in Geomechanics*, 35(14), 1507–1535.

Chester, F. and Chester, J. (1998). Ultracataclasite structure and friction processes of the punchbowl fault, San Andreas system, California. *Tectonophysics*, 199–221.

Chester, F.M., Evans, J.P., Biegel, R.L. (1993). Internal structure and weakening mechanisms of the San Andreas fault. *Journal of Geophysical Research: Solid Earth*, 98(B1), 771–786.

Chrysochoos, A. and Belmahjoub, F. (1992). Thermographic analysis of thermomechanical couplings. *Archives Mechanics*, 44(1), 55–68.

Colletini, C., Viti, C., Tesei, T., Mollo., S. (2013). Thermal decomposition along natural carbonate faults duing earthquakes. *Geology*, 41(8), 11508.

Cornet, F., Doan, M., Moretti, I., Borm, G. (2004). Drilling through the active Aigion fault: The AIG10 well observatory. *Comptes Rendus Geosciences*, 336(4–5), 395–406.

Einav, I. (2007a). Breakage mechanics Part I: Theory. *Journal of the Mechanics and Physics of Solids*, 55(6), 1274–1297.

Einav, I. (2007b). Breakage mechanics Part II: Modelling granular materials. *Journal of the Mechanics and Physics of Solids*, 55(6), 1298–1320.

Famin, V., Nakashima, S., Boullier, A.-M., Fujimoto, K., Hirono, T. (2008). Earthquakes produce carbon dioxide in crustal faults. *Earth. Plan. Sci. Let.*, 265, 487–497.

Faulkner, D.R., Lewis, A.C., Rutter, E.H. (2003). On the internal structure and mechanics of large strike-slip fault zones: Field observations of the carboneras fault in southeastern Spain. *Tectonophysics*, (367), 235–251.

Ferri, F., DiToro, G., Hirose, T., Shimamoto, T. (2010). Evidence of thermal pressurization in high-velocity friction experiments on smectite-rich gouges. *Terra Nova*, 22(5), 347–353.

Fowler, A., (ed.). (1997). *Mathematical Models in the Applied Sciences*, 2nd edition. Cambridge University Press.

Fujita, C. (1969). On the non-linear equations $du + e^u = 0$ and $v_t = dv + e^v$. *Bull. Am. Math. Soc.*, 75, 132–135.

Goren, L. and Aharonov, E. (2007). Long runout landslides: The role of frictional heating and hydraulic diffusivity. *Geophys. Res. Lett.*, 34, L07301.

Goren, L., Aharonov, E., Anders, M.H. (2010). The long runout of the heart mountain landslide: Heating, pressurization, and carbonate decomposition. *Journal of Geophysical Research: Solid Earth*, 115(B10).

Han, R., Shimamoto, T., Hirose, T., Ree, J., Ando, J. (2007). Ultralow friction of carbonate faults caused by thermal decomposition. *Science*, 316, 878–881.

Herwegh, M., Hurzeler, J., Pfiffner, O., Schmid, S., Abart, R., Ebert, A. (2008). The glarus thrust: Excursion guide and report of a field trip of the Swiss tectonic studies groups. *Swiss Journal of Geosciences*, 101(2), 323–340.

Hill, R. (1962). Acceleration waves in solids. *J. Mech. Phys. Solids*, 10, 1–16.

Hirono, T., Yokoyama, T., Hamada, Y., Tanikawa, W., Mishima, T., Ikehara, M., Famin, V., Tanimizu, M., Lin, W., Soh, W., Song, S.-R. (2007). A chemical kinetic approach to estimate dynamic shear stress during the 1999 Taiwan Chi-Chi earthquake. *Geophys. Res. Lett.*, 34, L19308.

Holdsworth, R., van Diggelen, E., Spiers, C., de Bresser, J., Walker, R., Bowen, L. (2011). Fault rocks from the {SAFOD} core samples: Implications for weakening at shallow depths along the San Andreas fault, California. *Journal of Structural Geology*, 33(2), 132–144.

Kennedy, L. and Logan, J. (1997). The role of veining and dissolution in the evolution of fine-grained ylonites: The McConnell thrust, Alberta. *J. Struct. Geology*, 19(6), 785–797.

Lachenbruch, A. (1980). Frictional heating, fluid pressure and the resistance to fault motion. *J. Geophys. Res.*, 85, 6097–6112.

Lyakhovsky, V. and Ben-Zion, Y. (2014a). A continuum damagebreakage faulting model and solid-granular transitions. *Pure Appl. Geophysics*, 171, 3099–3123 [Online]. Available at: https://link.springer.com/article/10.1007.

Lyakhovsky, V. and Ben-Zion, Y. (2014b). Damage-breakage rheology model and solid-granular transition near brittle instability. *J. Mech. Phys. Solids*, (64), 184–197.

Mandel, J. (1966). Conditions de stabilité et postulat de Drucker. In *Rheology and Soil Mechanics / Rhéologie et Mécanique des Sols. International Union of Theoretical and Applied Mechanics*, Kravtchenko, J., Sirieys, P.M. (eds). Springer, Berlin, Heidelberg.

Mühlhaus, H. and Vardoulakis, I. (1987). Thickness of shear bands in granular materials. *Geotechnique*, 37(3), 271–283.

Noda, H. and Shimamoto, T. (2005). Thermal pressurization and slip-weakening distance of a fault: An example of the Hanaore fault, southwest Japan. *Bull. Seism. Soc. Am.*, 95(4), 1224–1233.

Paola, N.D., Hirose, T., Mitchell, T., Toro, G.D., Togo, T., Shimamoto, T. (2011). Fault lubrication and earthquake propagation in thermally unstable rocks. *Geology*, 39(1), 35–38.

Papanicolopulos, S. and Veveakis, E. (2011). Sliding and rolling dissipation in Cosserat plasticity. *Granular Matter*, 13(3), 197–204.

Perzyna, P. (1966). Fundamental problems in viscoplasticity. *Adv. Appl. Mech.*, 9, 243–377.

Poulet, T., Veveakis, E., Regenauer-Lieb, K., Yuen, D.A. (2014a). Thermo-poro-mechanics of chemically active creeping faults 3: The role of serpentinite in episodic tremor and slip sequences, and transition to chaos. *Journal of Geophysical Research: Solid Earth*, 119(6), 4606–4625.

Poulet, T., Veveakis, M., Herwegh, M., Buckingham, T., Regenauer-Lieb, K. (2014b). Modeling episodic fluid-release events in the ductile carbonates of the glarus thrust. *Geophysical Research Letters*, 41(20), 7121–7128.

Rattez, H., Stefanou, I., Sulem, J. (2017a). The importance of Thermo-Hydro-Mechanical couplings and microstructure to strain localization in 3D continua with application to seismic faults. Part I: Theory and linear stability analysis, *Journal of the Mechanics and Physics of Solids*, 115, 54–76. [Online]. Available at: https://doi.org/10.1016/j.jmps.2018.03.004.

Rattez, H., Stefanou, I., Sulem, J., Veveakis, M., Poulet, T. (2017b). The importance of Thermo-Hydro-Mechanical couplings and microstructure to strain localization in 3D continua with application to seismic faults. Part II: Numerical implementation and post-bifurcation analysis. *Journal of the Mechanics and Physics of Solids*, 115, 1–29. [Online]. Available at: https://doi.org/10.1016/j.jmps.2018.03.003.

Rattez, H., Stefanou, I., Sulem, J., Veveakis, M., Poulet, T. (2017c). Numerical analysis of strain localization in rocks with thermo-hydro-mechanical couplings using Cosserat continuum, *Rock Mech. Rock Eng.*, 51, 3295331 [Online]. Available at: https://link.springer.com/article/10.1007% 2Fs00603-018-1529-7.

Regenauer-Lieb, K., Yuen, D., Fusseis, F. (2009). Landslides, ice quakes, earthquakes: A thermodynamic approach to surface instabilities. *Pure. Appl. Geophys*, 166(10–11), 1885–1908.

Regenauer-Lieb, K., Veveakis, M., Poulet, T., Wellmann, F., Karrech, A., Liu, J., Hauser, J., Schrank, C., Gaede, O., Trefry, M. (2013a). Multiscale coupling and multiphysics approaches in Earth Sciences: Applications. *Journal of Coupled Systems and Multiscale Dynamics*, 1(3).

Regenauer-Lieb, K., Veveakis, M., Poulet, T., Wellmann, F., Karrech, A., Liu, J., Hauser, J., Schrank, C., Gaede, O., Trefry, M. (2013b). Multiscale coupling and multiphysics approaches in Earth Sciences: Theory. *Journal of Coupled Systems and Multiscale Dynamics*, 1(1), 49–73.

Reiner, M. (1964). The Deborah number. *Physics Today*, 17(1), 152–153.

Rice, J.R. (2006). Heating and weakening of faults during earthquake slip. *J. Geophys. Res.*, 111, B05311.

Rosakis, P., Rosakis, A., Ravichandran, G., Hodowany, J. (2000). A thermodynamic internal variable model for the partition of plastic work into heat and stored energy in metals. *J. Mech. Phys. Solids*, 48, 581–607.

Rowe, C., Fagereng, A., Miller, J., Mapani, B. (2012). Signature of coseismic decarbonation in dolomitic fault rocks of the Naukluft Thrust, Namibia. *Earth Plan. Sci. Let.*, 333–334, 200–210.

Rudnicki J.W. and James, R.R. (1975). Conditions for the localization of deformation in pressure sensitive dilatant materials. *J. Mech. Phys. Solids*, 23, 371–394.

Sibson, R. (1973). Interaction between temperature and pore-fluid pressure during earthquake faulting – a mechanism for partial or total stress relief. *Nature*, 243, 66–68.

Sibson, R. (2003). Thickness of the seismic slip zone. *Bull. Seism. Soc. Am.*, 93, 1169–1178.

Stefanou, I., Sulem, J. (2014). Chemically induced compaction bands: Triggering conditions and band thickness. *Journal of Geophysical Research: Solid Earth*, 119(2), 880–899.

Sulem, J. and Famin, V. (2009). Thermal decomposition of carbonates in fault zones: Slip-weakening and temperature-limiting effects. *J. Geophys. Res.*, 114, B03309.

Sulem J., Vardoulakis I.O.H., Perdikatsis, V. (2005). Thermo-poro-mechanical properties of the Aigion fault clayey gouge – application to the analysis of shear heating and fluid pressurization. *Soils and Foundations*, 45, 97–108.

Sulem, J., Lazar, P., Vardoulakis, I. (2007). Thermo-poro-mechanical properties of clayey gouge and application to rapid fault shearing. *Int. J. Num. Anal. Meth. Geomechanics*, 31(3), 523–540.

Sulem, J., Stefanou, I., Veveakis, E. (2011). Stability analysis of undrained adiabatic shearing of a rock layer with cosserat microstructure. *Granular Matter*, 13(3), 261–268.

Taylor, G. and Quinney, H. (1934). The latent energy remaining in a metal after cold working. *Proc. R. Soc., Ser. A.*, 143, 307–326.

Tobolsky, A. and Andrews, R. (1945). Systems manifesting superposed elastic and viscous behaviour. *J.Chem. Phys.*, 13(1), 3–27.

Toro, G.D., Han, R., Hirose, T., De Paola, N., Nielsen, S., Mizoguchi, K., Ferri, F., Cocco, M., Shimamoto, T. (2011). Fault lubrication during earthquakes. *Nature*, 471, 494–498.

Townend, J. (2009). Drilling, sampling, and monitoring the alpine fault: Deep fault drilling project—alpine fault, New Zealand; Franz Josef, New Zealand, 22–28 March 2009. *Eos, Transactions American Geophysical Union*, 90(36), 312–312.

Vardoulakis, I. (2002a). Dynamic thermo-poro-mechanical analysis of catastrophic landslides. *Geotechnique*, 52, 157–171.

Vardoulakis, I. and Sulem, J. (eds). (1995). *Bifurcation Analysis in Geomechanics*. Blankie Acc. and Professional, London.

Veveakis, E. and Regenauer-Lieb, K. (2014). The fluid dynamics of solid mechanical shear zones. *Pure Appl. Geophys.*, 171, 3159–3174.

Veveakis, E. and Regenauer-Lieb., (2015). Cnoidal waves in solids. *Journal of Mechanics and Physics of Solids*, 78, 231–248.

Veveakis, E., Vardoulakis, I., Di Toro., G. (2007). Thermoporomechanics of creeping landslides: The 1963 vaiont slide, northern Italy. *J. Geophys. Res.*, 112, F03026.

Veveakis, E., Alevizos, S., Vardoulakis., I. (2010). Chemical reaction capping of thermal instabilities during shear of frictional faults. *J. Mech. Phys. Solids*, 58, 1175–1194.

Veveakis, E., Sulem, J., Stefanou, I. (2012). Modeling of fault gouges with cosserat continuum mechanics: Influence of thermal pressurization and chemical decomposition as coseismic weakening mechanisms. *J. Struct. Geology*, 38, 254–264.

Veveakis, E., Stefanou, I., Sulem, J. (2013). Failure in shear bands for granular materials: Thermo-hydro-chemo-mechanical effects. *Geotechnique Let.*, 3(2), 31–36.

Veveakis, E., Poulet, T., Alevizos, S. (2014a). Thermo-poro-mechanics of chemically active creeping faults 2: Transient considerations. *Journal of Geophysical Research: Solid Earth*, 119(6), 4583–4605.

Veveakis, E. and Regenauer-Lieb, K., Weinberg, R. (2014b). Ductile compaction of partially molten rocks: The effect of non-linear viscous rheology on instability and segregation. *Geophysical Journal International*, 200(1), 519–523.

Wibberley, C.A.J. and Shimamoto, T. (2003). Internal structure and permeability of major strike-slip fault zones: The median tectonic line in mid prefecture, southwest Japan. *J. Struct. Geol.*, 25, 59–78.

Wibberley, C.A.J. and Shimamoto, T. (2005). Earthquake slip weakening and asperities explained by thermal pressurization. *Nature*, 426(4), 689–692.

Yuen, D. and Schubert, G. (1977). Asthenospheric shear flow: Thermally stable or unstable? *Geophys. Res. Lett*, 4(11), 503–506.

10

Analysis of Instabilities in Faults

Hadrien RATTEZ[1], Ioannis STEFANOU[2], Jean SULEM[3], Manolis VEVEAKIS[4] and Thomas POULET[5]

[1] *Duke University, Durham, USA*

[2] *GeM, Ecole Centrale de Nantes, France*

[3] *Navier, Ecole des Ponts ParisTech, Université Paris-Est, Marne-la-Vallée, France*

[4] *Duke University, Durham, USA*

[5] *CSIRO, Kensington, Australia*

In this chapter, we are interested in modeling the creation of shear bands within a fault zone by taking into account the influence of the characteristic size of the microstructure through Cosserat continuum theory together with thermo-hydro mechanical (THM) couplings. The introduction of a characteristic length in the constitutive laws of the material in Cosserat's theory not only enables us to regularize the problem of localization of deformations, but also enables us to take into account the effect of the size of the microstructure.

Two approaches are considered to study the system of coupled nonlinear partial differential equations: linear stability analysis (LSA) and the finite element method. The LSA allows us to examine the conditions of the occurrence of instabilities (loss of homogeneity for the field of deformation) for a mechanical system with multiphysical couplings. In addition, considerations on the perturbation applied to the system in these analyses make it possible to estimate the thickness of the shear zone – a key parameter for understanding the fault mechanism. For deformations remaining within a given range, these estimates are confirmed by numerical integration. They are then compared to experimental and *in situ* observations and show a good correlation. On the other hand, numerical simulations make it possible to obtain the mechanical response

Instabilities Modeling in Geomechanics,
coordinated by Ioannis STEFANOU and Jean SULEM. © ISTE Editions 2020.

of the fault core, called a gouge, and thus to obtain information on the influence of the different couplings on the energy budget of an earthquake.

10.1. Introduction

Waves that propagate within the Earth's crust and cause dramatic tremors at the surface are generally the result of frictional instability along a fault (Scholz 2002). This instability leads to a sudden and rapid sliding along a preexisting fault. Indeed, the movement of tectonic plates generates an accumulation of elastic energy at their borders. When the shear strength of a fault is reached, the fault begins to slip. However, this slip can be either continuous and quasi-static or intermittent and dynamic. The difference is due to the evolution of the resistance along the fault during the slip. This mechanism can be illustrated by a simple spring-slider model (Rattez 2017). This model demonstrates the importance of the stress and strain response of the fault in triggering earthquakes.

In the spring-slider model, the fault is modelled as an interface. However, observations of samples from drilling cores or outcrops show that fault zones are very heterogeneous and present a significant thickness. In addition, further observations *in situ* or in the laboratory show the importance of strain localization within the fault zone in the triggering of seismic instability (see Figure 10.1).

This localization of deformations involves multiple physical couplings. First, the presence of a fluid inside the porous space affects the behavior of the material (Coussy 2004). Then, the velocity reached during a seismic event (about 1 m/s) causes a significant temperature increase. This can induce a thermal pressurization of the fluid contained in the pores, a decrease in friction at the contact microasperities, or even in some cases a melting of the rock, which is revealed by the presence of pseudotachylyte (Di Toro *et al.* 2009). It can also cause the activation of endothermic chemical reactions such as the decomposition of carbonates, which limits the temperature increase (Veveakis *et al.* 2010). Many of these phenomena are affected by the size of the microstructure, which decreases during sliding due to the crushing of the grains in zones where the deformations are localized (Sammis *et al.* 1986).

Strain localization is a phenomenon that can be modeled as a bifurcation (Rudnicki and Rice 1975). The homogeneous state of deformation becomes unstable in the Lyapunov sense. A LSA enables us to determine the conditions under which the localization is triggered (Benallal and Comi 2003).

However, the classical Cauchy continuum theory does not allow us to describe a finite thickness of the localization zone, as it predicts an infinitely small localized zone (Vardoulakis 1985), unless a viscoplastic behavior and THM couplings are considered (Platt *et al.* 2014; Rice *et al.* 2014). This can be explained by the fact that the constitutive law does not include any characteristic length. One of the ways that enables us to regularize the problem of strain localization and obtain a finite band thickness is to consider continua with microstructure, such as Cosserat continua (Sulem *et al.* 2011). The latter are particularly interesting because they also allow the size of the microstructure to be taken into account explicitly in the model and it has been shown that this parameter plays an important role in the behavior of faults (Anthony and Marone 2005; Cashman *et al.* 2007).

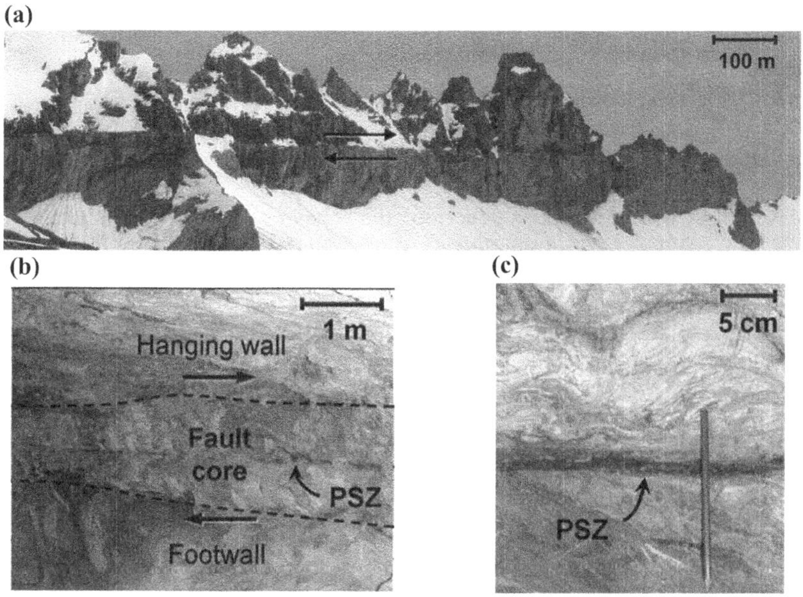

Figure 10.1. *Photos of the outcrop of the Glarus Thrust (Switzerland) showing the structure of a fault zone at three different scales: a) the fault has slipped along a very thin interface at kilometric scale, b) a closer look reveals a 1 m thick fault core and c) at the center of which a centimetric deformation localization zone (PSZ) appears (Poulet* et al. *2014). For a color version of this figure, see www.iste.co.uk/stefanou/instabilities.zip*

Considerations on the perturbations applied to the system in the LSA make it possible to estimate the evolution of the thickness of the localized zone (Sulem *et al.* 2011). Another way to obtain this evolution is to numerically integrate the system of partial differential equations describing the evolution of the structure using the finite element method (Rattez *et al.* 2018b). Moreover, numerical simulations make it possible to obtain the stress-strain response of a fault core and to explore its dependency on various parameters such the grain size and the applied shear rate. The precise evaluation of the stress-strain response of the fault zone under THM couplings is of key importance for studying earthquake nucleation as the softening rate controls the transition from aseismic to seismic slip (Scholz 2002).

In this chapter, a THM model for elasto-plastic Cosserat continua is presented in section 10.2. A LSA is applied to this model in section 10.3 in a general manner, and then used for the particular case of the problem of slip in a fault zone. The influence of the main parameters of the model is investigated as far it concerns the onset of localization and the shear band thickness evolution. Finally, the model is numerically integrated in section 10.4 in order to compare the evolution of the thickness between the two methods and to study the influence of model parameters on the stress-strain behavior.

10.2. Description of the model

A summary of the equations governing the behavior of the system is given in this section. A more detailed presentation, as well as the derivation of these equations, can be found in (Rattez *et al.* 2018c; Stefanou *et al.* 2017).

10.2.1. *Cosserat continuum theory*

Unlike the classical theory of Cauchy's (or Boltzmann's) continuum, each point presents three rotational degrees of freedom ω^c_k, in addition to the translations u_k for Cosserat continua in three dimensions. These rotations are related to the tensor of rotation of the microstructure by:

$$\omega^c_{ij} = -e_{ijk}\,\omega^c_k \qquad [10.1]$$

where e_{ijk} is the Levi-Civita symbol and the indices are equal to 1, 2, 3.

Several kinematic fields are introduced: the deformation tensor γ_{ij} – which is split into its symmetric $\gamma_{(ij))]} = \varepsilon_{ij}$ and antisymmetric part $\gamma_{[ij]}$ – and the curvature tensor κ_{ij} – also split into split its symmetric $\kappa_{(ij)}$ and antisymmetric part $\kappa_{[ij]}$.

$$\varepsilon_{ij} = \frac{1}{2}(u_{i,j} + u_{j,i}) \qquad \Omega_{ij} = \frac{1}{2}(u_{i,j} - u_{j,i})$$

$$\gamma_{[ij]} = \Omega_{ij} - \omega^c_{ij} \qquad \kappa_{ij} = \omega^c_{i,j}$$

$$\gamma_{ij} = \varepsilon_{ij} + \gamma_{[ij]} = u_{i,j} - \omega^c_{ij} = u_{i,j} + e_{ijk}\,\omega^c_k \qquad [10.2]$$

The macroscopic strain and rotation tensors (ε_{ij} and Ω_{ij}) are obtained as the symmetric and antisymmetric part of the displacement gradient, as in the classical Cauchy continuum. $\gamma_{[ij]}$ is the difference between the global rotation Ω_{ij} and the rotation of the microstructure ω^c_{ij}. The curvature κ_{ij} is defined as the gradient of the vector of Cosserat rotations and represents the gradient of microscopic deformations.

Correspondingly the stress tensor τ_{ij} is also divided into its symmetric σ_{ij} and antisymmetric part $\tau_{[ij]}$. The symmetric part corresponds to the macroscopic stresses (the ones that are considered in Cauchy continuum). Besides the stress tensor, a couple stress tensor μ_{ij} is introduced which is energy conjugate to the curvature (see Figure 10.2).

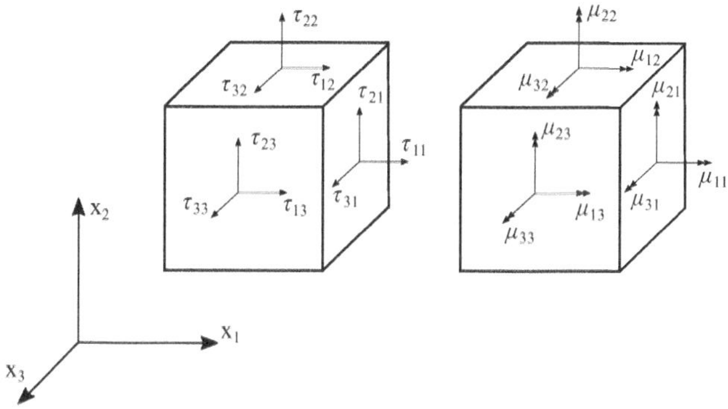

Figure 10.2. *Representation of the components of the stress and couple-stress tensors*

Without considering the volumetric forces acting on the medium and the inertia terms, the momentum balance equations are:

$$\tau_{ij,j} = 0 \qquad [10.3]$$

$$\mu_{ij,j} - e_{ijk}\,\tau_{jk} = 0 \qquad [10.4]$$

10.2.2. *Constitutive equations for a Cosserat continuum*

The general constitutive equations for a linear isotropic elastic Cosserat continuum are defined by six coefficients: the two classical deformation moduli, K and G, and four new coefficients L, M, M_c and G_c (Mindlin 1964). The effective stress tensor τ'_{ij} and couple stress tensor μ_{ij} can be expressed as:

$$\tau'_{ij} = C^e_{ijkl}\,\gamma^e_{kl} \qquad [10.5]$$

$$\mu_{ij} = M^e_{ijkl}\,\kappa^e_{kl} \qquad [10.6]$$

The effective stress tensor is linked to the total stress tensor by $\tau'_{ij} = \tau_{ij} + p\delta_{ij}$. C^e_{ijkl} is the elastic stiffness tensor and M^e_{ijkl} the elastic flexural bending rigidity tensor:

$$M^e_{ijkl} = (L - \frac{2}{3}M)\delta_{ij}\delta_{kl} + (M + M_c)\delta_{ik}\delta_{jl} + (M - M_c)\delta_{il}\delta_{jk} \qquad [10.7]$$

$$C^e_{ijkl} = (K - \frac{2}{3}G)\delta_{ij}\delta_{kl} + (G + G_c)\delta_{ik}\delta_{jl} + (G - G_c)\delta_{il}\delta_{jk} \qquad [10.8]$$

Of the six parameters introduced into the elastic constitutive law, three are measured by the same units as stress (K,G and G_c) and the other three, the units of a length squared times stress (L, M and M_c). Thus, any ratio of L, M or M_c to K, G or G_c will introduce a material parameter with dimension of length squared (Cowin 1970). For geomaterials, we usually consider $R = \sqrt{\frac{M}{G}}$ as a characteristic length related to the microstructure of the material (Unterreiner 1994). Moreover, we assume that $M = M_c$ to ensure that no out-of-plane couple stresses are produced during the shearing of a layer (invariance in x_1 and x_3 directions).

A flow theory of plasticity for granular media with Cosserat microstructure can be derived by keeping the same definitions for the yield surface and the plastic potential as in the classical theory, but by generalizing the stress and strain invariants to Cosserat continua (Mühlhaus and Vardoulakis 1987). We decompose the deformation and curvature rate tensors into elastic, plastic and thermal parts (small strain hypothesis):

$$\dot{\gamma}_{ij} = \dot{\gamma}^e_{ij} + \dot{\gamma}^p_{ij} + \dot{\gamma}^{th}_{ij} \qquad [10.9]$$

$$\dot{\kappa}_{ij} = \dot{\kappa}^e_{ij} + \dot{\kappa}^p_{ij} \qquad [10.10]$$

The thermal strain rate is $\dot{\gamma}^{th}_{ij} = \alpha\,\dot{T}\,\delta_{ij}$ with α the coefficient of thermal expansion. The effect of temperature on the curvature tensor is neglected herein.

The generalized deviatoric strain and stress second invariants are generalized as:

$$\tau = \sqrt{h_1\, s_{ij}\, s_{ij} + h_2\, s_{ij}\, s_{ji} + \frac{1}{R^2}(h_3\, m_{ij}\, m_{ij} + h_4\, m_{ij}\, m_{ji})}$$

$$\dot{\gamma}_p = \sqrt{g_1\, \dot{e}^p_{ij}\, \dot{e}^p_{ij} + g_2\, \dot{e}^p_{ij}\, \dot{e}^p_{ji} + R^2(g_3\, \dot{k}^p_{ij}\, \dot{k}^p_{ij} + g_4\, \dot{k}^p_{ij}\, \dot{k}^p_{ji})} \qquad [10.11]$$

where s_{ij}, m_{ij}, e_{ij} and k_{ij} are the deviatoric parts of the effective stress, couple-stress, strain and curvature respectively. h_i and g_i are coefficients determined by micro-mechanical considerations (Rattez *et al.* 2018c; Vardoulakis and Sulem 1995). With the above definition of the second deviatoric stress invariants, a Drucker-Prager plasticity model can be derived as in Mühlhaus and Vardoulakis (1987), or a von Mises plasticity model as in de Borst (1991). In the following, we consider only the case of a Drucker-Prager yield surface in order to represent the behavior of granular media such as fault gouges. The yield surface and the plastic potential are then defined by:

$$F = \tau + \mu\,\sigma \quad \text{and} \quad Q = \tau + \beta\,\sigma \tag{10.12}$$

where μ is the friction coefficient of the granular medium and β the dilantancy. σ is expressed as:

$$\sigma = \frac{\tau_{kk}}{3} \tag{10.13}$$

The strain-hardening modulus that describes the evolution of the material's behavior in the plastic regime is defined by:

$$-H_s = \frac{\partial F}{\partial \gamma^p} = \frac{\partial \mu}{\partial \gamma^p}\,\sigma \tag{10.14}$$

with a hardening coefficient:

$$h_s = \frac{\partial \mu}{\partial \gamma^p} \tag{10.15}$$

10.2.3. *Mass balance equation*

We consider a porous medium comprising two phases, i.e. the solid skeleton and the fluid contained in the pores of the material. By taking into account the conservation of the fluid mass, we can deduce the following form for the evolution of the pore pressure p:

$$\frac{\partial p}{\partial t} = c_{hy}\,p_{,ii} + \frac{\lambda^*}{\beta^*}\frac{\partial T}{\partial t} - \frac{1}{\beta^*}\frac{\partial \varepsilon_v}{\partial t} \tag{10.16}$$

where $c_{hy} = \chi/(\eta^f\beta^*)$ is the hydraulic diffusivity, $\beta^* = n\beta^f + (1-n)\beta^s$ is the compressibility of the skeleton and fluid mixture, $\lambda^* = n\lambda^f + (1-n)\lambda^s$ is the thermal expansion coefficient of the mixture (Vardoulakis 1986). χ is the intrinsic permeability of the porous medium and η^f is the viscosity of the fluid contained in the pores. β^f and β^s are the compressibilities per unit volume of the fluid and solid, respectively, and λ^f, λ^s are the thermal expansions per unit volume. n represents the Eulerian porosity of the medium.

10.2.4. *Energy balance equation*

The energy balance in a quasi-static transformation enables us to obtain the following equation for the evolution of temperature T:

$$\rho C\left(\frac{\partial T}{\partial t} - c_{th}T_{,ii}\right) = \tau'_{ij}\dot{\gamma}^p_{ij} + \mu_{ij}\dot{\kappa}^p_{ij} \tag{10.17}$$

where $c_{th} = \frac{k_T}{\rho C}$ is defined as the thermal diffusivity of the medium, k_T represents the thermal conductivity and ρC is the thermal capacity per unit volume of the material in its reference state.

10.3. Bifurcation analysis

10.3.1. *LSA for a Cosserat continuum with THM couplings*

We consider a problem whose evolution is described by equations [10.3], [10.4], [10.17] and [10.16]. This system of equations has a homogeneous steady state if adiabatic, undrained and total stresses boundary conditions are applied to the system. This steady state is defined by $T = T_s$, $p = p_s$, $\tau_{ij} = \tau_{ij}^0$, $\gamma_{ij} = \gamma_{ij}^0$, $\mu_{ij} = 0$ and $\kappa_{ij} = 0$, where T_s and p_s are the reference temperature and pressure, respectively, τ_{ij}^0 and γ_{ij}^0 are homogeneous tensors of stresses and deformations that depend on the geometry and boundary conditions. All fields corresponding to this steady state are marked with an exponent 0.

The phenomenon of localization of deformations is the result of an instability of the state of homogeneous deformations in a Lyapunov sense (Lyapunov 1892; Stefanou and Alevizos 2016). To determine the conditions under which this phenomenon occurs, the relevant variables of the problem are expressed as follows:

$$
\begin{aligned}
&T(x_i,t) = T^0 + T^*(x_i,t) \qquad && p(x_i,t) = p^0 + p^*(x_i,t) \\
&\tau'_{kl}(x_i,t) = \tau'^0_{kl} + \tau'^*_{kl}(x_i,t) \qquad && u_k(x_i,t) = u_k^0(x_i) + u_k^*(x_i,t) \\
&\omega_k^c(x_i,t) = \omega_k^{c0} + \omega_k^{c*}(x_i,t) &&
\end{aligned} \qquad [10.18]
$$

where the exponent * represents perturbed fields from the homogeneous reference state. The constitutive equations for the elastoplastic material can be linearized around the steady state to obtain the following relationships:

$$
\begin{aligned}
\tau'^*_{kl} &= C^{ep}_{klmn}\,\gamma^*_{mn} + D^{ep}_{klmn}\,\kappa^*_{mn} + E^{ep}_{klmn}\,T^*\,\delta_{mn} \\
\mu^*_{kl} &= M^{ep}_{klmn}\,\kappa^*_{mn} + L^{ep}_{klmn}\,\gamma^*_{mn} + N^{ep}_{klmn}\,T^*\,\delta_{mn}
\end{aligned} \qquad [10.19]
$$

where C^{ep}_{klmn}, D^{ep}_{klmn}, E^{ep}_{klmn}, M^{ep}_{klmn}, L^{ep}_{klmn} are N^{ep}_{klmn} are the fourth-order tensors of the linearized elastoplastic incremental law.

We then obtain a linear system of eight equations with eight unknowns (the perturbed fields). It is convenient to apply Fourier analysis to study it. The solutions of the system can be written in the following form:

$$
X_k^*(x_l,t) = \overline{X}_k(t).exp(i\,\xi\,x_l\,n_l) \qquad [10.20]
$$

where ξ is the inverse of the wavelength $\lambda = \frac{h}{2\pi N}$ (N is an integer enabling us to satisfy the boundary conditions).

In the following, the height of the layer h is an open parameter, so that all possible values of λ are examined. n_j is the polarization direction of the localization band, $i^2 = -1$ and $X_i^*(x_j, t)$ is a vector containing the eight unknown of the perturbed fields as follows:

$$X_k^*(x_l, t) = [u_k^*(x_l, t) \quad \omega_k^{c*}(x_l, t) \quad T^*(x_l, t) \quad p^*(x_l, t)]^t \qquad [10.21]$$

By inserting equation [10.20] into the system, we can obtain a system of first-order ordinary differential equations, for which solutions have the form:

$$\overline{X}_k(t) = \overline{X}_k.exp(s\ t) \qquad [10.22]$$

where s is the growth rate of the perturbation and $\overline{X}_k$ is vector composed of algebraic components. Finally, by inserting the expression of $\overline{X}_k(t)$ into the system, we get the following system of linear equations written in the matrix form:

$$\begin{bmatrix} \Gamma_{km} - \rho s^2 \delta_{km} & \Delta_{km} & E^{ep}_{klmn} i\xi n_l \delta_{mn} & -i\xi n_k \\ \Xi_{km} & \Pi_{km} - \rho I s^2 \delta_{km} & -e_{klm} E^{ep}_{lmnq} \delta_{nq} + N^{ep}_{klmn} \delta_{mn} & 0 \\ -\tau'^0_{kl} s i\xi n_l \delta_{km} & -\tau'^0_{kl} s i e_{klm} - \mu^0_{kl} s i \xi n_l \delta_{km} & \rho C(s + c_{th}\xi^2) & 0 \\ \frac{1}{\beta^*} s i \xi n_k & 0 & -\Lambda s & s + c_{hy}\xi^2 \end{bmatrix}$$

$$\begin{bmatrix} \overline{u}_m \\ \overline{\omega}^c_m \\ \overline{T} \\ \overline{p} \end{bmatrix} = 0 \qquad [10.23]$$

where,

$$\begin{aligned} \Gamma_{km} &= C^{ep}_{klmn} \xi^2 n_n n_l \\ \Delta_{km} &= C^{ep}_{klqn} e_{qnm} i\xi n_l - D^{ep}_{klmn} \xi^2 n_n n_l \\ \Xi_{km} &= -L^{ep}_{klmn} \xi^2 n_n n_l - e_{klr} C^{ep}_{lrmq} i\xi n_q \\ \Pi_{km} &= -M^{ep}_{klmn} \xi^2 n_n n_l + L^{ep}_{klrn} e_{rnm} i\xi n_l - e_{klr} C^{ep}_{lrnq} e_{nqm} - e_{klr} D^{ep}_{lrmq} i\xi n_q \end{aligned} \qquad [10.24]$$

In order to obtain non-trivial solutions, the matrix determinant in equation [10.23] must be zero. This condition leads to a polynomial equation in s. The roots of this equation then provide us information about the stability of the system. Indeed, if one of the roots has a strictly positive real part, the homogeneous state considered is unstable. If all roots have a strictly negative real part, the homogeneous state is stable. However, it is not possible to conclude in cases where at least one of the roots has a zero real part and all the others have a strictly negative real part.

10.3.2. *Localization conditions for a fault zone*

The LSA is then applied to the particular case of rapid shear in a fault gouge. The fault is modeled by an infinite saturated sheared layer (see Figure 10.3).

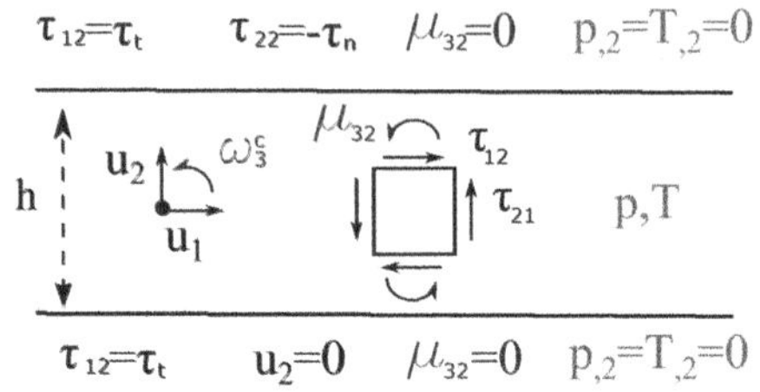

Figure 10.3. *Boundary conditions for the infinite sheared layer modeling a saturated fault gouge. For a color version of this figure, see www.iste.co.uk/stefanou/instabilities.zip*

The values of the different parameters of the model are chosen to represent a fault at a depth of 7 km, corresponding to a central depth of the fragile part of the lithosphere (see Table 10.1). These values are based on experiments on samples retrieved from the Aegean fault in Greece and the Median tectonic in Japan.

Parameters	Values	Units	Parameters	Values	Units
K	16.67×10^3	MPa	μ	0.5	
G	10×10^3	MPa	β	0	
Gc	5×10^3	MPa	λ^*	7.4×10^{-5}	/°C
R	0.01	mm	ρC	2.8	MPa/°C
ρ	2500	kg/m^3	c_{th}	1	mm^2/s
τ_n	200	MPa	c_{hy}	12	mm^2/s
τ'^0	67	MPa	α_s	2.5×10^{-5}	/°C
β^*	8.2×10^{-5}	MPa^{-1}	n	0.04	

Table 10.1. *Model parameter values for a fault gouge (Rice 2006; Sulem* et al. *2011)*

For this example, a bifurcation is observed for a critical hardening modulus $H_{cr} = 2\,\text{MPa}$ (corresponding to $h_s = 0.015$). Indeed, for $H_s > H_{cr}$, all the real roots of the characteristic polynomial are negative and some complex roots have a positive real part (see Figure 10.4). The dashed lines correspond to the real parts of the roots that have a non-zero imaginary part. For these roots, the system diverges by oscillating, which coincides with a flutter instability. As noted in Benallal and Comi (2003), our system is not differentiable because of the Kuhn-Tucker condition in plasticity, which allows us to have different constitutive laws for plastic loading and elastic unloading. Thus, to develop the above equations, we have assumed a monotonic loading to stay in the plastic regime. Therefore, we cannot conclude on the physical relevance of these instabilities.

On the other hand, for $H_s < H_{cr}$, some real roots of the polynomial are positive at certain wavelengths (Figure 10.5), which causes the triggering of strain localization. From these diagrams, the critical value of the hardening can be determined for different values of the model parameters.

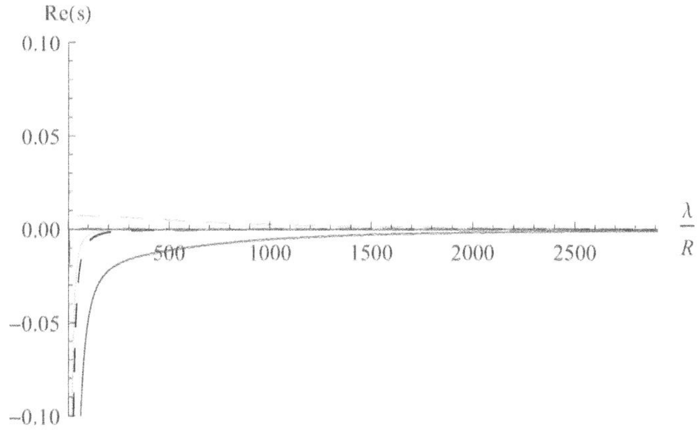

Figure 10.4. *Real part of the growth coefficient s as a function of the length of the perturbation for $H_s = 2.1$ MPa $> H_{cr}$. The colors represent different roots. The dotted lines represent roots with a non-zero imaginary part, whereas the solid lines represent real roots. For a color version of this figure, see www.iste.co.uk/stefanou/instabilities.zip*

In granular media, inelastic deformations can induce volume changes. For materials with low porosity, dilatancy is observed due to the rearrangement of grains by shearing or lifting during sliding along an asperity. The opposite effect, compaction, can also be observed for materials with greater porosity and is the result of the decrease of porous space or crushing of the grains. To illustrate the influence of these effects on the stability of our system, in Figure 10.6 we plot the value of the critical hardening modulus as a function of the dilatancy coefficient β, taking into account different couplings.

The purely mechanical system presents a linear evolution of the critical hardening modulus with β . For a dilatant material ($\beta > 0$), bifurcation occurs in the softening regime ($H_s < 0$), whereas for a compacting material ($\beta < 0$), it occurs in the hardening regime ($H_s > 0$).

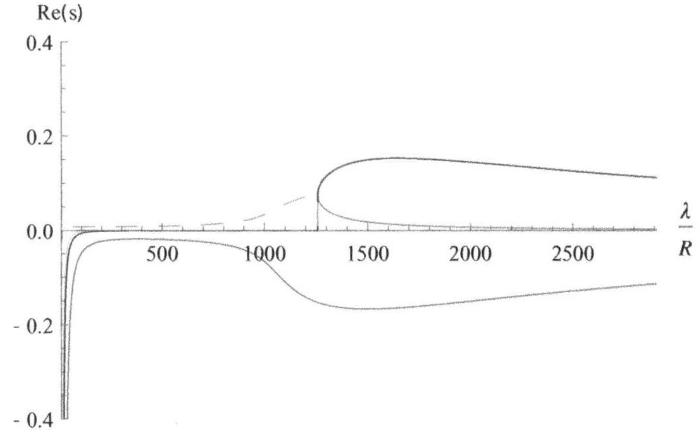

Figure 10.5. *Real part of the growth coefficient s as a function of the length of the perturbation for $H_s = 1.9$ MPa $< H_{cr}$. The colors represent different roots. The dotted lines represent roots with a non-zero imaginary part, whereas the solid lines represent real roots. For a color version of this figure, see www.iste.co.uk/stefanou/instabilities.zip*

In the case of hydromechanical couplings, the evolution is bilinear and the two lines cross in $\beta = 0$. The bifurcation is the same as for the mechanical system in the dilatant regime, whereas it is obtained for higher values of H_s in the compacting regime. The formation of this second line is due to the undrained behavior that becomes unstable before the drained behavior in the compacting regime (Rice 1975). Indeed, if the medium tends to compact, the size of the porous space decreases and causes an increase in fluid pressure in the pores, which triggers instability for larger values of the hardening modulus.

The introduction of thermal pressurization in the THM model has the effect of shifting the curve in the compacting regime to even higher values of H_{cr}, but the slope remains the same. Thus, the undrained behavior is affected by thermal pressurization but not the drained behavior.

10.3.3. *Shear band thickness evolution in a fault zone*

The introduction of an internal length within the constitutive law in Cosserat continua allows us, in addition to the condition of triggering the localization, to obtain a value of the shear band thickness.

For a given value of the hardening modulus below the critical value $H_s < H_{cr}$, we can observe in Figure 10.5 that the actual roots of the polynomial equation exhibit a maximum. This finite maximum value of the Lyapunov coefficient (or growth coefficient) is obtained by taking into account the inertia terms (Sulem *et al.* 2011).

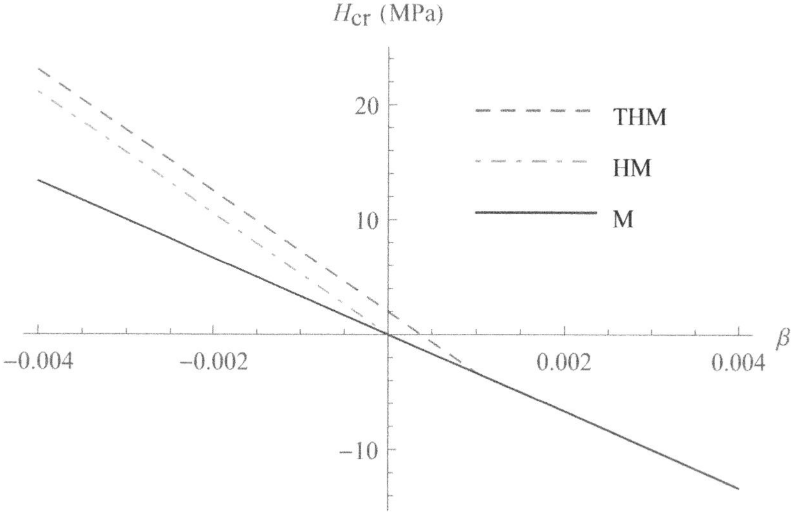

Figure 10.6. *Bifurcation parameter H_{cr} (hardening modulus for which a bifurcation is obtained) represented as a function of dilatancy β by considering a mechanical (M), hydro-mechanical (HM) or thermo-hydro-mechanical (THM) behavior. For a color version of this figure, see www.iste.co.uk/stefanou/instabilities.zip*

This maximum corresponds to the mode of instability with the fastest growth (see equation [10.22]). If we consider that the wavelength with the fastest growth represents the thickness of the localized zone, then we can plot, in Figure 10.7, this selected wavelength (λ_{max} normalized by the Cosserat characteristic length R) as a function of the hardening modulus and interpret it as the evolution of the shear band size.

The selected wavelength tends toward infinity for $H_s \to H_{cr}^-$ and decreases for a decreasing hardening modulus. This trend is similar to that observed in Mühlhaus and Vardoulakis (1987).

Hydraulic parameters are not well-constrained and depend a lot on the material considered, pressure, temperature, porosity, etc. Their influence on the evolution of the selected wavelength is studied in detail in (Rattez *et al.* 2018c). Among them, the one that has the most influence on the stability of the system is the thermal pressurization parameter $\Lambda = \lambda^*/\beta^*$ that controls the amplitude of the pore pressure increase for a given temperature increase (see Figure 10.8). Considering the different values for the thermal expansion and compressibility coefficients for the solid and fluid phase given in Rice (2006), this parameter can vary between 0.59 and 1.62 MPa/°C.

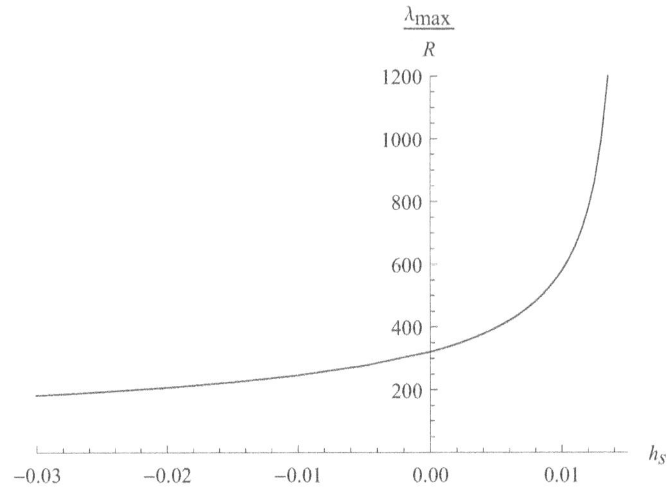

Figure 10.7. *Wavelength of the perturbation with the fastest growth λ_{max} as a function of the hardening coefficient h_s.*

10.4. Numerical analysis

The model presented in section 10.2 has been implemented and validated in a finite element code (Rattez *et al.* 2018a) and allows us to perform numerical simulations to study the problem of strain localization in a fault core and to complete the results obtained by LSA. Finite element analysis, unlike LSA, allows us to study the influence of all parameters on the stress-strain response of a fault. This response is of primary importance for the transition from an aseismic to seismic slip. It also gives us information on the energy budget of an earthquake.

10.4.1. *Regularization of the mesh dependency*

The presence of an internal length in the constitutive law of Cosserat continuum enables us to regularize the problem of strain localization and thus to solve the problem of mesh dependence encountered for Cauchy continuum.

To illustrate this effect, an infinite shear layer is considered in this section with a geometry and parameters coherent with a gouge in a fault core. Only the equations for an elastoplastic Cossserat medium are considered (without coupling).

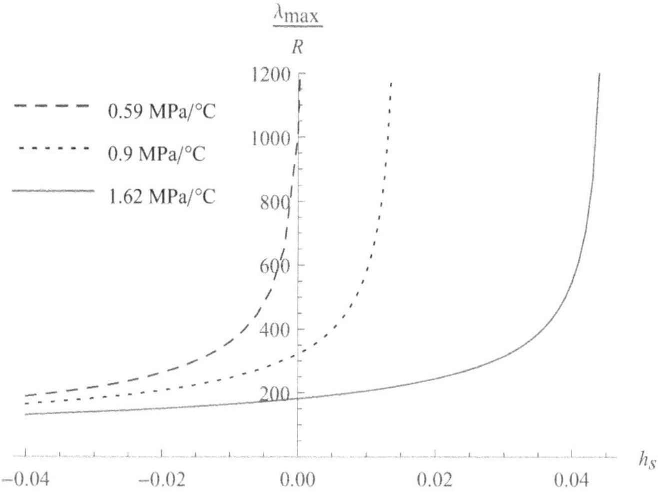

Figure 10.8. *Selected wavelength λ_{max} as a function of the hardening coefficient h_s for different values of the thermal pressurization. For a color version of this figure, see www.iste.co.uk/stefanou/instabilities.zip*

The fault core is modeled by an infinite layer of thickness $h = 1$mm, subjected to a constant shear rate $V = 1$m/s and a normal stress $\tau_n = 133$ MPa. Rotations are prevented at the upper and lower boundaries of the layer ($\omega_3^c = 0$), which leads to the development of a single band at the center of the layer and allows for a systematic comparison between the stress-strain responses of the different simulations (Rattez *et al.* 2018b).

The mechanical parameters of the system are the same as for the LSA (see Table 10.1). A linear softening law is considered with a hardening coefficient equal to 0.5.

A mesh convergence study is first carried out for a Cauchy continuum. A 3D geometry is considered with periodic boundary conditions on both sides of the sample, which results in a one-dimensional problem. A regular mesh with hexahedral elements is chosen with only one element in the x_1 and x_3 directions. Given the periodic conditions and the choice of shape functions (first-order Lagrange polynomials), the invariance in directions x_1 and x_3 is maintained. To simulate the case of a Cauchy continuum, the Cosserat shear modulus, as well as the internal length, are taken to be zero as explained in Iordache and William (1998). In Figure 10.9, the shear stress is plotted as a function of the horizontal displacement applied to the upper layer boundary.

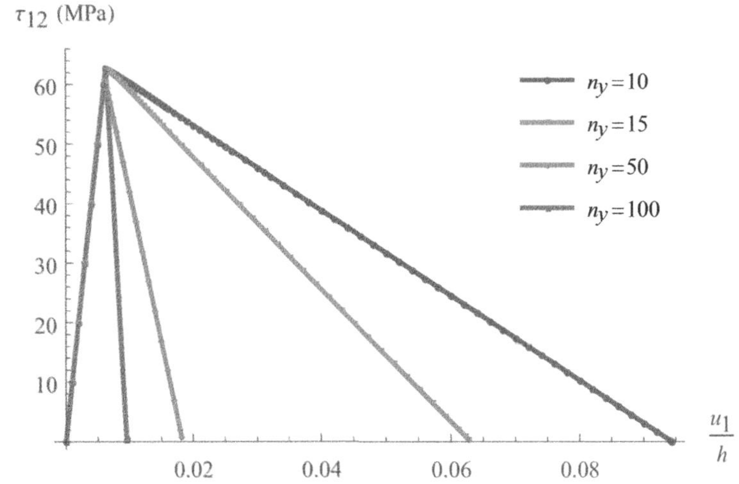

Figure 10.9. *Stress–strain response of an infinite sheared layer for Cauchy continuum and different numbers of mesh elements in the vertical direction. For a color version of this figure, see www.iste.co.uk/stefanou/instabilities.zip*

This figure exhibits the dependence of the softening branch on the mesh size. Plastic deformations localize in a single element of the hexahedral mesh (see Figure 10.10) and the finer the mesh, the steeper the softening.

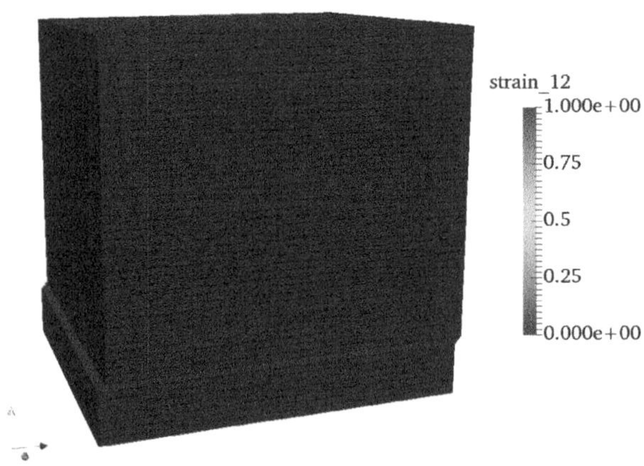

Figure 10.10. *Shear deformation plotted on the deformed mesh with 50 elements and $\tau_{12} = 1MPa$ for a Cauchy continuum. For a color version of this figure, see www.iste.co.uk/stefanou/instabilities.zip*

In Figure 10.11, the same analysis is repeated for a Cosserat continuum (G_c and R not zero this time). The results obtained for 160 and 240 elements in the vertical direction do not show any visible difference, indicating a mesh convergence.

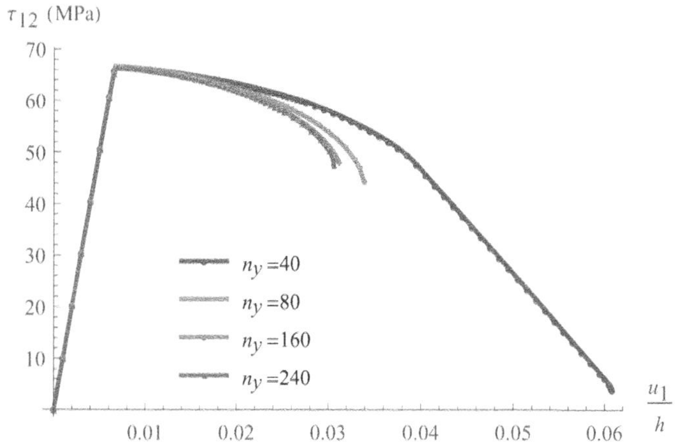

Figure 10.11. *Mesh convergence obtained for the stress-strain response of an infinite sheared layer modeled by a Cosserat continuum for different numbers of mesh elements in the vertical direction. For a color version of this figure, see www.iste.co.uk/stefanou/instabilities.zip*

In Figure 10.12, the Cosserat rotation ω_3^c is plotted on a deformed mesh with 80 elements in the vertical direction (not finer, for clearer visualization) at the last time step of the simulation. The amplitude of rotations is greater within the localization zone, as observed experimentally for granular materials (Hall *et al.* 2010).

This mesh dependency of the stress–strain response also results in a mesh dependency of the energy dissipated by the system for Cauchy continuum and leads to results that are not unphysical. This is all the more problematic when thermo-mechanical couplings are included in the model because the amount of heat produced depends on the plastic dissipation.

10.4.2. *Response and shear band thickness of a fault gouge*

In this section, we consider a saturated sheared layer modeled by a Cosserat continuum with THM couplings. We are interested in the effect of pore pressure and temperature change on the process of strain localization. A strain hardening represented by an exponential evolution toward a non-zero finite value of the friction coefficient is considered and allows for the modeling of granular materials such as fault cores (Rattez *et al.* 2018b).

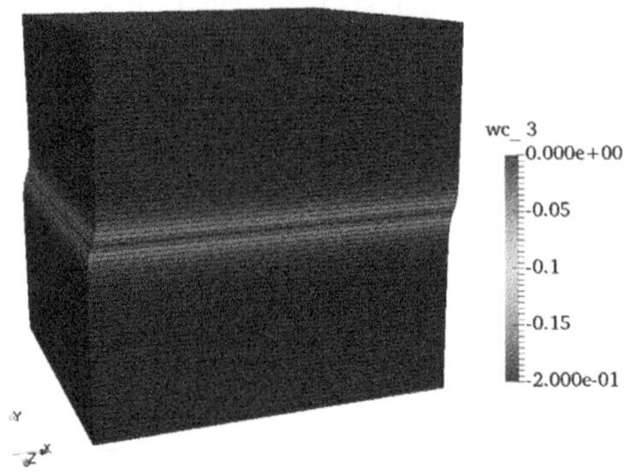

Figure 10.12. *Cosserat rotation plotted on the deformed mesh with 80 elements and $\tau_{12} = 48$MPa. For a color version of this figure, see www.iste.co.uk/stefanou/instabilities.zip*

In Figure 10.13, the stress–strain curve of the system is represented for different values of the microstructure size and the response of the purely mechanical system (in blue) is compared to the response of the THM system (in red dotted line). The size of the microstructure does not affect the asymptotic value reached by the shear stress, but it changes the slope of the softening. Thus, a gouge with finer grains is more likely to slip unsteadily and create earthquakes.

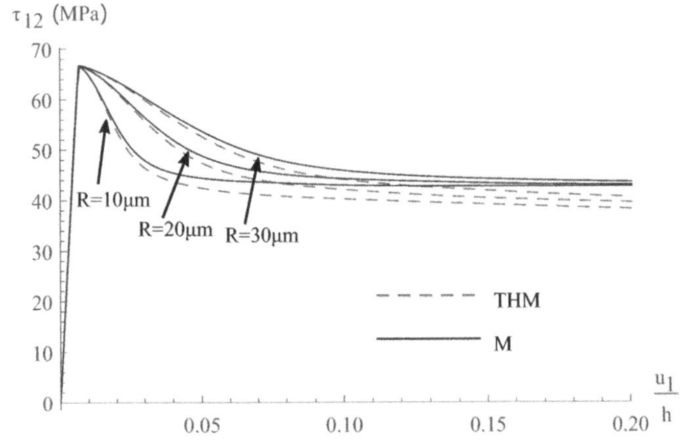

Figure 10.13. *Effect of THM couplings on the stress–strain response of the system for an exponential softening and different values of microstructure size (R = 10, 20 and 30 μm) considering a shear rate of 1 m/s. For a color version of this figure, see www.iste.co.uk/stefanou/instabilities.zip*

It can be observed that when THM couplings are considered, the shear stress τ_{12} does not reach a plateau, but continues to decrease, as the heat produced by the plastic dissipation continues to reduce the resistance of the fault due to thermal pressurization. However, the mechanical and THM models follow the same trend at the beginning of softening, and then diverge when plastic deformations become more important. Thermal pressurization does not play a significant role when instability is triggered, but rather on the evolution of slip and energy dissipation (Segall and Rice 2006).

The evolution of the shear band thickness is shown in Figure 10.14. The band size decreases rapidly at the beginning of the postbifurcation to reach a minimum and then increases slightly, before moving toward a residual value. THM couplings result in a thinner band and affect the residual thickness more than the minimum thickness. It can be noted that even if the shear stress does not reach a stationary value for THM couplings, the band size reaches an asymptotic value. Thus, thermal pressurization has little influence on the localization thickness if mechanical softening alone is already high. However, in the case of perfectly plastic behavior, localization is only observed if THM couplings are considered. In this case, thermal pressurization has a crucial role in controlling the size of the principal shear zone (PSZ).

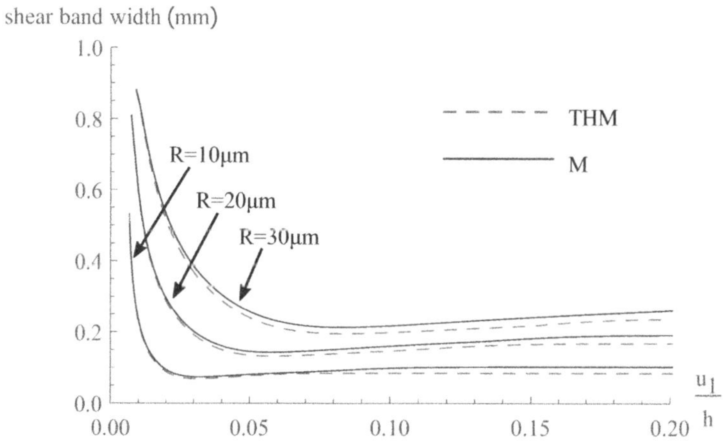

Figure 10.14. *Effect of THM couplings on the evolution of the shear band thickness for an exponential softening and different values of the internal length (R = 10, 20 and 30 μm). For a color version of this figure, see www.iste.co.uk/stefanou/instabilities.zip*

A significant result obtained in these simulations is the dependence of postpeak behavior and shear band size with the shear rate, despite the use of a rate independent constitutive law. A perfectly plastic behavior is considered in order to illustrate the effect of rate dependency on softening due only to couplings (see Figure 10.15). It can be noted that no softening occurs at low speeds (0.01m/s), because in this case the effect of thermal pressurization is negligible. This effect can be explained by the diffusion of pore pressure and temperature, which tend to widen the shear band thickness at lower speeds.

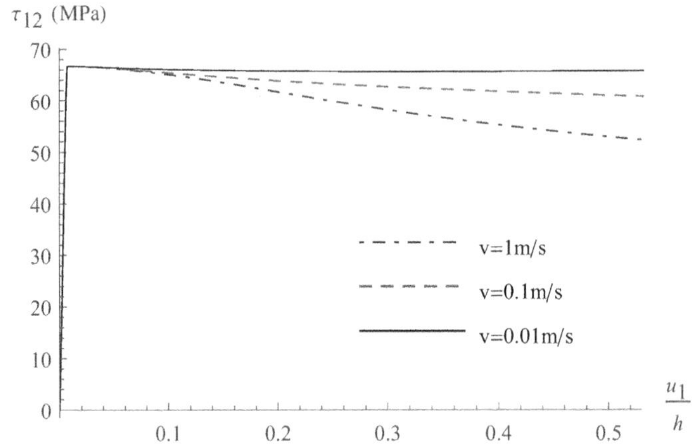

Figure 10.15. *Effect of shear rate on the stress–strain diagram for a perfectly plastic behavior. For a color version of this figure, see www.iste.co.uk/stefanou/instabilities.zip*

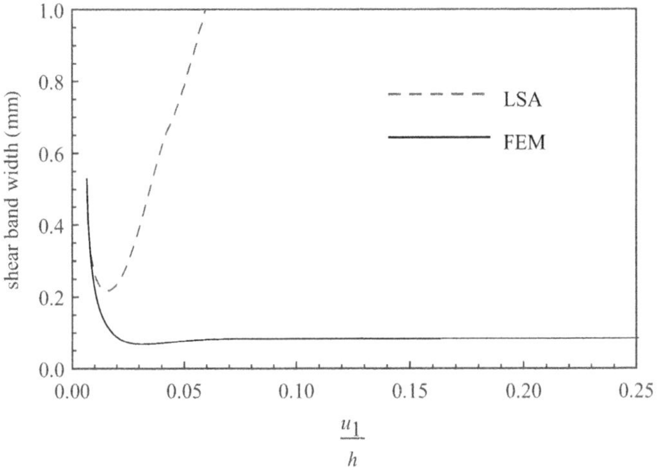

Figure 10.16. *Comparison of the evolution of the shear band thickness obtained by LSA or FEM considering THM couplings, an internal length of* $R = 10\mu m$ *and an exponential evolution of the softening. For a color version of this figure, see www.iste.co.uk/stefanou/instabilities.zip*

These evolutions in shear band sizes are then compared to those obtained by LSA. For this purpose, at each time step, an associated homogeneous problem is defined to which a LSA is applied (Rattez *et al.* 2018b). This analysis allows us to understand how the loss of system homogeneity influences the evolution of the localization process. In Figure 10.16, we plot the evolution of the localization thickness for a THM model with an exponential hardening law. The initiation and decrease of the band size are well-reproduced by LSA. As the deformation field increases and the system becomes more inhomogeneous, the band size estimates obtained by LSA diverge from the numerical results.

The size of the principal slip zone is then compared to observations on outcrops of the Punchbowl fault, located in California (Chester *et al.* 2005; Rice 2006). Inside the fault gouge, a shear band with a thickness of 100–300 μm was observed. The internal length of Cosserat required to reproduce these observations is about 10 μm (see Figure 10.17), an order of magnitude larger than the median diameter of the granular distribution D_{50} and which rather corresponds to D_{70} (sieve size for which 70% – in terms of mass – of the grains pass through). However, most laboratory experiments and discrete element simulations were performed with a narrow granular distribution and, in this case, the consideration of D_{50} as the Cosserat internal length allows the results to be reproduced (Mühlhaus and Vardoulakis 1987).

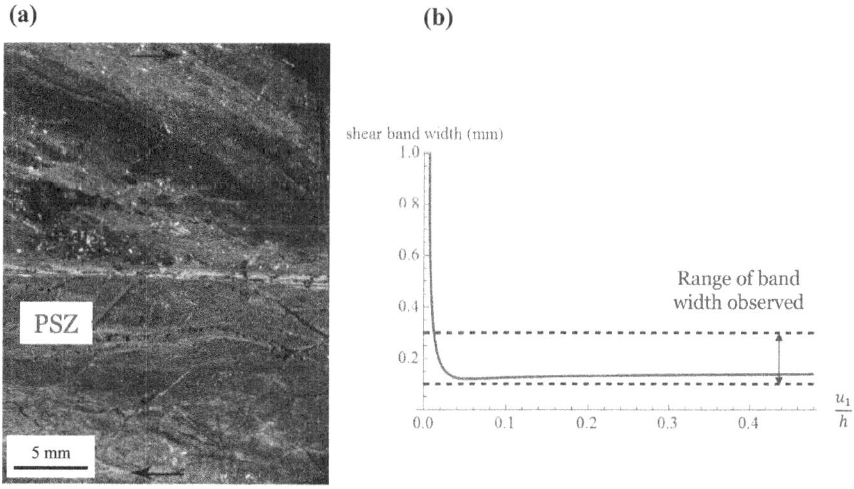

Figure 10.17. *a) Observation of the principal shear zone (PSZ) for the punchbowl fault from Chester* et al. *(2005) and Rice (2006). b) Comparison between the observed thicknesses and the evolution of the thickness obtained by numerical simulation. For a color version of this figure, see www.iste.co.uk/stefanou/instabilities.zip*

10.5. Conclusion

In this chapter, we have seen that the use of the Cosserat continuum allows us to model the size of the microstructure in fault gouges and to obtain a finite thickness of the strain localization zone. Thus, the energy dissipation of the mechanical system is correctly quantified. The system of partial differential equations for a Cosserat continuum with THM couplings is presented. It is first studied using a LSA to determine the onset of the localization of deformations and to qualitatively follow the evolution of the shear band size. It is then numerically integrated using finite element software. Moreover, numerical simulations provide the stress–strain response of the material, which is essential to determine the potential of a fault to slide seismically.

We have observed that finer microstructure size and higher thermal pressurization tend to decrease the size of the shear band and increase the slope of the softening branch. Furthermore, the localization band thicknesses obtained by the two methods are comparable as long as the assumptions of linearization of the constitutive equations around a homogeneous state remain valid.

10.6. Bibliography

Anthony, J.L. and Marone, C.J. (2005). Influence of particle characteristics on granular friction. *Journal of Geophysical Research B: Solid Earth*, 110(8), 1–14.

Benallal, A. and Comi, C. (2003). Perturbation growth and localization in fluid-saturated inelastic porous media under quasi-static loadings. *Journal of the Mechanics and Physics of Solids*, 51(5), 851–899.

Cashman, S.M., Baldwin, J.N., Cashman, K.V., Swanson, K., Crawford, R. (2007). Microstructures developed by coseismic and aseismic faulting in near-surface sediments. San Andres fault. California. *Geology*, 35(7), 611–614.

Chester, J.S., Chester, F.M., Kronenberg, A.K. (2005). Fracture surface energy of the Punchbowl fault. San Andreas system. *Nature*, 437(7055), 133–136.

Coussy, O. (2004). *Poromechanics*. Wiley, Hoboken, USA.

Cowin, S.C. (1970). Stress functions for Cosserat elasticity. *International Journal of Solids and Structures*, 6(4), 389–398.

de Borst, R. (1991). Simulation of strain localization: A reappraisal of the Cosserat continuum. *Engineering Computations*, 8(4), 317–332.

Di Toro, G., Pennacchioni, G., Nielsen, S. (2009). Pseudotachylytes and earthquake source mechanics. In *Fault-Zone Properties and Earthquake Rupture Dynamics, International Geophysics Series*, Elsevier, Cambridge, USA, 94, 87–133.

Hall, S.A., Bornert, M., Desrues, J., Lenoir, N., Pannier, Y., Viggiani, G., Bésuelle, P. (2010). Discrete and continuum analysis of localised deformation in sand using X-ray μCT and volumetric digital image correlation. *Géotechnique*, 60(5), 315–322.

Iordache, M.-M. and William, K. (1998). Localized failure analysis in elastoplastic Cosserat continua. *Computer Methods in Applied Mechanics and Engineering*, 151(3–4), 559–586 [Online]. Available at: http://www.sciencedirect.com/science/article/pii/S0045782597001667.

Lyapunov, A.M. (1892). The general problem of the stability of motion (in Russian). PhD thesis, University of Kharkov, Kharkov Mathematical Society, Ukraine.

Mindlin, R. (1964). Micro-structure in linear elasticity. *Archive for Rational Mechanics and Analysis*, 16(1), 51–78.

Mühlhaus, H.-B. and Vardoulakis, I. (1987). The thickness of shear bands in granular materials. *Géotechnique*, 37(3), 271–283.

Platt, J.D., Rudnicki, J.W., Rice, J.R. (2014). Stability and localization of rapid shear in fluid-saturated fault gouge 2: Localized zone width and strength evolution. *Journal of Geophysical Research: Solid Earth*, 119(5).

Poulet, T., Veveakis, M., Herwegh, M., Buckingham, T., Regenauer-Lieb, K. (2014). Modeling episodic fluid-release events in the ductile carbonates of the Glarus thrust. *Geophysical Research Letters*, 41(20), 7121–7128.

Rattez, H. (2017). Thermo-Hydro-Mechanical couplings and strain localization in Cosserat continuum: Application to stability analysis of rapid shear in faults. PhD thesis, Université Paris-Est, France.

Rattez, H., Stefanou, I., Sulem, J., Veveakis, M., Poulet, T. (2018a). Numerical analysis of strain localization in rocks with thermo-hydro-mechanical couplings using Cosserat continuum. *Rock Mechanics and Rock Engineering*, 51(10), 3295–3311.

Rattez, H., Stefanou, I., Sulem, J., Veveakis, M., Poulet, T. (2018b). The importance of Thermo-Hydro-Mechanical couplings and microstructure to strain localization in 3D continua with application to seismic faults. Part II: Numerical implementation and post-bifurcation analysis. *Journal of the Mechanics and Physics of Solids*, 115, 1–29.

Rattez, H., Stefanou, I., Sulem, J. (2018c). The importance of Thermo-Hydro-Mechanical couplings and microstructure to strain localization in 3D continua with application to seismic faults. Part I: Theory and linear stability analysis. *Journal of the Mechanics and Physics of Solids*, 115, 54–76.

Rice, J.R. (1975). On the stability of dilatant hardening for saturated rock masses. *Journal of Geophysical Research*, 80(11), 1531–1536 [Online]. Available at: http://doi.wiley.com/10.1029/JB080i011p01531.

Rice, J.R. (2006). Heating and weakening of faults during earthquake slip. *Journal of Geophysical Research: Solid Earth*, 111(5), 1–29.

Rice, J.R., Rudnicki, J.W., Platt, J.D. (2014). Stability and localization of rapid shear in fluid-saturated fault gouge 1: Linearized stability analysis. *Journal of Geophysical Research*, 119(5), 1–23.

Rudnicki, J.W. and Rice, J.R. (1975). Conditions for the localization of deformation in pressure-sensitive dilatant materials. *Journal of the Mechanics and Physics of Solids*, 23(6), 371–394.

Sammis, C.G., Osborne, R.H., Anderson, J.L., Banerdt, M., White, P. (1986). Self-similar cataclasis in the formation of fault gouge. *Pure and Applied Geophysics PAGEOPH*, 124(1–2), 53–78 [Online]. Available at: http://link.springer.com/10.1007/BF00875719.

Scholz, C.H. (2002). *The Mechanics of Earthquakes and Faulting*. Cambridge University Press, UK.

Segall, P. and Rice, J.R. (2006). Does shear heating of pore fluid contribute to earthquake nucleation? *Journal of Geophysical Research: Solid Earth*, 111(9), 1–17.

Stefanou, I. and Alevizos, S. (2016). Fundamentals of bifurcation theory and stability analysis. In *ALERT: Doctoral School 2016: Modelling of instabilities and bifurcation in Geomechanics*, Sulem, J., Stefanou, I., Papamichos, E., Veveakis, E. (eds). ALERT Geomaterials, Aussois, France, 29–68 [Online]. Available at: http://alertgeomaterials.eu/data/school/2016/2016_ALERT_schoolbook.pdf.

Stefanou, I., Sulem, J., Rattez, H. (2017). Cosserat approach to localization in geomaterials. In *Handbook of Nonlocal Continuum Mechanics for Materials and Structures*, Voyiadjis, G.Z. (ed.). Springer, Basel, Switzerland, 1–25.

Sulem, J., Stefanou, I., Veveakis, M. (2011). Stability analysis of undrained adiabatic shearing of a rock layer with Cosserat microstructure. *Granular Matter*, 13(3), 261–268.

Unterreiner, P. (1994). Contribution à l'étude et à la modélisation numérique des sols cloués: application au calcul en déformation des ouvrages de soutènement. PhD thesis, École Nationale des Ponts et Chaussées, France.

Vardoulakis, I. (1985). Stability and bifurcation of undrained, plane rectilinear deformations on water-saturated granular soils. *International Journal for Numerical and Analytical Methods in Geomechanics*, 9(5), 399–414 [Online]. Available at: http://doi.wiley.com/10.1002/nag.1610090502.

Vardoulakis, I. (1986). Dynamic stability analysis of undrained simple shear on water-saturated granular soils. *International Journal for Numerical and Analytical Methods in Geomechanics*, 10(2), 177–190 [Online]. Available at: http://doi.wiley.com/10.1002/nag.1610 100206.

Vardoulakis, I. and Sulem, J. (1995). *Bifurcation Analysis in Geomechanics*. Blackie, Glascow.

List of Authors

Sotiris ALEVIZOS
University of New South Wales
Sydney
Australia

Pierre BÉSUELLE
3SR
Université Grenoble Alpes
and
CNRS
France

Frédéric COLLIN
ArGEnCo
University of Liège
Belgium

Tomasz HUECKEL
Duke University
Durham
USA

Panagiotis KOTRONIS
GeM
Ecole Centrale de Nantes
France

Euripides PAPAMICHOS
Aristotle University of Thessaloniki
Greece
and
SINTEF Petroleum Research
Trondheim
Norway

Panos PAPANASTASIOU
Department of Civil and
Environmental Engineering
University of Cyprus
Nicosia
Cyprus

Benoît PARDOEN
iMMC
Université catholique de Louvain
Belgium

Thomas POULET
CSIRO
Kensington
Australia

Hadrien RATTEZ
Duke University
Durham
USA

Ioannis STEFANOU
GeM
Ecole Centrale de Nantes
France

Jean SULEM
Navier
Ecole des Ponts ParisTech
Université Paris-Est
Marne-la-Vallée
France

Manolis VEVEAKIS
Duke University
Durham
USA

Antonis ZERVOS
Faculty of Engineering and
Physical Sciences
University of Southampton
UK

Index

E

F, G

H, I

L

M

N, O, P

S

T, U

W, X

Printed and bound by CPI Group (UK) Ltd, Croydon, CR0 4YY

13/09/2026

14961043-0008